THE IISS
ARMED CONFLICT SURVEY
2017

published by

for
The International Institute for Strategic Studies

The International Institute for Strategic Studies
Arundel House | 13–15 Arundel Street | Temple Place | London | WC2R 3DX | UK

THE IISS ARMED CONFLICT SURVEY 2017

First published May 2017 by **Routledge**
4 Park Square, Milton Park, Abingdon, Oxon, OX14 4RN

for **The International Institute for Strategic Studies**
Arundel House, 13–15 Arundel Street, Temple Place, London, WC2R 3DX, UK

Simultaneously published in the USA and Canada by **Routledge**
711 Third Avenue, New York, NY 10017

Routledge is an imprint of Taylor & Francis, an Informa business

© 2017 The International Institute for Strategic Studies

DIRECTOR-GENERAL AND CHIEF EXECUTIVE Dr John Chipman
EDITOR Dr Anastasia Voronkova

CONTRIBUTING EDITOR Dr Nicholas Redman
ASSOCIATE EDITOR Chris Raggett
EDITORIAL Alice Aveson, Emma Champion, Holly Marriott Webb, Jessica Watson, Carolyn West
DESIGN AND PRODUCTION John Buck, Kelly Verity
RESEARCH ASSOCIATE Jens Wardenaer
CONFLICT AND MAP RESEARCH Raimah Amevor, Rojan Bolling, Henry Boyd, Emma Boyle, Ally Chandler, Nicholas Crawford, Diana Felix da Costa, Alexandra Dzero, Ali Gokpinar, Andrew Kelly, Janosch Kullenberg, Nishank Motwani, Antonio Sampaio, Philipp Schweers, Christopher Shay, Hebatalla Taha, Caitlin Vito, Alex Waterman
COVER IMAGES AFP/Getty; Ahmad Al-Rubaye/AFP/Getty; Yunus Keles/Anadolu Agency/Getty; Bullen Chol/Anadolu Agency/Getty

All rights reserved. No part of this book may be reprinted or reproduced or utilised in any form or by any electronic, mechanical, or other means, now known or hereafter invented, including photocopying and recording, or in any information storage or retrieval system, without permission in writing from the publisher.

British Library Cataloguing in Publication Data
A catalogue record for this book is available from the British Library

Library of Congress Cataloguing in Publication Data

ISBN 978-1-85743-914-4
ISSN 2374-0973

Contents

	Editor's Introduction	**5**
Chapter One	**Thematic Essays**	**13**
	Whither UN Peacekeeping?	13
	Conflict-related Sexual Violence	25
	The Islamic State's Shifting Narrative	39
	The Changing Foundations of Governance by Armed Groups	50
	Rebel-to-party Transitions	62
Chapter Two	**Maps, Graphics and Data**	**73**
	Territory lost by ISIS and operations against the group in 2016	
	Ten years of Mexico's 'war on drugs'	
	Distribution of highest reported level of rape during civil war	
	Refugee movements to selected non-Western countries	
	Global conflict fatalities 2016	
	Myanmar's newest insurgency	
Chapter Three	**Middle East**	**83**
	Egypt	83
	Iraq	91
	Israel–Palestine	100
	Lebanon–Hizbullah–Syria	107
	Libya	113
	Mali (The Sahel)	121
	Syria	128
	Turkey (PKK)	145
	Yemen	153
Chapter Four	**Sub-Saharan Africa**	**163**
	Central African Republic	163
	Democratic Republic of the Congo	169
	Ethiopia	180
	Nigeria (Boko Haram)	181
	Nigeria (Delta Region)	188
	Somalia	196
	South Sudan	202
	Sudan (Blue Nile, Darfur and South Kordofan)	209
Chapter Five	**South Asia**	**217**
	Afghanistan	217
	India (Assam)	232
	India (CPI–Maoist)	237
	India (Manipur)	244
	India (Nagaland)	248
	India–Pakistan (Kashmir)	253
	Pakistan	262

Chapter Six	**Asia-Pacific**	**279**
	China (Xinjiang)	279
	Myanmar	285
	Philippines (ASG)	292
	Philippines (MILF)	298
	Philippines (NPA)	304
	Southern Thailand	308
Chapter Seven	**Europe and Eurasia**	**315**
	Armenia–Azerbaijan (Nagorno-Karabakh)	315
	Russia (North Caucasus)	319
	Ukraine	321
Chapter Eight	**Latin America**	**327**
	Central America (Northern Triangle)	327
	Colombia	335
	Mexico	344
Chapter Nine	**Explanatory Notes**	**351**
	Index	**355**

Editor's Introduction

Fatalities in the world's conflicts declined for a second successive year in 2016, to 157,000, from 167,000 in 2015 and 180,000 in 2014. The war in Syria remained the world's most lethal, with a further 50,000 deaths there bringing the total since 2011 to around 290,000 – more than twice the number recorded in Bosnia's four-year fratricidal conflict in the 1990s. The wars in Iraq and Afghanistan claimed 17,000 and 16,000 lives respectively in 2016, although in lethality they were surpassed by conflicts in Mexico and Central America, which have received much less attention from the media and the international community. Mexico had the world's second-most-lethal conflict in 2016, with 23,000 fatalities. The number of homicides rose in 22 of Mexico's 32 states. The spike was linked to several factors. It is noteworthy that the largest rises in fatalities were registered in states that were key battlegrounds for control between competing, increasingly fragmented cartels. The violence grew worse as the cartels expanded the territorial reach of their campaigns, seeking to 'cleanse' areas of rivals in their efforts to secure a monopoly on drug-trafficking routes and other criminal assets. Amid growing security and governance vacuums, clashes among the cartels and between the cartels and state security forces became increasingly fierce and aggressive. Violence and repeated attacks resulting from such clashes and security voids greatly contributed to destabilisation across the country. Mexico's National Human Rights Commission highlighted the impact of crime on the population in the past decade, reporting in May 2016 that 35,433 people had been forcibly displaced nationwide since 2007. Around 90% of these people had fled their homes because of violence.

The combined total for Honduras, El Salvador and Guatemala was almost 16,000 – with El Salvador experiencing its second-most-violent year since 1999 – despite these countries establishing new agencies and approaches to reduce criminal groups' influence. The high number of fatalities reflected the significant presence, firepower and organisational capacity of rival gangs Mara Salvatrucha and Barrio 18. Collectively, these gangs, alongside smaller ones, had between 54,000 and 85,000 members spread across urban areas in Honduras, Guatemala and El Salvador, according to estimates made by the UN Office on Drugs and Crime and the US State Department in 2012. Regional cooperation was belatedly strengthened with the launch in November 2016 of a Tri-National Force against transnational organised crime, comprising around 1,500 personnel from the police and militaries, as well as the border and customs agencies, of the three countries. Yet it is unclear whether the force will be able to cope with patrolling 600 kilometres of shared land borders, especially given the long list of criminal activities they are tasked with suppressing: extortion, kidnapping, money laundering, gang violence and smuggling.

Although global conflict fatalities edged down in 2016, civilians caught amid conflict arguably suffered more than in the preceding years. Between January and August 2016, 900,000 people were internally displaced in Syria. The number of new internally displaced persons (IDPs) in Iraq and Afghanistan during the same period reached 234,000 and 260,000 respectively.

The latest figures on refugee movements to non-Western countries demonstrate that the long-term trend of refugees settling in urban centres is now being amplified by mass displacement due to the Syrian conflict. Around 90% of Syrian refugees have settled in urban and peri-urban centres in neighbouring countries, according to data published by the UN High Commissioner for Refugees. This trend marks an important shift away from the more traditional pattern of hosting refugees in designated refugee camps, frequently placed along border regions. It creates new challenges by placing unprecedented pressure on socio-economic infrastructure while increasing uncertainty for host communities, aid agencies and governments. Moreover, the trend complicates the provision of timely and equitable assistance, given that the majority of refugee communities are now more geographically dispersed.

Displacement rates have also been high in Sudan, where the UN Office for the Coordination of Humanitarian Affairs reported in June that 192,000 people had been displaced by fighting since the start of 2016 alone. Estimates suggested that by the end of 2016, there were 3 million registered and 5m–7m unregistered IDPs in Nigeria.

The capacity of international peacekeeping forces to effectively protect civilians seems to have diminished as a result of growing operational challenges, as well as threats to the security of their personnel. Indeed, 35 members of the UN Multidimensional Integrated Stabilization Mission in Mali (MINUSMA) were killed in 2016, making MINUSMA the most dangerous ongoing peacekeeping mission over the preceding three years, with 87 fatalities since the start of 2014. By comparison, the UN–African Union Mission in Darfur had incurred 27 fatalities since 2014. In the Central African Republic, there are signs that the public's attitude has turned against peacekeeping forces.

As displaced people have moved into cities, so too has conflict. Approximately half of the 36 conflicts featured in the *Armed Conflict Survey 2017*, and all of the high-intensity ones, have a significant urban component. If in the past the typical insurgent fought in the mountains, forest or jungle, today she or he is as likely to be found in an urban setting.

Turkey's conflict with the Kurdistan Workers' Party (PKK), for instance, became both more lethal and more urban in 2016. The number of recorded fatalities in the conflict reached 3,000, the highest level since 1997. The PKK and the government engaged in trench warfare in southern population centres. A surge in attacks on cities and public places put civilians in greater danger. Likewise, the conflict in South Sudan touched urban areas in the southern regions of the country. In Afghanistan, the Taliban extended its strategy of carrying out suicide attacks in urban centres. The group conducted one of the deadliest of these operations in

Kabul in April 2016, killing at least 64 people and injuring 347 others. Similarly, most of the highest-casualty attacks in Pakistan, many of them sectarian in nature, occurred in urban areas.

Siege warfare also played a more prominent role in 2016, as the Iraqi state and its allies recaptured cities from the Islamic State, also known as ISIS or ISIL, and took steps towards the objective of freeing Mosul. In Libya, the operation to drive ISIS out of Sirte led to prolonged urban warfare. And, in Syria, the siege of Aleppo took a decisive turn in favour of the forces of the president, Bashar al-Assad. The tactic proved cruel but effective, taking a heavy toll on civilians. According to an estimate by independent monitor Siege Watch, by the end of October 2016 at least 1.3m people were trapped in 39 besieged areas in Syria.

Lessons of the decline of ISIS

In 2016 ISIS lost nearly one-quarter of the territory that it controlled in Iraq and Syria. Seemingly, the group lost fighters at a still-faster rate. According to one estimate, in July 2016 the group had around 12,000 members, compared to the US Department of Defense's headcount of 31,500 in 2014. The group's efforts to build a 'caliphate' with the attributes of statehood appear doomed.

The losses suffered by ISIS underline the difficulties that any insurgent group faces in challenging a state conventionally, unless it has mass support and/or powerful external backers (such as diaspora communities or other states). In the last few decades, only a few insurgent groups stand out as victorious over a state adversary: those in Kosovo (and then only because of Western military intervention), South Sudan and Eritrea – although even they had a degree of external support. As a group with a defined territory and in possession of tanks and other conventional military platforms, ISIS presented an extensive target set for the coalitions ranged against it.

Coexisting and frequently competing with a state to provide services, exercise control and inspire loyalty among the population is also challenging. Sri Lanka's Liberation Tigers of Tamil Eelam are a well-documented example of a non-state armed group that comprehensively engaged in state-like governance for a limited time period. Yet despite running welfare, taxation and other services effectively, even this project cannot be considered fully successful as the group has not survived. In addition, securing the loyalty of the population requires robust social functions – a task ISIS has been and will be unable to perform in the long term, given its financial, administrative and other losses.

The shrinking of the caliphate does not, however, mean that the group is an irrelevance. Its losses and likely defeat have underscored the utility of insurgent and terrorist tactics for groups ranged against states. Furthermore, as groups fragment and multiply, the threat of brutality towards civilians increases. Wilayat Sina (Sinai Province), the Egyptian affiliate of ISIS, was the dominant anti-state force in the Sinai Peninsula in 2014–16, but in 2016 two new groups emerged with goals and rhetoric that differed quite radically from those of ISIS and its affiliates. New and particularly brutal militias have emerged in the Democratic Republic of the Congo and the Central African Republic. Islamist groups in the Sahel branched out in 2016, with one of them killing 19 people in an attack in the Ivorian town of Grand Bassam in March. In September, Burkina Faso became the site of the first attack by the Islamic State in the Greater Sahara. A new group that pledged allegiance to ISIS emerged in Somalia, carrying out its first attack in 2016. A Rohingya insurgent group emerged in Myanmar, carrying out several night-time raids. The frequent emergence of new militant groups in the Niger Delta region proved a major challenge for the military, largely due to a severe lack of intelligence on their capabilities and agendas.

Africa's mixed picture

Conflict fatalities in sub-Saharan Africa fell markedly in 2016, to 15,000 from 24,000 one year earlier. The decline was so sharp that the region recorded a lower fatality count than Central and Southern Asia for the first year since 2012. The main driver of the decline was a near-70% fall in fatalities relating to Nigeria-based insurgent group Boko Haram, which previously accounted for almost half of all conflict fatalities in sub-Saharan Africa. Among the areas affected by Boko Haram, Borno State continued to experience high-intensity conflict. Conflict intensity in other Nigerian states diminished, while the group's capacity to launch large-scale attacks in major cities decreased. Nevertheless, the group was able to perpetrate a series of attacks in Cameroon and the wider Lake Chad region, resulting in approximately 800 fatalities outside Nigeria. Despite the quantitative fall in fatalities, Boko Haram still had significant manpower; it continued to conduct hit-and-run assaults, raids and suicide attacks even in and around areas that had purportedly been cleared by the military, albeit at a significantly lower rate than in 2015. Militant cells remained in these territories, as remote terrain and a lack of personnel, materiel and surveillance capacity prevented the armed forces from holding and building there.

Nigeria's Delta region, however, saw a proliferation of attacks, predominantly against oil and gas pipelines, on a scale unseen since 2009. While the more than 100 attacks there in 2016 did not result in significant casualties, civilians frequently suffered human-rights abuses in the Nigerian military's security operations.

Conflict developments in eastern Africa were less positive. Fatalities rose in Sudan's internal conflicts, to around 3,350. As the under-reported war in South Kordofan and Blue Nile entered its fifth year, continued aerial attacks on civilians and severe fighting between the Sudan Armed Forces and armed groups caused hundreds of fatalities. Entering its thirteenth year, the conflict in Darfur generated persistent reports of killings, rapes and aerial attacks, especially in the Jebel Marra region. The violence followed an established pattern: fighting intensified in the dry season and lapsed in the rainy season, when it became difficult to travel on roads and in rural areas.

Meanwhile, the situation in South Sudan deteriorated significantly in the second half of 2016. Any cautious optimism about the implementation of the peace agreement signed in 2015 disappeared. There was a progressive polarisation among the population, largely incited by political leaders. This exacerbated insecurity and strengthened the pre-existing, if less visible, ethnic dimension of the conflict. State and non-state armed groups increasingly targeted civilians based on their ethnicity.

Al-Shabaab's battle for territory continued, particularly in the southern part of Somalia. Although the Somali government was largely able to maintain the pattern it had established in 2015 of driving al-Shabaab out of towns and cities while holding on to territory, progress was slow and sometimes only temporary. Even weakened, the group remained capable of conducting effective attacks and did not capitulate. Several times in the year, reports of al-Shabaab's imminent defeat circulated only for the government or the African Union Mission in Somalia to request thousands of additional troops to help in the fight shortly thereafter.

Many of the world's active conflicts have been running for decades. Colombia's conflict began in 1964, the Philippines' (with the New People's Army) in 1969, southern Sudan's in 1983 and Turkey's in 1984. India's internal conflicts, as well as Myanmar's and Eritrea's conflicts, have been running for several decades too. Some more recent conflicts, including those in Mali and Ukraine, seem to be developing the characteristics of protracted struggles. In Mali, Ukraine and the north of India, the failure to implement peace deals has not resulted in a significant increase in violence, but has weakened the prospects for the agreements to deliver lasting peace and stability. Although these conflicts underwent no major escalation in 2016, they have failed to move towards resolution due to the growing inability and, sometimes, apparent unwillingness of political actors to substantively address persistent rule-of-law and security problems. Moreover, they are best understood as 'simmering conflicts' rather than 'frozen conflicts', with potentially explosive characteristics such that violence could escalate at any moment.

Chapter One: **Thematic Essays**

Whither UN Peacekeeping?

Mats Berdal

United Nations peacekeeping – the deployment of military and police contingents, drawn from member states and authorised by the Security Council, to mitigate, contain and help create the conditions for overcoming violent and protracted conflict within the international system – has long been viewed as the UN's flagship activity, its most concrete and visible contribution to international peace and security. Measured in terms of overall troop numbers and money expended, that has not changed. Even as relations among the permanent members of the Security Council have deteriorated sharply in recent years, the demand for UN peacekeeping shows few signs of diminishing. The number of 'blue helmets' currently deployed worldwide – around 117,000, spread over 16 missions – remains close to an all-time high. With more than 125 countries contributing personnel, and an annual peacekeeping budget of almost US$8 billion (roughly triple the organisation's regular budget for 2017), UN peacekeeping appears firmly set to remain a core activity of the organisation. Lofty statements by government ministers have only reinforced that perception. At the conclusion of a 'UN Peacekeeping Defence Ministerial' held in London in September 2016, around 60 governments solemnly reaffirmed the 'indispensable part' and 'critical role that peacekeeping missions play to address today's international peace and security challenges'.[1]

Yet official statements and headline figures, while superficially impressive, conceal a much more troubling reality. Indeed, as António Guterres began his tenure as the ninth secretary-general of the UN in January 2017, the organisation's peacekeeping operations were faced with an unprecedented combination of mission-specific and wider systemic challenges. These problems were set against a backdrop of mounting geopolitical tension, rising nationalisms and growing disdain for multilateralism and values-driven diplomacy among key powers on the Security Council.

Peacekeeping malaise

Under-resourced, overextended and chronically short of specialist capacities in crucial areas such as engineering, communications, intelligence, aviation and logistics, UN peacekeepers are finding it progressively more difficult to meet rising expectations and reconcile multiple, frequently conflicting mandates handed down by the Security Council.

Along with other deficiencies that have historically bedevilled operations – from an absence of ready reserves and the highly uneven quality of troops provided by member states to antiquated and dysfunctional systems for financial and human-resources management – these impediments to operational effectiveness help explain why UN missions have always suffered from limited force cohesion, and have lacked anything resembling true unity of command. Over the past decade and a half, however, these deficiencies have had a more debilitating impact on mission performance.

In the past, as long as the conditions under which UN troops operated were relatively benign, it was possible to live with structural weaknesses. This was especially true when the deployment of peacekeepers followed rather than preceded a political settlement – or, at the very least, followed agreement on a workable ceasefire among belligerents, around which a settlement could then be negotiated. These kinds of conditions are now rare. Today, the majority of UN peacekeepers operate in what are euphemistically termed 'non-permissive' or 'high-tempo and insecure' environments.[2] The five largest and most troubled of the UN's missions – in the Democratic Republic of the Congo (DRC), Mali, Darfur, the Central African Republic (CAR) and South Sudan – are deployed in conditions of civil war, where power, resources and governmental authority are contested and, as a consequence, there is often very little peace to keep. Taken together, the five missions account for around 85,000 of the personnel currently serving under the UN flag. This growth of operations in 'situations of sustained armed conflict' also helps to explain why more than half of all UN peacekeeping fatalities since 1948 occurred in 2000–15.[3]

Amid such strains and stresses, Ban Ki-moon, then UN secretary-general, appointed in October 2014 a High-Level Independent Panel on UN Peace Operations (HIPPO) 'to take stock of evolving expectations of UN peacekeeping', producing the first report on the topic since Lakhdar Brahimi oversaw a similar effort in 2000.[4] Hard-hitting and constructive in equal measure, the final report paints a sobering, at times deeply dispiriting, picture of the multiple challenges facing ongoing operations. As with previous such reports, notably that overseen by Brahimi, the analysis of what ails the UN machinery for peacekeeping is hard to fault, and many of the concrete recommendations proposed by the panel – 166 in all – will, if implemented, undoubtedly enhance the UN's peacekeeping performance. As always, however, the main obstacles to UN reform are political in nature – and rarely stem from a paucity of good ideas. While good ideas should be, and are being, pursued, it is the HIPPO's larger and more important message that is in greatest need of attention from the UN's member states and Secretariat. Reflecting the experience of peacekeeping in civil-war-like situations over the past decade and a half, this message may be summed up as: peacekeeping in the absence of a viable political process is doomed to growing ineffectuality, and will over time contribute to a steady weakening of the UN's authority and leverage in a conflict. To understand how and why this dynamic has come about, and its implications for the future direction of UN field operations, it is necessary to analyse the impact of important changes in both the normative ambitions and practice of UN peacekeeping.

Rise of civilian protection and 'robust' peacekeeping
It might reasonably have been expected that the ignominious record of peacekeepers in Somalia, the former Yugoslavia and Rwanda in the early 1990s would

have permanently discredited UN peacekeeping as an instrument of third-party intervention in violent conflict. Although there was a period of retrenchment following the failures of the early post-Cold War era, 1999 saw the beginning of a dramatic growth in the quantity and scope of operations, with the number of blue helmets deployed rising from 20,000 to a new normal of well over 100,000. Nonetheless, these early disasters had an important legacy, one that has shaped nearly all of the 20 operations launched since 1999. The genocide in Rwanda in 1994 and, especially, the fall of the 'UN-protected' enclave of Srebrenica in 1995 fostered a 'never again' attitude among member states of the organisation. This, in turn, prompted a gradual but significant change in the normative expectations surrounding UN peacekeeping, while also stimulating a rethink about the long-established peacekeeping principles of consent, impartiality and the minimal use of force (except in self-defence).

When demand for UN operations picked up again after a hiatus of four years, the change in normative context was evident in the new importance that the Security Council formally attached to the protection of civilians (POC) in armed conflict. The first peacekeeping mission to be given an explicit mandate to protect civilians 'under imminent threat of physical violence' was established for Sierra Leone in October 1999.[5] Since then, POC has evolved from being an ancillary practice to a core task, and is now the *raison d'être* for almost all of the UN's current operations – including the missions in the DRC, Darfur, Mali, the CAR and South Sudan.[6]

The growing importance attached to the POC responsibilities of peacekeepers is inextricably linked to a second key feature of post-Brahimi Report peacekeeping: the shift towards more 'muscular' or 'robust' peacekeeping. Operating in insecure environments and charged with POC tasks, the Security Council has routinely given UN peacekeepers the authority – under the enforcement provisions of Chapter VII of the UN Charter – to 'use all necessary means' or 'take the necessary action' to accomplish their mission. In places such as Haiti and, above all, the DRC, UN forces have actively exercised that authority by undertaking high-intensity military operations – sometimes on their own, but more often in support of local security partners. The shift towards more robust peacekeeping culminated in March 2013, when the Security Council, following the humiliating fall of Goma in eastern DRC to the Rwanda-backed rebel group M23, strengthened the UN Stabilization Mission in the DRC (MONUSCO) by creating a Force Intervention Brigade (FIB). The FIB's mandate was 'to carry out targeted offensive operations … in a robust, highly mobile and versatile manner'.⁷ Ban described the decision as a 'milestone' in the evolution of UN peacekeeping.⁸ A similar, though yet to be deployed, Regional Protection Force (RPF) has been mandated for South Sudan – there, too, in an effort to strengthen civilian protection against the backdrop of an ongoing civil war.

The protection of civilians by UN peacekeepers, authorised to act robustly to this end by the Security Council, has proved a daunting task; the UN's record in the area is both bleak and discouraging, particularly in the DRC and South Sudan.

The FIB's initial successes in 2013, which led to the removal of M23 from eastern DRC, were soon nullified by the lack of a sustained and coordinated national or international effort to address the political issues at the heart of the conflict. At the same time, military action against the more than 50 other armed groups in the eastern part of the DRC quickly petered out following the apparent defeat of M23. Together, these shortfalls help explain why the number of internally displaced persons in the DRC has reached 1.7 million and is rising once again. They also explain why a Security Council delegation visiting eastern DRC in late 2016 was presented with 'an alarming picture of rampant victimisation of the civilian population by armed groups and human rights violations, including killings, sexual violence, destruction of schools and hospitals, and displacement'.⁹

Since the outbreak of civil war in December 2013, the situation in South Sudan has been bleaker still. The UN Mission in South Sudan (UNMISS) – which, even by UN standards, is critically hamstrung by acute logistical challenges and limited capabilities – has faced repeated protection crises since the conflict began. Indeed, in February and June 2016, government forces and government-backed militias deliberately targeted POC sites and UN compounds in Malakal and Juba respectively. In both these cases, UN-appointed inquiries harshly criticised the 'risk-averse' posture of peacekeepers in response to the attacks but, paradoxically, also insisted that 'protecting the POC sites … is beyond the capability of UNMISS or any peacekeeping mission, and a task that raises unreasonable expectations'.[10] The incoherence of the UN's mission and mandate in South Sudan – the adverse operational consequences of which have been aggravated by a stalled peace process – has resulted in a steadily worsening situation on the ground. In late 2016, the Special Commission on Human Rights in South Sudan even raised the possibility that the conflict there could lead to a 'Rwanda-like' genocide.[11]

While the UN's attempts to protect civilians in the DRC and South Sudan have been particularly problematic, its experiences there only serve to highlight more fundamental challenges encountered in the last decade and a half of peacekeeping in civil-war-like situations. As the UN assesses the future direction of peacekeeping operations under a new secretary-general – especially POC responsibilities and the role of force in discharging those responsibilities – two broad lessons will need to be absorbed.

Firstly, forces under UN command have proved structurally and politically ill-suited to conducting robust, let alone offensive, military operations in civil-war-like settings. As the HIPPO rightly emphasises and as recent experience affirms, 'there are outer limits for UN peacekeeping operations defined by their composition, character and inherent capability limitations'.[12] There is a place for forceful action by UN peacekeepers in some circumstances, such as when they are confronted with the prospect of mass atrocities and humanitarian disaster. But on the few occasions when the decisive use of force resulted in what might be described as a tactical success – as seen when the French-led Interim Multinational Emergency Force (IMEF) intervened to avert mass atrocities in Bunia, in eastern DRC, in 2003, or when Brazilian-led troops dismantled criminal-gang structures in Haiti in 2006–07 – the troops involved were well trained, properly resourced and subject to effective command and control.

However, such successes have been, and will likely continue to be, the exception rather than the norm. Since the drawdown of Western troops from Afghanistan in 2014, several traditional troops-contributing countries (TCCs) – including Canada, the United Kingdom, Norway, Sweden, the Netherlands and Denmark – have hinted at their return to UN peacekeeping. This has raised the hope that together with the US and other countries, they would offer 'niche' capabilities to plug gaps in peacekeeping performance.[13] Yet such promises have yet to move beyond tentative, largely symbolic, first steps. And the arrival of a new administration in Washington has firmly put paid to any hopes for a more active US role in UN peacekeeping along the lines suggested by then-president Barack Obama at a Leaders' Summit on Peacekeeping in September 2015. These factors ensure that the bulk of UN peacekeepers will for the foreseeable future continue to be provided by TCCs from developing countries – most prominently Pakistan, India, Bangladesh and Ethiopia.[14] The individual and unit-level soldiering skills of peacekeepers provided by these countries have often been very good, but their continuing lack of enabling assets (especially in air mobility) limits their effectiveness. Moreover, most of these now-dominant TCCs have also been reluctant, for principled and prudential reasons, to use force more robustly in peacekeeping, including in the protection of civilians. At the Leaders' Summit on Peacekeeping in 2015, China, which deployed an infantry battalion to South Sudan in 2014, announced that it was creating an 8,000-strong 'standby force' for UN peacekeeping operations. The offer, however, was noticeably short on detail, and it made no mention of the conditions that would attach to the deployment of Chinese personnel. Member states have a long history of offering 'standby forces' for UN peacekeeping; making troops available when they are actually needed has always proved a greater challenge.

Yet the quality of troops, risk aversion among TCCs and a lack of capabilities reflect only one aspect of the challenges that have been exposed by current operations. Another factor has been equally, if not more, important: even where well-prepared peacekeeping forces defeated much weaker opponents, the utility of force could not compensate for the lack of a political strategy that addressed the underlying drivers of conflict. The reason for this stems, in part, from the inherent difficulties encountered by the UN's intergovernmental machinery and bureaucratically fragmented structures when asked to act strategically – in this case, by linking military action on the ground to wider political purposes.

This speaks to a deeper issue revealed by post-Brahimi Report peacekeeping. When UN peacekeeping missions have deployed in civil-war-like situations, they have been unable to distance themselves from the messy and frequently violent politics of the conflict. The problem is especially acute where they are entrusted with a broad, intrusive and partly conflicting mandate, ranging from disarmament, demobilisation, reintegration, security-sector reform and civilian protection to monitoring the human-rights record of the government and its opponents. A UN peacekeeping mission must deal with – and through its presence is meant to bolster – the recognised government, which has given its consent to the mission (however grudgingly). But it is in the nature of civil-war-like situations that the legitimacy of the government is often contested and tenuous – a fact that inevitably shapes perceptions of the UN's role and that of its peacekeepers on the ground. When host governments and their security forces directly threaten civilians – as occurs in South Sudan and the DRC – they undermine the legitimacy of the UN presence in their countries. Such activity increasingly compromises the UN's ability to steer the conflict towards a lasting settlement through mediation and political engagement with all belligerents, damaging the organisation's power to cajole, convene and influence. This is critical because, in the long term, true protection of civilians can only emerge from an agreement on a political settlement. The continuing absence of such a settlement drives violence and insecurity, and often wrong-foots and compromises overextended and under-resourced blue helmets left to keep the peace in a political vacuum.

Of course, this does not mean that UN peacekeepers will be left without challenges or that they should always assume a passive posture. Judgements about the use of force will always prove difficult, and no amount of time spent on refining doctrine or improving tactical guidance will change that. Thus, in any peacekeeping operation, the quality of mission leadership is vital to its success. Even so, enforcement and war fighting can never be the main business of blue helmets. As Jean-Marie Guéhenno, head of UN peacekeeping from 2000 to 2008, has perceptively observed:

> The decisive use of force may help prevent marginal actors from derailing a peace process, and may contribute to the credibility of an international deployment by creating leverage that can then be used to influence and even shape a political process. But in the end, the critical element is the political foundation of peace.[15]

UN peacekeeping under Guterres

Any assessment of the prospects for UN peacekeeping under Guterres must factor in the likely impact of geopolitical change, rising nationalisms and the reassertion of power politics on the workings of the UN – specifically on the Security Council, the intergovernmental body that authorises, oversees and, in theory, provides strategic direction for UN operations.

The steady deterioration of relations between Russia and the Western powers on the Security Council has been apparent since Kofi Annan's 2001–06 term as secretary-general, if not earlier. However, it accelerated markedly under Ban, particularly following Russia's illegal annexation of Crimea in March 2014. Although this rising tension has not prevented the launch of new peacekeeping missions – a 12,000-strong deployment to the CAR was unanimously agreed on less than a month after the takeover of Crimea – Security Council politics are having an increasingly negative impact on field operations. Given that, historically, the relative success of UN peacekeeping operations correlated closely with the degree of Security Council unity and support, there are strong grounds for pessimism about the future effectiveness of the missions.

Furthermore, there continues to be a wide gap between declared and actual commitments to strengthen peacekeeping (with money, troops and specialist capabilities) among traditional troop-contributors and self-professed supporters of UN peacekeeping. As one long-time, perceptive observer of the UN scene has pointedly noted, the European Union's much-vaunted 'Battle Groups' – once touted as an obvious force multiplier for UN operations – have shown themselves to be a 'phantom strategic reserve … [of] rapid-reaction forces that are meant to aid the UN but never go into action'.[16] Concern about diminishing commitment from key UN member states will only have been strengthened by US President Donald Trump's overt disavowal of multilateralism, as well as his penchant for deal-making and transactional diplomacy.

In view of the challenges and uncertainties surrounding UN peacekeeping, what should be the priorities of the UN Secretariat and its new, much-respected secretary-general?

Thant Myint-U, grandson of former secretary-general U Thant and himself an experienced UN hand, has argued that, 'though peacekeeping has been a core Secretariat focus since the early 1990s … there is nothing to say that this needs to remain the case'.[17] This might seem a heretical thought to practitioners, analysts and long-term observers who have invested in the growth of UN peacekeeping after the Cold War, focusing on how to improve its effectiveness but never questioning its status as the organisation's central activity in peace and security. Reflecting on the political and structural barriers facing the UN as it operates in modern conflict zones, Myint-U contends further that the Secretariat's 'future advantages … may be in the mediation field, an area of work that has been relatively neglected'.[18] It is a thought that merits careful consideration. Former secretary-general Dag Hammarskjöld, who first gave intellectual coherence to the idea of peacekeeping, always stressed the limitations of the practice, seeing it as the UN's distinctive response to the 'ideological conflicts and the conflicts of power' dividing the world of his day.[19]

Today, addressing the limits of peacekeeping – highlighted by recurring protection crises and deep rifts among Security Council members – requires a sustained, systematic focus on the ways in which the activities of peacekeepers on the ground can reinforce the search for a political settlement. Blue helmets cannot impose political solutions to complex and deep-seated conflicts, but they can and should underpin and support progress towards solutions. In the words of the HIPPO, 'any peacekeeping operation must be a part of a robust political process in which the UN is deeply involved, and must continuously seek to build consent to the UN role and presence through an impartial posture'.[20] This will require greater realism about the tasks that it is appropriate to give peacekeepers, avoiding the current 'Christmas tree' approach to mandating peacekeeping operations – that is, the tendency to entrust missions with a laundry list of tasks without regard to prioritisation, sequencing or resource implications. It will also require the secretary-general to strengthen – through personal diplomacy, investment and prioritisation within the Secretariat – the organisation's capacity for political engagement and mediation in what remains a depressingly wide range of armed conflicts.

Mats Berdal is Professor of Security and Development and Director of the Conflict, Security and Development Research Group at King's College London. He was Director of Studies at the IISS from 2000 to 2003.

Notes

[1] UN, 'UN Peacekeeping Defence Ministerial: London Communiqué', 8 September 2016.

[2] High-Level Independent Panel on United Nations Operations, 'Uniting Our Strengths for Peace – Politics, Partnerships and People', 16 June 2015, paragraph 210.

[3] *Ibid.*, paragraph 115.

[4] UN, 'Report of the Panel on Peace Operations', A/55/305-S/2000/809, 21 August 2000.

[5] UN, S/RES/1270, 22 October 1999.

[6] The 20,000-strong mission in Darfur is a joint African Union/UN hybrid operation (UNAMID), and has POC as its core mandate.

[7] UN, S/RES/2098, 28 March 2013.

[8] UN, 'Secretary-General's Remarks at Security Council Debate', 11 June 2014.

[9] UN, 'Dispatches from the Field: Council Meetings in Beni', What's in Blue?, 14 November 2016.

[10] UN, 'Statement Attributable to the Spokesman for the Secretary-General on South Sudan', 1 November 2016, New York.

[11] UN, 'Statement by Yasmin Sooka, Chair of the Commission on Human Rights in South Sudan', 26th Special Session of the UN Human Rights Council, 14 December 2016.

[12] High-Level Independent Panel on United Nations Operations, 'Uniting Our Strengths for Peace', p. x.

[13] UN, 'Declaration of Leaders' Summit on Peacekeeping', 28 September 2015.

[14] As of December 2016, these four countries provided roughly 30,000 of the military and police personnel on peacekeeping duty.

15 Jean-Marie Guéhenno, *The Fog of Peace: A Memoir of Internal Peacekeeping in the 21st Century* (Washington DC: Brookings Institution Press, 2015), p. 262.

16 Richard Gowan, 'Red China's New Blue Helmets', 30 September 2015, Brookings Institution, https://www.brookings.edu/blog/order-from-chaos/2015/09/30/red-chinas-new-blue-helmets/.

17 Thant Myint-U, 'The Next Secretary-General, Secretariat Reform and the Vexed Question of Senior Appointments', Center on International Cooperation, April 2016.

18 *Ibid*.

19 UN, 'Introduction to the Annual Report, June 1959–June 1960', A/4390/Add.1.

20 High-Level Independent Panel on United Nations Operations, 'Uniting Our Strengths for Peace', paragraph 124.

Conflict-related Sexual Violence

Elisabeth Jean Wood and Julia Bleckner

Scholars, advocates and policymakers now understand that sexual violence against civilians is not an inevitable by-product of war. Rather, it varies widely in form, frequency and targeting – not only across conflicts but also among armed actors within conflicts.[1] Some armed actors effectively prohibit their members from engaging in sexual violence against civilians; some adopt rape or different forms of sexual violence as organisational policy (sometimes for military purposes); others simply tolerate its occurrence.

This essay assesses recent research on conflict-related sexual violence (CRSV), defined here as sexual violence by armed actors during conflict.[2] The first section focuses on the well-documented variation in its form (such as forced abortion, sexual slavery and forced marriage), frequency (including the near-absence of rape by some organisations) and targeting, which may include boys and men as well as women and girls, and is often based on ethnic or political identity.

The second section of the essay discusses new approaches to understanding the complex variation in patterns of CRSV. There is evidence that some armed actors purposefully adopt rape or other forms of sexual violence for military purposes. But others adopt sexual slavery, forced marriage or forced abortion as a policy for other reasons – often to manage the sexual and reproductive lives of their members. When a form of sexual violence is adopted as organisational policy, it may be authorised under some conditions but not explicitly ordered. Moreover, sexual violence can be frequent without having been explicitly adopted by the organisation.

The essay concludes by identifying some implications for policy both during and after conflict.

Variation and complexity in conflict-related sexual violence

Frequency

In some conflicts such as the ongoing war in South Sudan, sexual violence is common, widespread and perpetrated by almost all conflict actors. Since December 2013, factions led by President Salva Kiir and former vice-president Riek Machar have conducted increasingly violent attacks on civilians as they vie for control of the South Sudanese government. Both forces often engage in rape, much of which is ethnically targeted.[3] During April–September 2015, the United Nations recorded more than 1,300 cases of rape in Unity State alone;[4] according to a UN survey carried out in late 2016, 70% of women in South Sudan's capital, Juba, had been sexually assaulted since the beginning of the conflict.[5]

While many armed organisations engage in extensive sexual violence, not all have been documented as doing so: more than half of the armed actors in the civil wars that took place in 20 African countries between 2000 and 2009 were not reported to have engaged in rape or other forms of sexual violence.[6] Of course, there is severe under-reporting of CRSV in many contexts; however, these data were collected after human-rights and women's organisations had begun to actively document rape and other forms of CRSV.[7] While under-reporting undoubtedly continues, there are significant documented differences across armed organisations.

Given the apparently low level of rape by some armed actors, scholars increasingly focus on specific armed organisations rather than the wider conflict, investigating which organisations engage in CRSV, and in what form. This approach has brought to light an asymmetry that has important implications for analysis and policy. In a study of all civil wars between 1980 and 2009, Dara Kay Cohen documented variation in rape across actors in civil conflicts. She found that in 38% of conflicts in which rape occurred moderately or very frequently, it was reportedly committed by one party but not the other. Furthermore, in 80% of these asymmetric cases, it was state forces, not rebels, who committed moderate to high levels of rape.[8]

Form

Armed actors engage in different forms of sexual violence – including rape, gang rape, sexual torture, sexual slavery, non-penetrating assault and sterilisation, as well as forced abortion, prostitution and pregnancy – to highly varying degrees. Some organisations, such as Serb forces in the Bosnian conflict in the 1990s, have participated in many forms of CRSV, while others have perpetrated a much smaller range of these crimes (though not necessarily less often).

When wartime rape is frequent, the proportion of rape carried out by multiple perpetrators appears to be significantly higher than that during peacetime.[9] In the conflicts in Sierra Leone and eastern Democratic Republic of the Congo (DRC) – two of the best-documented cases – around 75% of conflict-related rapes of women were carried out by more than one perpetrator, in sharp contrast to the 5–20% typical in peacetime.[10] The office of the UN High Commissioner for Human Rights reported that multiple-perpetrator rape is highly prevalent in South Sudan. One witness told investigators that 'if you looked young or good looking, about ten men would rape the woman; the older women were raped by about seven to nine men.'[11] Rebels that abduct recruits and states that press-gang recruits (rather than merely conscripting them) are particularly likely to engage in frequent gang rape.[12]

Some armed groups such as the Islamic State, also known as ISIS or ISIL, perpetrate CRSV in the form of sexual slavery and/or coerced or forced marriage. During the peak of the Lord's Resistance Army's (LRA's) control of northern Uganda, the group engaged in widespread violence against civilians to punish them for, or deter them from, supporting the government. The organisation also forcibly recruited children on a massive scale, sometimes coercing them to kill their families.[13] The organisation's repertoire of sexual violence was largely dominated by forced marriage or sexual enslavement of women and girls, many of whom had been forcibly recruited.[14] Some abducted girls were given to fighters as rewards; resistance was severely punished with rape, beatings or other forms of torture, or execution.[15] The LRA strictly controlled the reproductive lives of its sex slaves, pressing 'wives' to bear children for the LRA and inflicting 'medical treatment' on those who did not.[16] Some organisations, such as Colombian insurgent group FARC, force them to use contraception and to have an abortion if they do become pregnant.[17]

State actors are especially likely to engage in CRSV against detained enemy combatants and/or political prisoners. For instance, the US Senate's 2014 report on torture revealed that CIA officers had subjected male detainees to forced nudity, 'rectal feeding', sexual threats and other forms of sexual humiliation.[18] Human Rights Watch reported in 2012 that the Syrian government sexually abused and tortured some male and female detainees, often those held for their political activism. The techniques it used included rape, penetration with objects, sexual groping and prolonged forced nudity, as well as the delivery of electric shocks and beatings to genitals.[19] During and after the civil war in Sri Lanka, government forces detained Tamil men and women for allegedly supporting the Liberation Tigers of Tamil Eelam. Detainees reported rape, gang rape and sexual torture; some said that they had been hung naked from the ceiling as uniformed Sri Lankan military and police officers beat them and applied chilli powder to their genitals.[20]

In some conflicts, such as that between the state and various ethnic armed groups in Myanmar, sexual violence has occurred within the context of forced labour. Having escaped or been released by their captors, civilian women from Kachin, Shan, Mon and Chin states reported that government soldiers gang raped or repeatedly raped them while they were forced to transport goods or perform other work for the military.[21]

Targeting

Armed organisations may indiscriminately target their victims or choose them based on their behaviour, such as their support for a rival group, or their identity, such as their membership of an ethnic, religious, political or community group thought to represent or back an opponent. In South Sudan, Kiir's and Machar's armed factions are based on distinct ethnic identities (Dinka and Nuer respectively), and the groups attack each other's ethnic bases, targeting civilians for their supposed support of the rival force. In July 2016, Kiir's army 'purged' the city of Wau based on the belief that the Fertit people there were supporting Machar. Sexual violence, particularly gang rape, has been a feature of these sorts of purges.[22] In Myanmar's Mon State, some women reported being tortured and sexually abused for their alleged links to a rebel group.[23] ISIS engages in different forms of CRSV against different ethnic groups, targeting girls and women with sexual slavery if they are Yazidi and with coerced marriage if they are Sunni Muslims (as defined by the organisation).[24]

Some armed groups only target females, while others also target males – as was the case in the immediate aftermath of the war in Sri Lanka, when government forces specifically targeted and raped Tamil men as well as women.[25] In a population-based survey of conflict-affected regions in eastern DRC, around 15% of men and 30% of women reported having suffered some form of CRSV.[26]

Advances in understanding conflict-related sexual violence

The accumulating evidence on CRSV has formed the basis for a more nuanced understanding of how, when and why different patterns of sexual violence emerge during war. These advances have led both the academic and policy communities to reconsider previous assumptions about the nature of CRSV. Given the asymmetry in the frequency of rape in civil conflict, CRSV cannot simply reflect the gender relations of the patriarchal culture of the society in which a war occurs. Some armed organisations radically reshape the behaviour (and norms) of their recruits – sometimes towards more frequent rape of a wider range of targets, but sometimes towards significantly less rape.[27] Indeed, a few organisations effectively prohibit rape by civilians (including partners), as well as by their members, in areas they control. However, the pattern of CRSV may reflect peacetime gender relations if the organisation accepts, promotes or simply ignores existing patriarchal norms.

To account for the complex variation in patterns of CRSV, it is important to identify and analyse differences in policies across organisations. Firstly, scholars and policymakers increasingly recognise that frequent rape by members of an armed organisation is not necessarily part of a strategy.[28] Rather, where commanders do not order or authorise rape but nonetheless allow their subordinates to engage in it, the crime may occur frequently if these subordinates' private preferences or social dynamics (the norms and incentives governing the interactions of the group's members) support sexually abusive violence. In such cases, rape is best understood as a practice, not a strategy or other type of policy purposefully adopted by the organisation.[29] For example, US soldiers engaged in the murder and rape of civilians in Vietnam with near-complete impunity. Those who testified about atrocities at informal war-crimes tribunals claimed that they had been authorised to kill civilians but did not make the same claim of rape, indicating that rape occurred as a practice.[30]

Where the relevant norms and incentives of both commanders and their subordinates reflect society's highly unequal gender norms, these norms can foster a view among combatants and civilians that men are entitled to sexual access to girls and women, or that the rape of women and girls causes no harm because its victims are of little worth. Behaviour may also reflect specific organisational norms, such as perceiving sexual access to women and girls as the 'spoils of war'. For example, soldiers in the DRC military see the 'lust' rape of women – rape involving forced sexual intercourse born out of frustration (as opposed to 'evil' rape, which includes mutilation and other violence) – as the compensation they are owed for their service, in place of what they consider to be adequate pay to support a family.[31] In a UN report published in March 2016, investigators noted indications that government-allied militia groups in South Sudan were being 'allowed to rape women in lieu of wages',[32] suggesting that rape was evolving from a practice to a policy. Cohen argues that in many settings, frequent gang rape occurs not as a purposefully adopted policy but because it improves cohesion in organisations that rely on forced recruitment (and are thus composed of hostile, bewildered recruits).[33]

Secondly, an organisation may purposefully adopt a form of sexual violence as policy in pursuit of organisational objectives but not immediate military goals – as happens when it seeks to manage the sexual and reproductive lives of its members. For example, while still primarily operating in Uganda, the LRA regulated and monitored compliance with its rules of forced marriage,[34] strengthening its strict code of conduct as a systematic mechanism of control.[35] As argued by Christopher Blattman and Jeannie Annan, the organisation's rituals and beliefs were 'a clear attempt to create new social bonds and loyalty based on a shared cosmology (as well as fear)'.[36] In this context, forced marriage forms part of a policy with institutionalised guidelines for its implementation. Similarly, ISIS regulates the sexual slavery of Yazidi women with explicit proclamations that state in elaborate detail the conditions under which sexual relations with slaves are acceptable.[37] The organisation also occasionally punishes its members for raping Muslim women in a way that breaks its regulations.[38]

Thirdly, commanders may authorise sexual violence as part of organisational policy without directly ordering it (as in the case of ISIS and sexual slavery). States may also provide such authorisation without direct orders. For example, US leaders appear to have authorised but not directly ordered sexual torture and humiliation as part of the interrogation of detainees in the 'war on terror'. The 2005 Bradbury memos explicitly authorised forced nudity as an accepted form of so-called 'enhanced interrogation',[39] essentially authorising sexual humiliation as a torture technique, particularly when combined with other authorised methods of enhanced interrogation.[40] Sexual violence – including the rape of male detainees with broomsticks or rifles, and beatings to the genitals – also occurred with the knowledge of commanding officers.[41]

Fourthly, whether rape is adopted as a policy or emerges as a practice, perpetrators often benefit from institutionalised impunity. In the case of the Myanmar military discussed above, women are encouraged or even threatened not to report sexual assault. After two young ethnic Kachin teachers were raped and murdered at a dormitory in Kaung Kha in Shan State in January 2015, allegations that the principal suspects were soldiers prompted the military-run news outlet Myawaddy to publish a statement threatening legal action against anyone who suggested that army personnel were responsible.[42] Human-rights groups continue to advocate for amendments to the parts of Myanmar's 2008 constitution that essentially institutionalise the immunity of the military, even for violations of international law.[43]

Policy implications

That rape and other forms of sexual violence are not inevitable in war but perpetrated only by some armed actors strengthens the grounds for holding them accountable. Understanding variation in the frequency, form and targeting of CRSV has important implications for designing policy for prevention, prosecution, peacekeeping, peacebuilding and transitional justice.

As there is evidence that members of armed groups can frequently engage in rape while their commanders merely tolerate its occurrence, prosecution under command responsibility could be the most effective legal approach to tackling CRSV in such cases.[44] Under this concept, a commander can be held liable for a crime if she or he knew, or had reason to know, that subordinates over whom she or he had effective control had engaged (or would soon engage) in sexual crimes, but failed to take all necessary and reasonable measures in their power to punish the subordinates or prevent the crime.[45] The International Criminal Court recently used this legal instrument to convict Jean-Pierre Bemba Gombo for rape as a crime against humanity and a war crime, without the need for evidence that he had ordered or authorised rape.[46]

And where rape is organisational policy, it may be authorised rather than ordered. Prosecutors may find it easier to demonstrate that commanders authorised the crime than to show that they had ordered it.[47] Other forms of CRSV may also be organisational policy, such as when commanders authorise or order sexual slavery, forced marriage or sexual torture.

These recent advances in the understanding of CRSV also contribute to the development of policy to address the issue *during* war.[48] Policies should be informed by a nuanced conception of gender in which CRSV is not a 'women's issue' but one driven by an armed organisation's norms and beliefs concerning gender, which may not reflect the local society's gender relations. Organisational norms, beliefs and rules may licence sexual violence (including in novel forms) not only against women and girls but also against men, boys and sexual minorities. In other cases, those norms, beliefs and rules effectively prevent sexual violence by the organisation's forces.

Policies will be more effective if they are tailored to address an armed organisation's specific pattern of CRSV, particularly its repertoire and targeting. Measures to address CRSV that is part of organisational policy will generally differ from those to address rape that is tolerated but not authorised or ordered. More specifically, policymakers should be aware of settings in which CRSV is particularly likely to occur. For instance, it is important to acknowledge that there is an increased risk of gang rape by non-state actors who recruit by abduction and state forces that do so by press-ganging; sexual torture where other forms of torture and abuse (such as forced labour) occur; and forced contraception and abortion within organisations that are particularly dependent on female combatants. However, persuading commanders to prohibit forms of CRSV that had hitherto been tolerated will be effective only in organisations with strong institutions capable of enforcing prohibition through ongoing education or strong disciplinary measures.

Understanding the institutional contexts in which commanders tolerate CRSV should also help policymakers address sexual violence by other actors, such as UN peacekeepers – who have rarely been punished for sexual crimes, as part of a pattern of institutionalised impunity reflecting the sovereignty of troop-providing states. Documenting all forms of CRSV by all actors in a conflict and the social groups they target contributes to peacebuilding and transitional justice, as it helps ensure that victims are recognised and perpetrators held accountable. While girls and women often bear the brunt of such violence, it is important to recognise that men, boys and sexual minorities may also be targets. By building on these advances to improve the documentation and analysis of

CRSV, advocates and policymakers can more effectively deter, intervene against and prosecute such violence.

Elisabeth Jean Wood is Professor of Political Science, International and Area Studies at Yale University and a member of the External Faculty of the Santa Fe Institute. She is currently working on a book on sexual violence during war. More information can be found on her personal website at http://campuspress.yale.edu/elisabethwood/.

Julia Bleckner is a PhD student in comparative politics at Yale University. Previously, she worked as a Senior Asia Research Associate at Human Rights Watch, where she focused on torture and sexual violence, and as a Fulbright Research Fellow in Bangladesh, where she studied the efficacy of all-female peacekeeping units in deterring sexual violence in UN missions.

Notes

[1] Dara Kay Cohen, 'Explaining Rape during Civil War: Cross-National Evidence (1980–2009)', *American Political Science Review*, vol. 107, no. 3, 2013, pp. 461–77; Dara Kay Cohen, *Rape during Civil War* (Ithaca, NY: Cornell University Press, 2016); Elisabeth Jean Wood, 'Conflict-related Sexual Violence and the Policy Implications of Recent Research', *International Review of the Red Cross*, vol. 96, no. 894, 2015, pp. 457–78; Ragnhild Nordas, 'Sexual Violence in African Conflicts', PRIO, 2011.

[2] Sexual violence perpetrated by civilians usually continues, and may increase, during conflict. Indeed, in some settings, it appears to be more frequent than that by armed actors. See Wood, 'Conflict-related Sexual Violence and the Policy Implications of Recent Research'. Sexual violence is sometimes committed by actors who are not active parties to the conflict. For example, UN peacekeepers have engaged in sexual assault, rape, sex trafficking, organised prostitution and abduction. See Julia Bleckner, 'From Rhetoric to Reality: A Pragmatic Analysis of the Integration of Women into UN Peacekeeping Operations', *Journal of International Peacekeeping*, vol. 17, no. 3–4, 2013, pp. 337–60. The UN is currently investigating more than 100 cases of sexual violence by its peacekeepers deployed in Central African Republic's Kemo province in 2014 and 2015. See Sandra Laville, 'UN Inquiry into CAR Abuse Claims Identifies 41 Troops as Suspects', *Guardian*, 5 December 2016, https://www.theguardian.com/world/2016/dec/05/un-inquiry-into-car-abuse-claims-identifies-41-troops-as-suspects. For a discussion of the relationship between peacetime sexual violence and that during war, see Wood, 'Conflict-related Sexual Violence and the Policy Implications of Recent Research'; and Christopher Butler and Jessica Jones, 'Sexual Violence by Government Security Forces: Can Peacetime Levels of Sexual Violence Predict Levels of Sexual Violence in Civil Conflict?', *International Area Studies Review*, vol. 19, no. 3, 2016.

[3] UN Human Rights Council, 'UN Experts Call for UN Special Investigation into Epic

Levels of Sexual Violence in South Sudan', 2 December 2016, http://reliefweb.int/report/south-sudan/un-experts-call-un-special-investigation-epic-levels-sexual-violence-south-sudan.

4   Office of the United Nations High Commissioner for Human Rights, 'Assessment Mission by the Office of the United Nations High Commissioner for Human Rights to Improve Human Rights, Accountability, Reconciliation and Capacity in South Sudan', A/HRC/31/49, 10 March 2016.

5   UN Human Rights Council, 'UN Experts Call for UN Special Investigation into Epic Levels of Sexual Violence in South Sudan'; Office of the United Nations High Commissioner for Human Rights, 'Assessment Mission by the Office of the United Nations High Commissioner for Human Rights to Improve Human Rights, Accountability, Reconciliation and Capacity in South Sudan'.

6   Nordas, 'Sexual Violence in African Conflicts'.

7   Wood, 'Conflict-related Sexual Violence and the Policy Implications of Recent Research', pp. 457–78.

8   Cohen, 'Explaining Rape during Civil War', p. 472.

9   Elisabeth Wood, 'Multiple Perpetrator Rape during War', in Miranda A.H. Horvath and Jessica Woodhams (eds), *Handbook on the Study of Multiple Perpetrator Rape: A Multidisciplinary Response to an International Problem* (Abingdon: Routledge, 2013); Cohen, 'Explaining Rape during Civil War', pp. 461–77.

10 Wood, 'Multiple Perpetrator Rape during War'. One exception is South Africa, where the fraction of rapes by multiple perpetrators is significantly higher in peacetime. See Theresa da Silva, Leigh Harkins and Jessica Woodhams, 'Multiple Perpetrator Rape as an International Phenomenon', in Horvath and Woodhams (eds), *Handbook on the Study of Multiple Perpetrator Rape*.

11 Office of the United Nations High Commissioner for Human Rights, 'South Sudan: UN Report Contains "Searing" Account of Killings, Rapes and Destruction', 11 March 2016, https://www.ohchr.org/EN/NewsEvents/Pages/DisplayNews.aspx?NewsID=17207&angID=E.

12 Cohen, 'Explaining Rape during Civil War', pp. 461–77; Wood, 'Multiple Perpetrator Rape during War'; Cohen, *Rape during Civil War*.

13 Since the situation was referred to the International Criminal Court in 2004, LRA leader Joseph Kony has been indicted with 21 counts of war crimes and 12 counts of crimes against humanity, including sexual enslavement, rape and inducement of rape. See Sophie Kramer, 'Forced Marriage and the Absence of Gang Rape: Explaining Sexual Violence by the Lord's Resistance Army in Northern Uganda', *Journal of Politics and Society*, vol. 23, no. 1, 2012.

14 Some rapes were opportunistic, but these occurred at significantly lower levels. According to a representative 2005–07 survey of 881 young men and 857 young women from two districts in northern Uganda, 93.5% of women forced into marriage experienced rape, as did 6.9% of women and girls who were abducted but not forced into marriage, compared to 1.7% who were not abducted. See Jeannie Annan et al., 'Civil

War, Reintegration, and Gender in Northern Uganda', *Journal of Conflict Resolution*, vol. 55, no. 6, pp. 877–908. The LRA's repertoire of sexual violence after it was forced out of northern Uganda into neighbouring countries is not well documented.
15. Khristopher Carlson and Dyan Mazurana, 'Forced Marriage within the Lord's Resistance Army, Uganda', Feinstein International Center, Tufts University, 2008.
16. Ibid.
17. Centro Nacional de Memoria Histórica, *¡Basta Ya! Colombia: Memorias de Guerra y Dignidad* (Bogota: Centro Nacional de Memoria Histórica, 2012).
18. Senate Select Committee on Intelligence, 'Committee Study of the Central Intelligence Agency's Detention and Interrogation Program', 3 December 2014, https://www.gpo.gov/fdsys/pkg/CRPT-113srpt288/pdf/CRPT-113srpt288.pdf.
19. Human Rights Watch, 'Syria: Sexual Assault in Detention', 15 June 2012, https://www.hrw.org/news/2012/06/15/syria-sexual-assault-detention.
20. Human Rights Watch, 'We Will Teach You a Lesson: Sexual Violence against Tamils by Sri Lankan Security Forces', 26 February 2013, https://www.hrw.org/report/2013/02/26/we-will-teach-you-lesson/sexual-violence-against-tamils-sri-lankan-security-forces; Human Rights Watch, 'We Live in Constant Fear: Lack of Accountability for Police Abuse in Sri Lanka', 23 October 2015, https://www.hrw.org/report/2015/10/23/we-live-constant-fear/lack-accountability-police-abuse-sri-lanka.
21. Kachin Women's Association Thailand, 'Ongoing Impunity', June 2012, http://www.burmapartnership.org/2012/06/ongoing-impunity/; Karen Women's Organization, 'Shattering Silences: Karen Women speak out about the Burmese Military Regime's use of Rape as a Strategy of War in Karen State', April 2004, http://www.ibiblio.org/obl/docs/Shattering_Silences.pdf; Shan Human Rights Foundation and Shan Women's Action Network, *License to Rape*, May 2002, http://www.burmacampaign.org.uk/reports/License_to_rape.pdf; Woman and Child Rights Project and Human Rights Foundation of Monland, *Catwalk to the Barracks*, July 2005, http://www.ibiblio.org/obl/docs3/Catwalk_to_the_Barracks.htm; Women's League of Chinland, 'Unsafe State', March 2007, http://burmacampaign.org.uk/media/UnsafeState.pdf.
22. Obi Anyadike, 'South Sudan's Never-ending War', *Irin News*, 12 October 2016, http://www.irinnews.org/analysis/2016/10/12/south-sudans-never-ending-war.
23. Woman and Child Rights Project and Human Rights Foundation of Monland, *Catwalk to the Barracks*.
24. Mara Revkin, 'Patterns of ISIS Violence', unpublished paper, Yale University, 2016; Human Rights Watch, 'Iraq: ISIS Escapees Describe Systematic Rape', 14 April 2015, https://www.hrw.org/news/2015/04/14/iraq-isis-escapees-describe-systematic-rape.
25. Human Rights Watch, 'We Will Teach You a Lesson'.
26. Wood, 'Multiple Perpetrator Rape during War', p. 143.
27. Elisabeth Wood, 'Armed Groups and Sexual Violence: When Is Wartime Rape Rare?', *Politics and Society*, vol. 37, no. 1, March 2009, pp. 131–62; Amelia Hoover Green,

'The Commander's Dilemma: Creating and Controlling Armed Group Violence against Civilians', *Journal of Peace Research*, vol. 53, no. 5, 2016, pp. 619–32.

28. Xabier Agirre Aranburu, 'Beyond Dogma and Taboo: Criteria for the Effective Investigation of Sexual Violence', in Morten Bergsmo, Alf Butenschon Skre and Elisabeth Jean Wood (eds), *Understanding and Proving International Sex Crimes* (Oslo: Torkel Opsahl Academic EPublisher, 2012), pp. 267–94; Elisabeth Wood, 'Rape during War Is Not Inevitable: Variation in Wartime Sexual Violence', in Bergsmo, Skre and Wood (eds), *Understanding and Proving International Sex Crimes*, pp. 389–419; Maria Eriksson Baaz and Maria Stern, 'Why Do Soldiers Rape? Masculinity, Violence, and Sexuality in the Armed Forces in the Congo (DRC)', *International Studies Quarterly*, vol. 53, no. 2, 2009, pp. 495–518; Cohen, 'Explaining Rape during Civil War', pp. 461–77; Cohen, *Rape during Civil War*.

29. A practice differs from opportunistic violence in that it may be the product not of individual preferences but social interactions – for example, the strong social pressures during training and combat for a combatant to conform to the behaviour of others in the unit. See Wood, 'Conflict-related Sexual Violence and the Policy Implications of Recent Research', pp. 457–78; and Elisabeth Jean Wood, 'Rape as a Practice of War: Toward a Typology of Political Violence', unpublished paper, Yale University, 2016.

30. Wood, 'Conflict-related Sexual Violence and the Policy Implications of Recent Research', pp. 457–78.

31. Eriksson Baaz and Stern, 'Why Do Soldiers Rape?', pp. 495–518.

32. Office of the United Nations High Commissioner for Human Rights, 'Assessment Mission by the Office of the United Nations High Commissioner for Human Rights to Improve Human Rights, Accountability, Reconciliation and Capacity in South Sudan'.

33. Cohen, 'Explaining Rape during Civil War', pp. 461–77; Cohen, *Rape during Civil War*.

34. Carlson and Mazurana, 'Forced Marriage within the Lord's Resistance Army, Uganda'.

35. Annan et al., 'Civil War, Reintegration, and Gender in Northern Uganda'.

36. Christopher Blattman and Jeannie Annan, 'On the Nature and Causes of LRA Abduction: What the Abductees Say', in Tim Allen and Koen Vlassenroot (eds), *The Lord's Resistance Army: Myth and Reality* (Chicago, IL: Zed Books, 2010), pp. 132–55.

37. Amnesty International, 'Escape from Hell: Torture and Sexual Slavery in Islamic State Captivity in Iraq', December 2014, http://www.amnestyusa.org/sites/default/files/escape_from_hell_-_torture_and_sexual_slavery_in_islamic_state_captivity_in_iraq_mde_140212014_.pdf; Middle Eastern Media Research Institute, 'Islamic State (ISIS) Releases Pamphlet on Female Slaves', 4 December 2014.

38. Mara Revkin, 'Patterns of ISIS Violence', unpublished paper, Yale University, 2016.

39. US Department of Justice, 'Memorandum for John A Rizzo Senior Deputy General Council, Central Intelligence Agency', 30 May 2005, https://www.justice.gov/sites/default/files/olc/legacy/2013/10/21/memo-bradbury2005.pdf.

40 Physicians for Human Rights, 'Broken Laws, Broken Lives', June 2008, http://brokenlives.info/?page_id=69.

41 *Ibid.*, p. 6.

42 Lawi Weng, 'Army Statement Warns against Linking Teachers' Murders to Troops', 29 January 2015, http://www.irrawaddy.com/news/burma/army-statement-warns-linking-teachers-murders-troops.html; David Mathieson, 'Impunity for Sexual Violence in Burma's Kachin Conflict', 21 January 2016, https://www.hrw.org/news/2016/01/21/dispatches-impunity-sexual-violence-burmas-kachin-conflict.

43 Women's League of Burma, 'Same Impunity, Same Patterns: Sexual Abuses by the Burma Army Will Not Stop Until There is a Genuine Civilian Government', January 2014, http://womenofburma.org/wp-content/uploads/2014/01/SameImpunitySamePattern_English-final.pdf.

44 See Wood, 'Rape as a Practice of War'.

45 International Committee of the Red Cross, 'Rule 153, Command Responsibility for Failure to Prevent, Repress or Report War Crimes', https://www.icrc.org/customary-ihl/eng/docs/v1_cha_chapter43_rule153. Of course, military and political leaders commonly respond to accusations of strategic rape by their forces by claiming that the troops were not under their control, but this can be countered by other indicators of control. See Wood, 'Rape during War Is Not Inevitable', pp. 389–419.

46 However, the record is mixed, as other prosecutions for sex crimes under command responsibility did not lead to convictions, including those of Mathieu Ngudjolo Chui (DRC), Germain Katanga (DRC) and Callixte Mbarushimana (Rwanda).

47 In any case where rape occurs as part of a widespread or systematic attack on a civilian population, it is a crime against humanity; rape itself need not be widespread or systematic.

48 For more on this topic, see Wood, 'Conflict-related Sexual Violence and the Policy Implications of Recent Research'; and Elisabeth Jean Wood and Dara Kay Cohen, 'How to Counter Rape during War', *New York Times*, 28 October 2015, https://www.nytimes.com/2015/10/29/opinion/how-to-counter-rape-during-war.html.

The Islamic State's Shifting Narrative
Nelly Lahoud

In early 2014, cities and towns in Iraq and Syria were falling like dominoes into the hands of the Islamic State, also known as ISIS or ISIL. In June that year, the group proclaimed itself a 'caliphate', a global 'Islamic state', dropping 'Iraq' and 'the Levant' from its name. However, the size of the territory controlled by ISIS peaked before the group celebrated its first anniversary. During its winning streak, ISIS held to a narrative that claimed that its *tamkin* (territorial strength) was a product of *wa'du Allah* (God's Promise); its deeds were such that this narrative appeared plausible to many. Yet as it began to incur losses, the group modified its narrative, promising – or, perhaps, praying for – a return to *tamkin*. As such, it has struggled to deliver on the propaganda of the (divine) deed that marked its claim to legitimacy as it expanded. Moreover, territorial losses are having a negative impact on the flow of foreign fighters joining the group, while eroding its online empire. Although the jihadist rivals of ISIS are savouring the opportunity to ridicule its early claims of divine legitimacy, they are not gaining from the group's change of fortune. In Syria, these rival groups have failed to form a unified front against the regime: in early 2017, they began to turn their guns on one another, possibly paving their path towards self-destruction.

In April 2014, two months before the proclamation of the caliphate, ISIS spokesman Abu Muhammad al-Adnani concluded his public statement with a *mubahala*, a public supplication that echoes a verse in the Koran (3:61). He beseeched God to furnish His worshippers with proof concerning the legitimacy of ISIS. Adnani implored God that if ISIS is the true Islamic State, may He reward the group with victory against its enemies; and if it is not, he prayed that God should defeat it and kill its leaders.[1] At that time, ISIS was on its ascent and on its path to capture Mosul and link its territories in Iraq with some of those it had captured in Syria. When this happened, Adnani proclaimed the caliphate, the global Islamic State, in June 2014, claiming that the legitimacy of ISIS is premised on God's Promise, echoing the Koranic verse 24:55 in which God promises true believers *tamkin* so that they may implement His Law.

Yet the size of the global caliphate peaked before it celebrated its first anniversary, and by the end of 2016 it had reportedly lost a quarter of its territories.[2] Adnani was killed in August 2016, and several other ISIS leaders have met the same fate. If assessing this change of fortune through a religious prism, one would likely conclude not that God had changed His mind about ISIS, for divine wisdom is not vulnerable to the whims of mortals. Instead, a religious person would likely conclude that ISIS is not after all the state that God promised.

Winning narrative

When the caliphate was proclaimed, Adnani painted a picture of a complete and orderly state of the kind that God had promised the faithful. He spoke of Islamic courts serving the legal needs of the people, the *hudud* (deterrent penalties for certain crimes) being administered and religious taxes levied, among other things. Thus, for ISIS, the caliphate was not an aspiration but an existent and expanding reality.

As such, the leaders of ISIS saw themselves as fulfilling an Islamic legal obligation. Adnani used elements of the classical Islamic corpus to argue that when Muslims enjoyed *tamkin*, there would be a *wajib kifa'i* (legal obligation of the community) to proclaim a caliphate and appoint a caliph to lead it.[3] Within days of the proclamation, ISIS leader Abu Bakr al-Baghdadi made his first public appearance, leading the Friday prayer at the Grand Mosque of al-Nuri in Mosul. Baghdadi had an obligation to prove his existence – which until then had been shrouded in mystery – before Muslims could be expected to pledge allegiance to him.

At the time, the leaders of ISIS had reason to describe their state in miraculous terms. In Iraq, the group had captured most of Ninewa and Anbar provinces, as well as large parts of Kirkuk and Salahuddin provinces. In Syria, it had taken control of most of northern and eastern Rif Aleppo, most of Raqqa and Deir ez-Zor provinces, large parts of Hasakah province and a scattering of areas in Homs, Hama, Damascus and Badia provinces.

In November 2014, ISIS launched its *wilayat* (provinces) project to mark its global expansion. In this orchestrated effort, groups in Egypt, Algeria, Yemen, Libya and Saudi Arabia pledged *bayat* (oath of allegiance) to the caliph on the same day. Three days later, Baghdadi accepted their *bayat*, bestowing upon the groups the title of provinces. In time, the *wilayat* grew to span more than ten countries. Although they proved unable to hold and govern territory, except for small parts of Libya, all of the *wilayat* carried out militant operations (with varying degrees of success). While ISIS continued to capture large Iraqi and Syrian cities in the remainder of 2014 and the first half of 2015, the provinces appeared to have a propaganda effect greater than their actual reality.

Irrespective of the extent to which supporters and enemies of ISIS exaggerated its strength, the group's successes on the battlefield held what to some seemed like a divine promise. The seeming certitude of ISIS leaders about the destiny of their caliphate, combined with the group's battlefield victories, had a global appeal. According to the 'Worldwide Threat Assessment of the US Intelligence Community' published in February 2016, ISIS had inspired around 36,500 foreign fighters outside Iraq and Syria – 6,600 of them in Western countries – to travel to join the group.[4]

If providence was at play as the leaders of ISIS claimed, God's hand seemed to have extended to the virtual world. The group rapidly built an online army whose members produced magazines and other publications in many languages, including Arabic, English, Bosnian, Kurdish, French, German, Indonesian, Pashtun, Russian, Turkish and Uighur. Simultaneously, ISIS engaged in another state-building activity: the production of its own *anashid* (a capella songs/chants), videos and radio programmes.

Losing narrative

By 2016, the global state of ISIS seemed more like a struggling start-up than an enduring state. This is not to suggest that the group's end is imminent. Indeed, ISIS has fought on several fronts despite facing a 68-member coalition designed to, in the words of former US secretary of state John Kerry, 'support a rehabilitated Iraqi military, the Kurdish Peshmerga and other local partners to liberate territory once occupied by [the group]'.[5]

In early 2017, ISIS continued to hold several cities in Iraq, such as parts of Mosul, Tal Afar and Tal Kaif, in Ninewa province; Hawija in Kirkuk; and Ana, Rawa and Qaim, in Anbar province. In Syria, amid a proliferation of militant groups that increased competition for territory, the caliphate covered strategically important cities and many small, dispersed territories, including Raqqa and surrounding areas; Al-Bab and other areas north and northeast of Aleppo; most of Deir ez-Zor city and nearby areas close to the Iraqi border; recaptured Palmyra and surrounding areas; parts of southern Hasakah province bordering Iraq; areas east of Hama and Damascus; parts of Yarmouk refugee camp; areas in northern Suwayda province; and parts of southern Daraa province bordering Jordan.[6]

Nonetheless, ISIS had lost a considerable amount of territory. In Iraq, the group was driven out of Ramadi, Falluja, Rutba and Hit, in Anbar province; Sinjar, Zummar and Shirqat (in the lead-up to the battle to free Mosul); Tikrit, Baiji and many small villages in Salahuddin province; and Jalawla and Sadiya, in Diyala province. In Syria, the group lost Kobani, Jarabulus, Azaz, Manbij and Khanasir, in Aleppo province; Qaryatayn, Maheen and Palmyra (later recaptured), in Homs province; Shaddadi and parts of Hasakah city; and Tal Abyad, Ayn Issa and Suluk, in Raqqa province.

Adnani used his last public statement – issued in May 2016, three months before he was killed by a US airstrike in Syria – to argue that the United States, not ISIS, faced defeat. Denouncing Washington's role in Iraq, he asked:

> Where is the unified and free Iraq [which you had promised]; where is democracy?! Do you fool yourself and your people and the world, or are your recognising the [legitimacy/inevitability] of the Islamic State?
>
> ...
>
> America, are you lying or are you simply incapable of delivering on your promises?

Adnani was reminding Sunni Iraqis within the anti-ISIS coalition of the calamitous effects of *Operation Iraqi Freedom*, the 2003 US-led intervention in the country. But he was also attempting to shift the ISIS narrative away from reliance on tangible territorial victories. It would not be difficult for even a casual observer to point to the differences between the respective failures of the US and ISIS. After all, the US had failed to deliver on promises that were made by mortals and for mortals, whereas ISIS appeared to have fallen short of the divine promise that its leaders believed shaped their destiny. Indeed, during the group's ascent, their public statements were void of humility, holding to a narrative that lacked the ambiguity to accommodate any possibility of defeat.

If capturing territory reflected God's Promise of *tamkin* as a reward for believers, what might losing territory mean? Understandably, ISIS leaders chose not to meditate on this question in their public statements in 2016. Instead, they seemed to suffer from a sort of amnesia, forgetting their one-time promise that ISIS is but His Promise. As its territorial losses continued, ISIS replaced its narrative of *tamkin* with an emphasis on the centrality of Islamic beliefs, attempting to boost the morale of its fighters in Mosul, Raqqa and other areas of strategic importance.

In what ended up being his *adieu* speech, Adnani crafted this narrative shift, assuring ISIS fighters that they would not be defeated even if they were left to 'roam in the desert with no city and land' or lost 'Mosul or Sirte or Raqqa and all the [other cities we continue to hold], and [even] if we [are forced to] return to our prior [stateless] status'. In this version of the narrative, defeat is not about losing territories; rather, defeat comes when 'one loses the will and the desire to fight' and the US succeeds 'in removing the Koran from the hearts of Muslims'.[7] As a consequence, victory acquires a poetic dimension and can be achieved through new means – including defeat.

Perhaps realising the ISIS narrative's reliance on God's Promise, Baghdadi returned to the theme in his only public statement of 2016. 'Verily, this is God's Promise', he assured his supporters. The fact that the 'infidels', despite their differences, were united to defeat the group, he held, was a 'prelude to the enduring/firm victory' that would allow Muslims to 'reclaim *tamkin*'.[8] Furthermore, this statement suggests that he was worried about the loss of Mosul and feared that his fighters might flee the battlefield.[9] He called on them to be steadfast in the face of the 'planes of America and its allies', calling on all those who sought martyrdom to 'turn the infidels' nights into days, destroy their abode and [liberally shed] their blood so that it may [flood the streets] just like [water flows in] the rivers'.[10]

Abu Hasan al-Muhajir, Adnani's successor, echoed this concern about the desertion of ISIS fighters. In his first public statement, issued in December 2016, he appealed to them to battle fiercely, adding: 'you should not talk yourself into fleeing'. He assured them that this phase of territorial losses would pass, and 'this tribulation [*mihna*] is in reality a gift [*minha*]'.[11] If Baghdadi was worried about Mosul, Muhajir's focus was on the defence of Tal Afar, a strategically important city west of Mosul and close to the border with Syria. It is likely that ISIS wants to use Tal Afar as a transit point for moving its resources in Iraq to Syria, where the territory the group holds appears to be more secure.

Effects of territorial losses

The territorial losses ISIS experienced have had a significant impact on the flow of foreign fighters to the group, its online empire and the fortunes of its jihadist competitors. According to the *Washington Post*, the number of foreign fighters joining ISIS appears to be 'falling as precipitously as the terrorist group's fortunes'.[12] The effects of territorial losses and the resulting shift in narrative largely account for this decline but, as the paper reports, 'enhanced intelligence-sharing between Turkey and Western governments' has also made it difficult for fighters to join ISIS by crossing the Turkey–Syria border. As a consequence, the number of foreign fighters using this route has dropped from 2,000 per month to 50 per month.[13]

This fall-off prompted ISIS to intensify its campaign against Turkey. All the public statements released by ISIS leaders in 2016 called on their supporters to attack the country, and some of the faithful have responded – as seen in the high-casualty attack on a nightclub in Istanbul on New Year's Eve 2016. Beyond that, and perhaps anticipating the group's seemingly inevitable territorial losses in Iraq and Syria, Baghdadi is now highlighting the importance of *hijra* (emigration) to the 'blessed provinces' – not necessarily in Iraq or Syria – to establish an abode where God's Law will reign supreme.

The group's online empire is also undergoing a noticeable decline, although it continues to produce publications in many languages. However, it remains unclear whether the digital caliphate is more enduring than the physical one, particularly as online ISIS soldiers are under less pressure than their counterparts on the battlefield. Nonetheless, ISIS has stopped producing its most popular English magazine, *Dabiq*; instead, it is now producing a new magazine, *Rumiyah*, in English and other languages. Whereas *Dabiq* appeared to publish mostly original articles, *Rumiyah* seems to rely largely on translations of articles produced in the group's Arabic magazine, *Al-Naba*. This may suggest that online support for ISIS is diminishing.

The jihadist enemies of ISIS are not grieving over the group's change of fortune; they have been savouring the opportunity to ridicule the group's claims of divine legitimacy. One powerful ideological argument they have mounted against it concerns Adnani's *mubahala*, in which he implored God to kill ISIS leaders if the state they proclaimed was illegitimate. Thus, according to the argument used by rival jihadists, Adnani's death was God's verdict on the caliphate.

Bahrain-born Turki al-Binali, an ISIS ideologue, took on the unenviable task of refuting this claim. Drawing on a citation by the classical Muslim scholar Ibn

Hajar, Binali argued that he who lies in a *mubahala* will be punished by God within a year, stressing that Adnani's death occurred more than two years after his *mubahala*.[14] This was not a compelling defence, and Binali's jihadist critics were quick to highlight the pedestrian nature of his argument, not least since it lacked supporting evidence from Islamic foundational texts (the Koran and the Hadith).[15] In fairness to Binali, Adnani's *mubahala* made it impossible for anyone to come up with a convincing defence. It is also difficult to envisage a narrative that might give ISIS the compelling edge it was able to project during its winning streak.

But if ISIS is losing the ideological and territorial battles, its jihadist enemies are faring little better. Militants in Syria, home to the group's main jihadist opponents, have been unable to form a unified front against either it or the regime. The Syrian landscape is littered with militant factions, but the only groups that may be in a position to compete with ISIS ideologically are al-Qaeda affiliate Jabhat Fateh al-Sham (JFS) and Ahrar al-Sham, which has been fighting under an Islamic (but not a pure jihadist) umbrella.[16] For most of their history, the groups have been at least cordial with each other and have in some instances collaborated, seeing themselves as facing the same enemy, the Syrian regime. Yet conflict between them broke out in January 2017, as each attempted to absorb the other.

Al-Majlis al-Islami al-Suri, a council formed by Sunni scholars in Istanbul in 2014, released on 24 January 2017 a statement accusing the JFS of holding to the same extremist ideology as ISIS and indiscriminately shedding the blood of Muslims. The council called on other Islamist groups in the Syrian war to fight against ISIS and the JFS.[17] Two days later, 'responding to the calls made by the people of knowledge', Suqour al-Sham, Jaysh al-Islam (Idlib branch), Jaysh al-Mujahideen, Tajammu Fastaqim Kama Umirt and Jabhat al-Shamiya (western Aleppo branch) merged with Ahrar al-Sham.[18] It is reported that they all merged because the JFS 'attacked their positions in Idlib and Aleppo provinces', but it would not be implausible to surmise that al-Majlis al-Islami was seeking through its statement to strengthen support for Ahrar al-Sham.[19]

In contrast, the JFS argued that it sought to bring all Islamic factions under one umbrella, only to be thwarted. The group believes that it suffered the lion's share of the attacks in the Russian-led air campaign in Aleppo, implying that the campaign largely ignored its rivals because they would compromise on the principles of the revolution. In saying this, the JFS appeared to have in mind the Astana peace talks that Russia organised in January 2017. And, on 28 January, the group announced the formation of Tahrir al-Sham, merging with Harakat Nour al-Din al-Zenki, Liwa al-Haqq, Jabhat Ansar al-Din and Jaysh al-Sunna.[20]

Although it is difficult to ascertain what prompted these realignments, a telegram by an ISIS supporter indicates that the leadership of Tahrir al-Sham includes former leaders of both the JFS and Ahrar al-Sham. If this is true, it would suggest that the divisions between Syria's Islamist groups may well be ideological. In other words, it is possible that the divide is between those who wish to be part of the (democratic) political process and are prepared to stop fighting against the Syrian regime and those who want to maintain their jihadist path and establish an Islamic political entity, perhaps in the manner of the Afghan Taliban.

The great historian Ibn Khaldun remarked that those who lack *asabiyya* (group feeling) resort to religious propaganda as a justification for rebelling against entrenched rulers and dynasties. 'Most men who adopt such ideas', he held, 'will be found to be either deluded and crazy, or to be swindlers who, with the help of such claims, seek to obtain (political) leadership – which they crave and would be unable to obtain in the natural manner.'[21] His analysis may not entirely apply to the motivations and methods of ISIS, but his critique of the dangers of resorting to religious propaganda remains relevant. From the vantage point of 2017, a sober ISIS supporter would surely agree that claiming that the caliphate's territorial strength was nothing other than God's Promise was, in Ibn Khaldun's parlance, 'delusional and crazy'.

Nelly Lahoud is IISS Senior Fellow for Political Islamism.

Notes

1. Abu Muhammad al-Adnani, 'Ma Kana Hadha Manhajuna wa-lan Yakun', April 2014. My thanks to a colleague who drew my attention to the Koranic verse that is echoed in the *mubahala*, and to my research assistant for collecting some of the primary sources for this essay.
2. 'Islamic State Group "Lost Quarter of Territory" in 2016', BBC, 19 January 2017, http://www.bbc.com/news/world-middle-east-38641509.
3. Abu Muhammad al-Adnani, 'Hadha Wa'du Allah', 29 June 2014. Unless otherwise stated, Arabic translations are my own.
4. James R. Clapper, 'Worldwide Threat Assessment of the US Intelligence Community', statement for the record to the US Senate Armed Services Committee, 9 February 2016, http://www.armed-services.senate.gov/imo/media/doc/Clapper_02-09-16.pdf.
5. John Kerry, 'What We Got Right', *New York Times*, 19 January 2017, https://www.nytimes.com/2017/01/19/opinion/john-kerry-what-we-got-right.html.
6. My thanks to my research assistant for his assistance in identifying the geographical areas in this section.
7. Abu Muhammad al-Adnani, 'Wa-Yahya man Hayya an Bayyina', May 2016.
8. Abu Bakr al-Baghdadi, 'Hadha Ma Wa'adana Allahu wa-Rasuluhu', November 2016.
9. See Anne Speckhard and Ahmet S. Yayla, *Isis Defectors: Inside Stories of the Terrorist Caliphate*, Advances Press, 2016; and Aymenn Jawad Al-Tamimi, 'Review of ISIS Defectors: Inside Stories of the Terrorist Caliphate', https://medium.com/narrative-inquiry/review-of-isis-defectors-inside-stories-of-the-terrorist-caliphate-beb01c9a97c1#.xpcb12q1j.
10. *Ibid.*
11. Abi al-Hasan al-Muhajir, 'Fa-sa-Tadhkuruna ma Sa-Aqulu Lakum', December 2016. Note the play on words for effect (*mihna* and *minha*).
12. Griff Witte, Sudarsan Raghavan and James McAuley, 'Flow of Foreign Fighters Plummets as Islamic State Loses Its Edge', *Washington Post*, 9 September 2016, https://www.washingtonpost.com/world/europe/flow-of-foreign-fighters-plummets-as-isis-loses-its-edge/2016/09/09/ed3e0dda-751b-11e6-9781-49e591781754_story.html?utm_term=.9b77a483e7b8.
13. *Ibid.*
14. Turki al-Binali, 'Jawab al-Sheikh Turki al-Binali 'ala man Za'ama anna Maqtala al-Sheikh al-'Adnani Dalilun 'ala anna Khusumahu Zaharu 'alayhi bi-al-Mubahala', *Idha'at al-Bayan*, transcribed by *al-Nusra al-Maqdisiyya li-al-I'lam*, December 2016.
15. See 'Abd al-Rahman al-Jaza'iri, Turki al-Binali … Ahuwa Shar'i … Am Shurru', December 2016.
16. The group was initially called Jabhat al-Nusra, before it changed its name to Jabhat Fateh al-Sham to dissociate itself from al-Qaeda. However, the two groups closely coordinated the name change, suggesting that they had not completely severed their ties.

17 Al-Majlis al-Islami al-Suri, 24 January 2017. The JFS split from ISIS in April 2013, and the two have been staunch enemies since December that year.
18 'Bayan Mushtarak min Kubra al-Fasa'il al-Thawriyya fi al-Shamal', 26 January 2017.
19 Mariya Petkova, 'Syrian Opposition Factions Join Ahrar al-Sham', *Al-Jazeera*, 26 January 2017.
20 Tahrir al-Sham, 'Bayan Tashkil Hay'at Tahrir al-Sham', 28 January 2017.
21 Ibn Khaldun (trans. by Franz Rosenthal), *The Muqaddimah* (New York: Pantheon Books, 1958), vol. 1, chapter 3, p. 326.

The Changing Foundations of Governance by Armed Groups

William Reno

How and why armed groups govern civilians owes much to the particularities of specific conflicts. Nevertheless, there has been a systemic change in the past 50 years. Global political changes have altered who supports armed groups from outside conflict zones, and for what purpose. This shift has had important effects on armed groups' motivations and resources for civilian governance. Broad changes have also occurred in the character of the states in which armed groups fight, and this has had a corresponding impact on the kinds of social relationships and reactions that these groups encounter among civilians. In particular, many contemporary armed groups have formed in the wake of state failure and the collapse of authoritarian personalist governments. The fragmented social environments that this process creates play critical roles in shaping how, and with whom, armed groups negotiate in order to govern.

Broad systemic change does not influence all armed groups in equal measure. Some of these groups govern civilian areas in much the same way as their counterparts did 50 years ago. At its height in the 2000s, Sri Lanka's Liberation Tigers of Tamil Eelam stood out for its comprehensive governance of civilians.[1] The Islamic State, also known as ISIS or ISIL, and the Afghan Taliban exhibit more typical contemporary patterns of governance by armed groups. Both face difficulties in governing societies fragmented by earlier processes of state collapse, and both face an international system that has become much more hostile to armed groups. The fragmentation of contemporary armed groups also has precedent.[2]

Nevertheless, one can observe the broad contours of a fundamental change in the conditions and outcomes of armed groups' governance of civilians – a change that intensifies the challenges faced by these groups. This essay explains the causes of this change, and how it has affected the ways in which armed groups govern civilians, as well as the ways in which civilians have responded to their efforts. To do so, it begins by outlining the relevant transformations of the past half-century.

The big picture

Global politics casts a long shadow on armed groups. Fifty years ago, state backers typically provided support to armed groups on the condition that they at least appeared to govern civilians in what were often called 'liberated zones'. The decisions of US or Soviet officials about whether to support a particular armed group had much to do with expanding a sphere of political and economic influence in a competitive bipolar strategic environment. Armed groups recognised that their choice to side with one of the superpowers would affect their capacity to tap outside material and political support.[3] Leaders of armed groups also anticipated that a backer would press for commitments to particular organisational patterns and ideological perspectives, with significant consequences for their decisions on how to govern civilians.[4]

The decades after the Second World War also saw wide popular and government interest in how armed groups might create new kinds of societies – particularly as struggles against colonial rule and apartheid gained strength from the 1950s up until the end of the Cold War. Many armed groups fit a narrative in which they would bring an end to colonial and apartheid rule. In its Soviet version, governance by armed groups was central to creating socialist societies. US officials recognised a need to maintain contacts with, and occasionally aid, armed nationalist groups to counter Soviet influence. In all its versions, in addition to fighting, mastering the process of social change required that armed groups design systems of governance.

This global approach to governance by armed groups was reflected in international organisations that gave them degrees of formal recognition. The

Organisation of African Unity (OAU) established the Coordinating Committee for the Liberation of Africa (Liberation Committee) in 1963 to identify armed groups that were 'authentic' movements on the basis of demonstrated military success against foreign rule and control of communities on the ground. Determining whether an armed group was authentic involved diplomatic wrangling and the reports of teams sent to examine and certify its governance capabilities.[5] The Liberation Committee solicited contributions from OAU member states and donations from other sympathetic governments, organisations and individuals to pay for weapons and training for these authentic movements. Liberation Committee assistance to armed groups in Africa between 1963 and 1977 totalled US$27.8 million (equivalent to around US$100m today).[6] The Non-Aligned Movement, the UN Special Committee on Decolonization and many other organisations also provided public forums in which armed groups could appeal for more support.

Partial incorporation into the international system helped armed groups gain access to direct funding, such as that provided by Scandinavian governments and sympathetic citizens to groups that fought colonial and apartheid rule.[7] Gaining observer status in the United Nations was a significant step for these groups. The Palestine Liberation Organization (PLO) won UN observer status in 1974 and established diplomatic relations with more than 100 states.[8] In 1972 the UN General Assembly recognised the Southwest African People's Organization (SWAPO) as the 'sole legitimate representative of Namibia's people'. In 1974 Norway's government started direct aid to SWAPO of about US$50m over 15 years, for which SWAPO's UN status and coordination with the Liberation Committee were prerequisites.[9] In 1976 the UN opened the Institute for Namibia in Lusaka, in Zambia, in which foreign donors coordinated with SWAPO to train cadres in practical aspects of administration, in preparation for Namibia's independence. Assistance from the Soviet, Chinese, Cuban and other foreign governments owed much to their strategic calculations, but for the PLO the legitimacy it gained through recognition in international organisations eased this process and channelled resources to a distinct set of armed-group leaders.

The coordination of international recognition and support of armed groups was important in shaping their governance decisions. Firstly, to the degree that external actors coordinated their decisions (which was problematic in light of Cold War superpower rivalries and splits within the socialist camp), consolidating aid flows to selected armed-group leaders was critical to reducing factionalism within these groups. Detailed histories of armed groups almost always reveal intense conflict within them, but leaders of armed groups who could demonstrate their military and administrative skills had an advantage. Moreover, regardless of their political orientation, these leaders generally had to exhibit a commitment to a modern secular nation-state project along the lines of European models. This commitment was a standard feature of armed groups' rhetoric and civilian-governance practices.

While external assistance remains important in the fates of contemporary armed groups, they encounter nothing like the old system of partial incorporation into the international system. The OAU's Liberation Committee disbanded in 1994 – its mission accomplished with the end of the apartheid regime in South Africa. Armed groups still face external judgement, but more in relation to their violation of human-rights norms. UN sanctions committees and panels of experts, as well as non-governmental organisations such as Global Witness and the International Crisis Group (among many others), scrutinise and publicise connections to the criminal networks that have always played a role in supporting armed groups. Due to global concerns about terrorism and the pernicious effects of violent illicit commerce, armed groups are often depicted as being part of a criminal conspiracy to grab personal power and wealth.[10] Equally, concerns about the role that armed groups play in 'ungoverned spaces'[11] and their possible links to terrorist networks have landed many of them on government lists of terrorist organisations. Armed groups counter these narratives through social media and other mechanisms for communicating with their supporters. But these means are accessible to many factions and even individuals within armed groups, potentially undermining their leaders' efforts to coordinate members to fight and govern.

Situation on the ground

It has become difficult to find classic twentieth-century-style wars in which armed groups build liberated zones in pursuit of modern state-building projects with support from international organisations and solidarity networks. It is much more common to find wars of state failure that have a distinctly twenty-first-century character. In these conflicts, the army dissolves into multiple militias. Then, fragmented state forces confront an array of disparate rebel groups that often include key players in the pre-conflict regime. External non-state actors, such as those involved in criminal networks, are important. In contrast to the situation 50 years ago, these actors play enhanced roles in decentralised warfare that bring together armed groups and criminal networks. As a result, civilians are more likely to be exposed to violence from multiple sources, which results in mass displacement more often than it does where there is sustained governance by armed groups.

These conditions arise out of the distinct character of the states in which armed groups fight. Paramount among these conditions is the incumbent state leader's recognition of, and response to, risk. Between 1956 and 2001, all but six of Sub-Saharan Africa's 48 countries experienced military interventions in politics. *Coups d'états* succeeded in 30 (62.5%) of these countries, and multiple coups occurred in 18 (37.5%) of them. Many of the coups caused the death of the incumbent leader.[12] By 2001, nine of 16 Middle Eastern states (56%) had experienced coups since gaining independence, with six in Syria, four in Iraq and three in Algeria. Paul Collier and Anke Hoeffler have noted that coups tend to legitimate future coups, and that 'societies can collapse into political black holes of repeated regime change generated from within the army'.[13]

Leaders who face such risks often recognise that it is in their interest to undermine the formal institutions of their states to prevent other officials from acquiring power bases. Surviving a coup attempt or a major uprising often signals a decisive shift from building state institutions to undermining them. In surviving an uprising in Hama in 1982, the Assad regime discovered that a project of modernisation and state-building is risky. The regime responded by expanding the security forces, including quasi-official and illicit ones, and increasing citizens' reliance on powerful figures by cultivating dependence on business networks under the control of family members and regime insiders.[14] In sum, this push among leaders to undermine the foundations of their states, and of public order, in pursuit of regime survival has created a serious crisis.

Associates who are beholden to a leader due to family or other personal ties, as well as illicit business deals, present much less danger to regimes than citizens or government officials who are vital to state-building projects but may plot against them. The exercise of authority through these channels, behind the façade of formal state institutions, plays a central role in the fragmented patterns of violent competition when these states fail, and in how armed groups fight and govern. High-level officials from presidents down are typically involved in these networks of patronage, which they often fund through criminal activities that ease the strain on the national treasury. Economic reforms that privatise state firms and deregulate the economy simply provide more avenues for building and funding these networks. Members of the networks are meant to be unsure of the security of their positions and to compete for the leader's favour. This system is effective at undermining elite-level cooperation and works so long as the leader can maintain control over other people's access to resources.

In some instances, it is difficult to distinguish networks of state officials from criminal syndicates. For example, prior to the outbreak of civil war in South Sudan in 2013, the country's president controlled militias by drawing resources and personnel from official security agencies that had received foreign funding. Many South Sudanese politicians also enjoyed access to insider deals.[15] Mali's president armed ethnic and clan militias to exploit inter-group rivalries, and then exacerbated these rivalries by selectively permitting access to criminal networks – allegedly including those involved in drug trafficking. This technique of governance was important in shaping the conflict that intensified as the state collapsed in 2012.[16]

In a formal institutional sense, state failure occurs well before the international community recognises it as such. But, once the veneer of domestic order slips, the underlying fragmentation reshapes the political system as members of the establishment struggle with one another for control of resources. Those who are best positioned in this struggle may become warlords, using these resources to support armed groups loyal to them. Thus, some armed groups are products of a system of personalist rule and its networks of power. These networks also shape the environment in which all armed groups fight and govern.

By using these divide-and-rule tactics to manage risk, leaders enable their subordinates to employ violence for personal benefit – as occurs when politicians partner with criminal gangs, or local strongmen use vigilantes to defend a community and press its claims against neighbours. This violence makes other people less secure and herds them into these networks for protection, reinforcing a pattern of governance by armed groups based on rigid definitions of communal identity. Rulers can then turn purely local, and even personal, tensions among their subjects into tools to disrupt and forestall the organisation of effective opposition to their leadership. This tactic sweeps up otherwise private or intensely local rivalries and disputes, transforming them into points of tension that rulers can manipulate to divide communities. The decentralisation and privatisation of violence turns disorder into a political instrument.[17]

The experience of Sierra Leone shows how these divide-and-rule tactics shape the organisation of, and governance by, armed groups after a state collapses. In the 1970s, paramilitary groups that had presidential protection, under joint control of politicians and Lebanese merchants, mined diamonds in defiance of official regulations. The president used these militias to suppress a 1982 rebellion among supporters of an outlawed opposition party.[18] By that stage, members of the militias outnumbered troops in the national army by four to one.[19] The fragmented patterns of violence in Sierra Leone's 1991–2002 civil war featured diamond-mining gangs that fought alongside Revolutionary United Front rebels while continuing to pursue their old occupation. Elements of the army also mined diamonds, particularly after desperate politicians started recruiting fighters to put down the rebellion from political militias and mining gangs.[20] As in other failed states, this created a chaotic environment in which multiple armed groups and agendas undermined these groups' efforts to govern. The regime's divide-and-rule tactics reached into the periphery of the state, preventing rebel groups from governing liberated zones in the manner of classic twentieth-century guerrillas.

Social atomisation and governance challenges

In the past 50 years, the connections between fighting and governance have been highlighted by the changes that have occurred in the political environments in which armed groups operate. These changes reflect the importance of external actors in aiding or undermining leaders of armed groups who are attempting to organise a cohesive support base and to govern civilians. The quality of their social connections and the state's formal and informal institutions also has great bearing on these issues. In particular, a personalist regime can cause the state to fail by creating an environment of social fragmentation and antagonism among communities. In such an environment, multiple contending armed groups may organise around parochial disputes and the exploitation of the commercial opportunities of the old patronage system.

A close examination of the connection between social relationships in communities and governance by armed groups highlights how these broad shifts – especially in the impact of regimes' patronage politics and divide-and-rule tactics – have made such governance more difficult. In her study of FARC's governance of areas of Colombia it controlled, Ana Arjona shows how communities with intense bonds of solidarity kept the group at arm's length and limited its efforts at tax collection and the provision of basic security. She identifies true 'rebelocracy' in communities with strong social bonds. Armed groups are able to govern in these communities because they can identify local intermediaries who use personal authority to manage rebel directives and encourage civilians to accept a sort of social contract. Arjona finds that armed groups' efforts to govern result in the greatest disorder and failure – often accompanied by intensifying violence against civilians – in socially atomised communities.[21]

Ultimately, armed groups cannot directly govern every individual all the time. Writing on insurgency in Southeast Asia in the 1960s, James Scott noted that armed groups comprehensively govern only when there are local social bonds that that their leaders can co-opt. These bonds help them explain and interpret the rules of governance to civilians, and secure their complicity with the armed group's presence. This social embeddedness is essential for enabling armed groups to manage parochial issues, while insulating their members from direct involvement in civilians' personal or local problems.[22] If these social bonds are weak or absent, an armed group's efforts to govern lead to the disorder that Arjona describes in Colombia, as the group absorbs communal divisions.

The mechanisms of state failure – or, more precisely, the failure of patronage-based personalist regimes – break the bonds of dependence on a single leader, exacerbating and spreading the animosities that the regime had co-opted to retain power. This shift empowers those who view disorder as an opportunity for personal enrichment, and gives a platform to narrow-based groups of militants. The ferocity of the resulting environment often compels civilians to seek the protection of these same groups. Meanwhile, social degradation presents enormous obstacles to collective action, especially sustained governance by armed groups in the old image of a state-building project. In such circumstances, civilians struggle to comply with competent governance by armed groups.

In many failed states, armed groups have faced severe difficulties in establishing sustainable governance of civilians among divided and disorganised communities. Islamist insurgents in Somalia repeatedly succumb to the divisive politics of clan kinship. Foreign-military action has helped prevent al-Shabaab from sustaining its form of governance, but so too has the group's entanglement in complex parochial disputes that repeatedly poison it from within.[23] This is reflected in the tendency among members of al-Shabaab to engage in violence related to personal issues at the expense of its overall military or governance strategies. Likewise, Iraq's Mahdi Army encountered the greatest challenges to its attempts to govern eastern Baghdad in the neighbourhoods that had been most dependent upon the patronage networks of the Ba'athist state prior to the 2003 US invasion.[24]

Governance by armed groups continues to take many forms. But, as long as the international community withholds coordinated support for such efforts and an increasing proportion of wars involve state failure, the obstacles to the task of governance by armed groups will grow. For example, armed groups in the Middle East show little sign of governing civilians in ways that present a credible alternative to incumbent governments, regardless of how unpopular or incompetent the latter may be. Options for international action include tackling the underlying problem of state failure and devising a standard of appropriate governance by armed groups that is worthy of broad international support. Both options are exceedingly unlikely to receive significant political support. It is more likely that the international community's approach to governance by armed groups will be driven by the interests of individual states, which may seek to recruit these groups as proxies to pursue terrorists or create buffer zones. This less hopeful alternative is realistic in reflecting the growing, multisided geopolitical competition and the persistent problem of state failure that also shape governance by armed groups.

William Reno is Professor of Political Science and Director of the Program of African Studies at Northwestern University. He is the author of *Corruption and State Politics in Sierra Leone* (Cambridge University Press, 1995), *Warlord Politics and African States* (Lynne Rienner Publishers, 1999), *Warfare in Independent Africa* (Cambridge University Press, 2011) and numerous other publications on the politics of state failure and associated conflicts.

Notes

[1] Zachariah Mampilly, *Rebel Rulers: Insurgent Governance and Civilian Life during War* (Ithaca, NY: Cornell University Press, 2011).

[2] Stathis Kalyvas, '"New" and "Old" Civil Wars: A Valid Distinction?', *World Politics*, vol. 54, no. 1, January 2001, pp. 99–118; Stephen V. Ash, *When the Yankees Came: Chaos and Violence in the Occupied South* (Chapel Hill, NC: University of North Carolina Press, 1995), pp. 38–75.

[3] Forrest Colburn, *The Vogue of Revolution in Poor Countries* (Princeton, NJ: Princeton University Press, 1994).

[4] Stathis Kalyvas and Laia Balcells, 'International System and Technologies of Rebellion: How the End of the Cold War Shaped Internal Conflict', *American Political Science Review*, vol. 104, no. 3, August 2010, pp. 415–29.

[5] Emmanuel Dube, 'Relations between Liberation Movements and the OAU', in

Nelson Shamuyarira (ed.), *Essays on the Liberation of Southern Africa* (Dar es Salaam: Tanzania Publishing House, 1975), pp. 25–68.

6   Paul Shipale, 'The Role of the OAU Liberation Committee in the Southern African Liberation Struggle', SWAPO Party, http://www.swapoparty.org/the_role_of_the_oau_liberation_committee.html.

7   Tor Sellström, *Sweden and National Liberation in Southern Africa* (Uppsala: Nordiska Afrikainstitutet, 1999).

8   James Crawford, 'Israel (1948–1949) and Palestine (1998–1999): Two Studies in the Creation of States', in Guy Goodwin-Gill and Stefan Talmon (eds), *The Reality of International Law: Essays in Honour of Ian Brownlie* (New York: Oxford University Press, 1999), pp. 95–115.

9   Eva Helene Ostbye, 'The Namibian Liberation Struggle: Direct Norwegian Support to SWAPO', in Tore Linné Eriksen (ed.), *Norway and National Liberation in Southern Africa* (Uppsala: Nordiska Afrikainstitutet, 2000), pp. 91–3.

10 Paul Collier, 'Rebellion as a Quasi-Criminal Activity', *Journal of Conflict Resolution*, vol. 44, no. 6, December 2000, pp. 839–53.

11 Anne Clunan and Harold Trinkunas (eds), *Ungoverned Spaces: Alternatives to State Authority in an Era of Softened Sovereignty* (Palo Alto, CA: Stanford University Press, 2010).

12 Patrick McGowan, 'African Military Coups d'État, 1956–2001: Frequency, Trends, and Distribution', *Journal of Modern African Studies*, vol. 41, no. 3, September 2003, p. 345.

13 Paul Collier and Anke Hoeffler, 'Coup Traps: Why Does Africa Have So Many Coups d'Etat?', Centre for the Study of African Economies, Oxford University, 2005, p. 3.

14 Bassam Haddad, *Business Networks in Syria: The Political Economy of Authoritarian Resilience* (Palo Alto, CA: Stanford University Press, 2012).

15 United Nations Security Council, *Final Report of the Panel of Experts on South Sudan Established Pursuant to Security Council Resolution 2206 (2015)* (New York: UN, 22 January 2016), pp. 14–17, http://www.securitycouncilreport.org/atf/cf/%7B65BFCF9B-6D27-4E9C-8CD3-CF6E4FF96FF9%7D/s_2016_70.pdf.

16 Ivan Briscoe, *Crime after Jihad: Armed Groups, the State and Illicit Business in Post-Conflict Mali* (The Hague: Clingendael Institute, 2014), pp. 19–32.

17 Patrick Chabal and Jean-Pascal Daloz, *Africa Works: Disorder as Political Instrument* (Oxford: James Currey, 1999).

18 Jimmy Kandeh, 'Ransoming the State: Elite Origins of Subaltern Terror in Sierra Leone', *Review of African Political Economy*, vol. 26, no. 81, September 1999, pp. 349–66.

19 'Sierra Leone: The Unending Chaos', *Africa Confidential*, vol. 23, no. 21, 20 October 1982, p. 6.

20 Arthur Abraham, 'War and Transitions to Peace: A Study in State Conspiracy in Perpetuating Armed Conflict', *Africa Development*, vol. 22, 1997, pp. 3–4.

21 Ana Arjona, *Rebelocracy: Social Order in the Colombian Civil War* (New York: Cambridge University Press, 2016).

22 James Scott, *The Moral Economy of the Peasant: Rebellion and Subsistence in Southeast Asia* (New Haven, CT: Yale University Press, 1977).

23 Stig Jarle Hansen, *Al-Shabaab in Somalia: The History and Ideology of an Islamist Group, 2005–2012* (New York: Oxford University Press, 2013), pp. 73–103.

24 Nicholas Krohley, *The Death of the Mehdi Army: The Rise, Fall, and Revival of Iraq's Most Powerful Militia* (New York: Oxford University Press, 2015).

Rebel-to-party Transitions
Carrie Manning

Since the end of the Cold War in 1989, the overwhelming majority of negotiated peace agreements have based their political settlements on electoral politics. In more than half of these cases, rebel groups formed parties to participate in post-war politics.[1] Thus, rebel-to-party conversions have been a key feature of post-conflict peacebuilding for more than two decades. What challenges do rebel groups face in attempting to make this change? What conditions affect the success of these attempts? And to what extent do post-rebel parties affect outcomes such as peace and democratisation?

Political parties are defined here as organisations that are dynamic, non-unitary actors in which the bases of organisational power can respond to changing incentives. Parties comprise sub-groups whose membership and interests may shift over time. These sub-groups might be based on deeply rooted ideological differences or disagreements over tactics or strategy arising from sub-groups' functional roles.[2] In this essay, the term 'transition' refers to a rebel group's registration as a legal party eligible to compete in elections.[3] For such a group to successfully transform itself into a political party, it must adapt organisational routines, internal authority patterns and an identity forged in war to an environment in which competition occurs in the election booth, the legislature and the policy arena. Therefore, this essay defines 'transformation' as the process by which a rebel group becomes a party in practice, recruiting and running candidates for elected office, participating in institutions of governance and encountering the risks and opportunities that this work entails.

This essay focuses on transformation. While all rebel groups face a similar set of challenges upon entering post-war politics, the difficulty of transformation varies based on two sets of factors: the characteristics and history of a group; and the characteristics of its new environment. Key characteristics of a rebel group include its connections with a popular base, its organisational structure and coherence, and its prior experience with politics.

The environment shapes the expectations of the parties making the transformation in two important ways. Firstly, it influences party leaders' calculations of the relative costs of peaceful politics and violence in achieving their goals. Important factors in this include electoral rules, the competitiveness of the party system, the number of different avenues through which a party might gain access to the state (such as subnational elections), the regional security context and long-standing patterns of socio-economic inequality or political marginalisation.

Secondly, the environment shapes expectations through its perceived stability. The more stable the rules of the game, the longer the time horizons of the players and the more likely they are to invest in prevailing under these rules. As Dankwart Rustow noted in 1970, electoral politics launches a competitive process in which early adapters quickly gain the advantage.[4]

Contours of rebel-to-party transformation

It is relatively easy to understand why most rebel groups have formed post-war parties when given the opportunity to do so. Particularly where elections are overseen by multinational peacekeeping organisations or other influential international actors, parties often receive financial or logistical support and face little electoral competition – other than that from the ruling party and rival former rebel groups. Frequently, rebel groups have helped establish the rules governing electoral competition, allowing them to design systems that play to their strengths. Consequently, the opportunity costs of failing to participate in elections are often higher than the risks of participation. And there has often been ample international support for the major tasks of transition, such as disarmament, the reintegration of combatants and technical advice on electoral campaigning.

Yet the organisational challenges of repeatedly participating in elections and surviving between elections – the challenges of party transformation – are a different matter. While a growing literature addresses rebel groups' decisions to form parties after conflict ends, research on party transformation is still relatively scarce.

Participating in politics across multiple electoral cycles will require both attitudinal and behavioural changes. These changes will, in turn, require shifts in organisational structure – in the distribution of authority and resources within the organisation, and in the kinds of skills the organisation needs to succeed. Indeed, electoral success often requires such adjustments, as members of the party who are elected to public office may gain both motives and opportunities to challenge the position of their superiors within the party. Once they succeed in placing a candidate in public office, parties often find that their new representative begins to develop a stake in his or her personal success in office that outweighs loyalty to the party. This can sharpen existing internal divisions and empower new groups within the party to challenge its leadership.

Despite these problems, armed opposition groups have made this transformation in a range of dramatically different historical contexts covering, inter alia, Angola, Bosnia, Burundi, El Salvador, Ethiopia, Mozambique, Liberia, Nepal and Northern Ireland. More than half of the parties that participate in the first post-conflict election continue to participate in all subsequent elections.[5]

Parties composed of former rebels usually remain in opposition. But in a handful of countries, including Burundi and Ethiopia, former rebel groups have become the ruling party through elections or as part of a peace deal. For organisations such as Hizbullah in Lebanon or Hamas in Palestine, 'transformation' might not be a wholly accurate term. In these cases, participation in electoral politics implies not so much a break with the past as the acquisition of another tool for achieving organisational goals.⁶

A successful party must be capable of participating in elections and government. This requires mechanisms for acquiring and allocating material resources within the party; establishing organisational routines for decision-making and gathering and distributing information; and mobilising and sustaining a voter base through the use of collective incentives grounded in a shared identity or ideology.

Organisational characteristics

There is evidence that rebel groups that have experience as a political party are more likely to enter mainstream post-war politics than those that do not. They might also be expected to perform better. No large-scale comparative studies on this question have been completed, but careful case studies and 'small-n' comparisons suggest that parties with prior political experience may have retained some organisational structures that can facilitate the transition to party politics after a war, such as consensual decision-making mechanisms for gathering input from the party base.⁷ Furthermore, an organisation's characteristics tend to be 'sticky', outlasting changes in its environment. As argued by John Ishiyama and Anna Batta, post-communist successor parties with internal democratic features had generally developed these features prior to the end of communist rule.⁸

This insight also applies to organisational characteristics that might hinder successful adaptation to peacetime politics. Pointing to Burundi, Katrin Wittig notes that wartime command structures can have lasting impact – positive or negative – on a party's organisational structure and, by extension, its performance either in power or in opposition. Drawing on the experience of Ethiopia, Terrence Lyons contends that rebel groups that fought long wars in small territories and with less external support are more likely to develop strong, coherent (and authoritarian) party organisations after a war.

Strong and coherent leadership and institutionalised decision-making structures forged out of the need to govern territory in wartime can be useful in the transformation to peacetime politics. Regardless of whether they are deployed to support democratic politics, such leadership structures can also be used to secure autocratic rule once such parties win power, as occurred in Ethiopia, Rwanda and Burundi.

Parties in Bosnia-Herzegovina, El Salvador, Mozambique and Timor-Leste have experienced organisational problems arising from electoral success.[9] Subnational elections in which a party's candidate gains executive power – along with the visibility and resources that translate into political leverage at election time – have been particularly challenging to central party leaderships. As members of a party participate in legislatures, municipal government or state agencies, their perspectives and incentives tend to change. This results in incremental shifts in attitudes and behaviour within the sub-groups of a party across a range of institutions.

The reactions of a party's leaders determine the impact of these changes on its identity, commitment to electoral politics and wider approach to politics. In weakly institutionalised parties such as the Mozambique National Resistance Movement, party leaders find it relatively easy to quash challenges without repercussions. In the absence of internal party rules and routines for candidate selection and the adjudication of disputes, these leaders have few constraints in reacting to threats. At the same time, the absence of rules limits the party's mechanisms for adaptation to democratic politics.

In contrast, El Salvador's Farabundo Martí National Liberation Front entered post-war politics with the legacy of a political front – multiple factions loosely grouped for a common political purpose. This necessitated reliance on organisational procedures for leadership selection and the resolution of disputes. As procedural legitimacy was important for any member of the party, representatives who had been empowered by electoral politics – principally, newly elected mayors – used these procedures to challenge the leadership's approach to building popular support, which ultimately meant re-examining the party's identity. In this case, broad acceptance of existing rules facilitated the kinds of internal changes that lead to party adaptation and transformation.[10]

Environmental factors

A constellation of environmental factors – including formal institutions and the involvement of external actors who might act as guarantors of a peace deal – shape actors' willingness to commit and adapt to the post-war political settlement. These factors affect party leaders' calculus on the relative price of participation in politics versus a return to violence (or the maintenance of both capacities concurrently).

Formal institutions include peace agreements, which elaborate the terms of the settlement, and political entities such as electoral rules or subnational elected government. Specific provisions for rebel-to-party transformation may be important as well. According to Mimmi Söderberg Kovacs and Sophia Hatz, while such provisions increase the chances that a rebel group will form a party, they are neither necessary nor sufficient for this outcome.[11] Provisions such as the integration of rebel fighters into state administration and the security forces or other guarantees of economic and physical security also influence the durability of peace.[12]

Peace agreements may be important due to their frequent inclusion of external actors who can act as guarantors. These actors can be empowered to monitor various aspects of the peace agreement and, at a minimum, call out violations of the deal. Some external actors may even have a mandate to directly intervene in response to such violations.

Donors may also play an important role in post-conflict environments, even relative to peacekeepers. For example, the presence of a multilateral peace operation does not in itself significantly increase the chance that a rebel group will form a party. Instead, the impact of external intervention is significant only when it includes a committee of donors overseeing the peace agreement.

Mozambique's experience provides insight into this dynamic. There, bilateral donors who had provided humanitarian aid for years were able to leverage the relationships they had built in those years to strengthen compliance using incentives and disincentives, ranging from additional financial support for certain elements of the peace process to private diplomatic intervention with particular actors. Donors have also used their deep contextual knowledge to continuously refine the peace process, anticipating and working to remove diplomatic and financial obstacles. Moreover, they have ensured that the rules agreed to by both sides in the peace settlement are respected, and have served as a kind of court of public opinion to which either side could appeal when the rules are broken.[13]

Generally, post-rebel parties are more likely to adapt to new rules when they expect that those rules will not only be enforced but will endure. In relatively stable environments, the time horizons of political actors tend to expand, making them more likely to respond to rules and processes that can be backed by credible enforcement mechanisms. Although donors and other external actors can play a pivotal early role in enforcing the rules and convincing politicians that the postwar settlement based on electoral politics will hold, ultimately these actors must be replaced with new mechanisms for ensuring predictability and rule stability.

There is little evidence that specific electoral rules have a significant impact on party formation and political participation in the long term.[14] However, one factor that does appear to be influential is political inclusion, which may be defined in a variety of ways. For instance, Ishiyama and Michael Marshall argue that in conflicts settled between 1979 and 2014 that included at least one postwar election, 'including relevant rebel groups (and perhaps more significantly not excluding significant groups) [as political actors] reduces the likelihood of civil war and conflict resumption'.[15]

Charles Call suggests that a different aspect of political inclusion may be important. For him, inclusion extends beyond institutional arrangements such as power-sharing to the broader political process. The meaning of inclusion depends on the perceptions of the post-rebel party; its expectations of inclusion are set during the transition from war to peace. Exploring cases in which a conflict appeared for several years to have been 'settled' but then recurred, he contends that conflict becomes more likely when post-rebel parties' expectations of inclusion are not met. Call argues that '*political exclusion*, rather than economic or social factors, plays the decisive role in most cases of civil war recurrence: Political exclusion acts as a trigger for renewed armed conflict.'[16]

In any political system, actors and institutions interact in a dynamic process to shape political outcomes over time. A post-rebel party's organisational experience can provide it with the raw material that will shape its initial participation in politics. But this participation launches an endogenous process of organisational change within the groups involved. This process of organisational change, as well as broader political outcomes, is conditioned by the rules that constrain the behaviour of the post-rebel party, as well as that of other political actors. Thus, the key to a successful transformation from rebel to party seems to lie in the interaction between an organisation and its environment.

Colombia's FARC offers an opportunity to examine this interaction as it unfolds. After 50 years in existence, FARC is unusually well institutionalised and has a clear identity and ideological base. But the environmental context poses challenges for the group's political incarnation. Its initial foray into politics, in the 1980s, was marked first by electoral success and then by a wave of retaliatory violence and intimidation from paramilitary and state forces. The group withdrew from politics and the peace process unravelled. This early political experience provides little encouragement for the risk-taking that rebel-to-party transformation entails. In the last two years, FARC's peace negotiations with the Colombian government have shifted its calculus in favour of participating in politics, which offers ex-combatants protection from prosecution – except for the most serious war crimes – as well as opportunities for social reintegration. Complicating matters is FARC's reliance on a portfolio of criminal enterprises, from kidnapping and protection rackets to the drug trade. These economic engagements are likely to divide the party's leaders, with some reluctant to embrace politics fully if this entails relinquishing illicit income. Facing an uncertain future in politics, some FARC leaders might find the transition a tough sell. Finally, the group faces a complex political environment, with significant public opposition to its integration into politics. Still, the peace deal likely represents the best chance yet for FARC to make the transition from rebel group to party.

Carrie Manning is Professor and Chair, Department of Political Science, Georgia State University. She is the author of three books and dozens of journal articles on post-conflict politics.

Notes

[1] Carrie Manning and Ian Smith, 'Political Party Formation by Former Armed Opposition Groups after Civil War', *Democratization*, vol. 23, no. 6, October 2016.

[2] This view of parties draws on Giovanni Sartori, *Parties and Party Systems* (Colchester: ECPR Press, 2005); and Angelo Panebianco, *Political Parties: Organization and Power* (Cambridge: Cambridge University Press, 1988).

[3] For an introduction to the burgeoning literature on rebel groups in politics and related conceptual issues, see John Ishiyama (ed.), *Democratization*, vol. 23, no. 6, October 2016; and Gyda Maras Sindre and Johanna Söderström (eds), *Civil Wars*, vol. 18, no. 2, July 2016.

[4] Dankwart Rustow, 'Transitions to Democracy: Toward a Dynamic Mode', *Comparative Politics*, vol. 2, no. 3, April 1970.

[5] Carrie Manning and Ian Smith, 'Persistence in Party Politics by Former Armed Opposition Groups', unpublished manuscript.

[6] Benedetta Berti, 'Rebel Politics and the State: Between Conflict and Post-Conflict, Resistance and Co-Existence', *Civil Wars*, vol. 18, no. 2, July 2016.

[7] Valerie Alfieri, 'Political Parties and Citizen Political Involvement in Post-Conflict Burundi: Between Democratic Claims and Authoritarian Tendencies', *Civil Wars*, vol. 18, no. 2, July 2016; Carrie Manning, *The Making of Democrats: Elections and Party Development in Postwar Bosnia, El Salvador, and Mozambique* (New York: Palgrave Macmillan, 2008); Katrin Wittig, 'Politics in the Shadow of the Gun: Revisiting the Literature on "Rebel-to-party Transformations" through the Case of Burundi', *Civil Wars*, vol. 18, no. 2, July 2016.

[8] John Ishiyama and Anna Batta, 'Swords into Ploughshares: The Organizational Transformation of Rebel Groups into Political Parties', *Communist and Post-Communist Studies*, vol. 44, no. 4, December 2011.

[9] Carrie Manning, *The Politics of Peace in Mozambique* (Westport, CT: Praeger, 2002); Manning, *The Making of Democrats*; Gyda Maras Sindre, 'Internal Party Democracy in Former Rebel Parties', *Party Politics*, vol. 22, no. 4, July 2016.

[10] For a more detailed discussion of both of these cases, see Manning, *The Making of Democrats*.

[11] Mimmi Söderberg Kovacs and Sophia Hatz, 'Rebel-to-party Transformations in Civil War Peace Processes 1975–2011', *Democratization*, vol. 23, no. 6, October 2016.

[12] Charles T. Call, *Why Peace Fails: The Causes and Prevention of Civil War Recurrence* (Washington DC: Georgetown University Press, 2012); Söderberg Kovacs and Hatz, 'Rebel-to-party Transformations in Civil War Peace Processes 1975–2011'; Caroline Hartzell and Matthew Hoddie, 'Institutionalizing Peace: Power Sharing and Post-Civil War Conflict Management', *American Journal of Political Science*, vol. 47, no. 2, April 2003; Devon Curtis and Jeroen de Zeeuw, 'Rebel Movements and Political Party Development in Post-Conflict

Societies – A Short Literature Review', City University of New York, 2007.

[13] Carrie Manning and Monica Malbrough, 'Bilateral Donors and Peacebuilding in Mozambique', *Journal of Modern African Studies*, vol. 48, no. 1, March 2010.

[14] Manning and Smith, 'Political Party Formation by Former Armed Opposition Groups after Civil War'.

[15] Michael Christopher Marshall and John Ishiyama, 'Does Political Inclusion of Rebel Parties Promote Peace after Civil Conflict?', *Democratization*, vol. 23, no. 6, October 2016.

[16] Call, *Why Peace Fails*. Emphasis in original.

Chapter Two

Maps, Graphics and Data

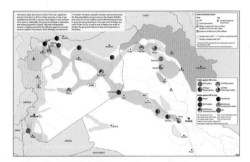

Territory lost by ISIS and operations against the group in 2016   **74**

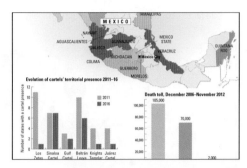

Ten years of Mexico's 'war on drugs'   **76**

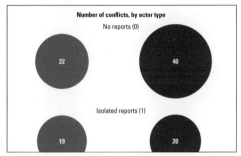

Distribution of highest reported level of rape during civil war   **77**

Refugee movements to selected non-Western countries   **78**

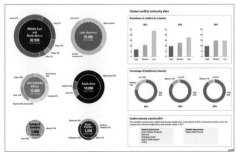

Global conflict fatalities 2016   **80**

Myanmar's newest insurgency   **82**

Territory lost by ISIS and operations against the group in 2016

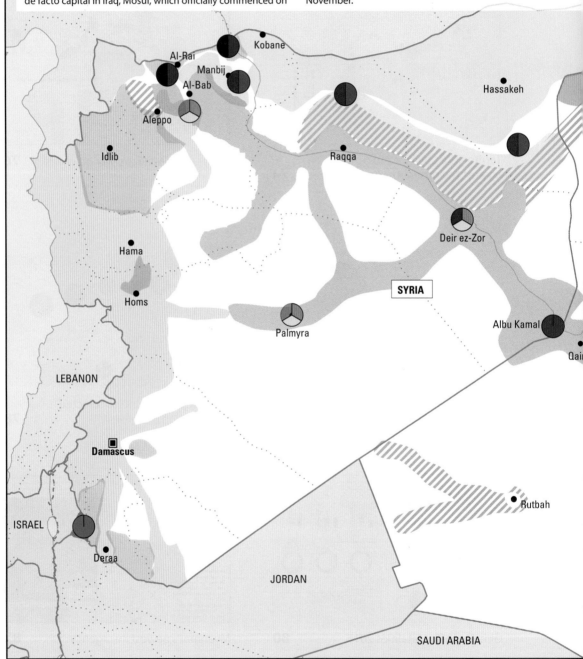

The Islamic State, also known as ISIS or ISIL, lost a significant amount of territory in 2016. In Anbar province, in Iraq, it was expelled from Ramadi in January, from Falluja in June and from Hit in April. In Salahuddin, the group lost Shirqat in September, after losing Qayyarah in August. Qayyarah subsequently became a staging ground for the offensive against the group's de facto capital in Iraq, Mosul, which officially commenced on 17 October. The Mosul operation includes Iraqi security forces, the Shia paramilitary groups known as the Popular Mobilisation Units, the US-led coalition and Kurdish Peshmerga forces. In Syria, ISIS was pushed back in the area north of Raqqa and north of Deir ez-Zor, as well as east of Aleppo and south of Manbij. The group lost Palmyra in May, but recaptured it in November.

Sources: IISS; Institute for the Study of War

Maps, Graphics and Data

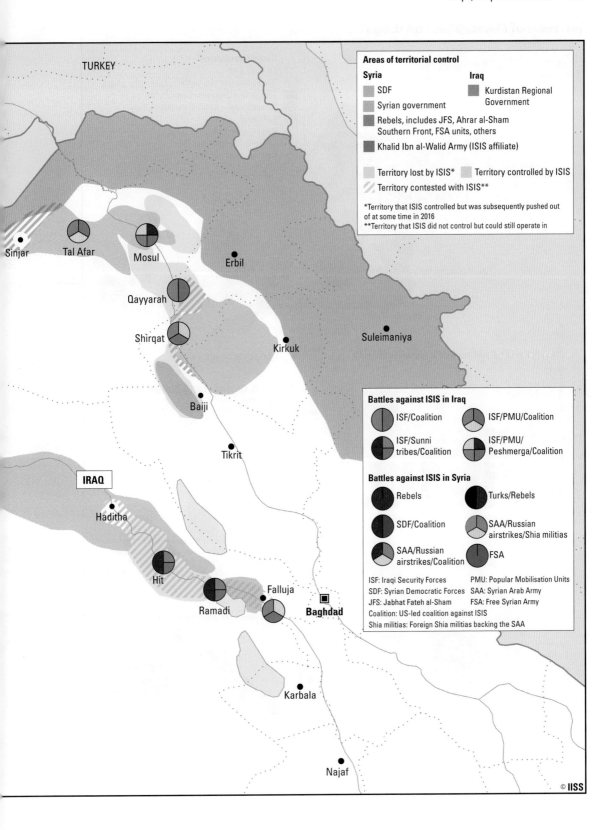

Ten years of Mexico's 'war on drugs'

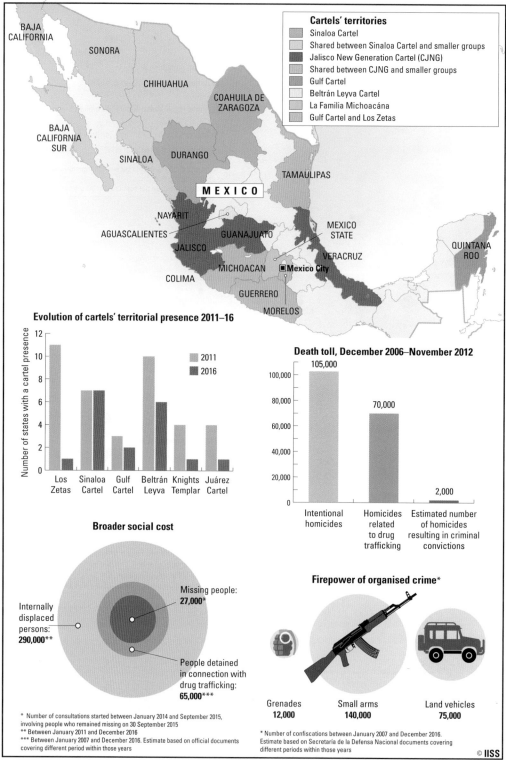

Distribution of highest reported level of rape during civil war

Each figure depicts the highest reported level of rape by actor type (state or rebel forces) over the course of a conflict for 91 civil wars between 1980 and 2012. For each conflict, there are two observations: the highest reported level of rape perpetrated by state forces, and the highest reported level of rape perpetrated by rebel forces. While the reported level of rape may vary over the course of a conflict from 0 reports (coded as 0) to isolated reports (coded as 1) and numerous or widespread reports (coded as 2), the levels shown here represent the highest reported level during the entire span of the conflict. The figures highlight the considerable variation between perpetrators of rape, even within one conflict. Overall, state forces are more likely than rebel forces to be reported to have engaged in numerous or widespread levels of rape.

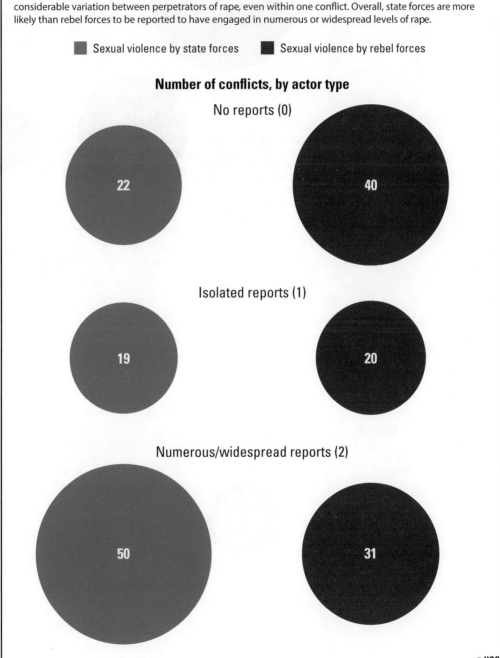

Sources: Dara Kay Cohen, *Rape During Civil War* (Ithaca, NY: Cornell University Press, 2016)

Refugee movements to selected non-Western countries

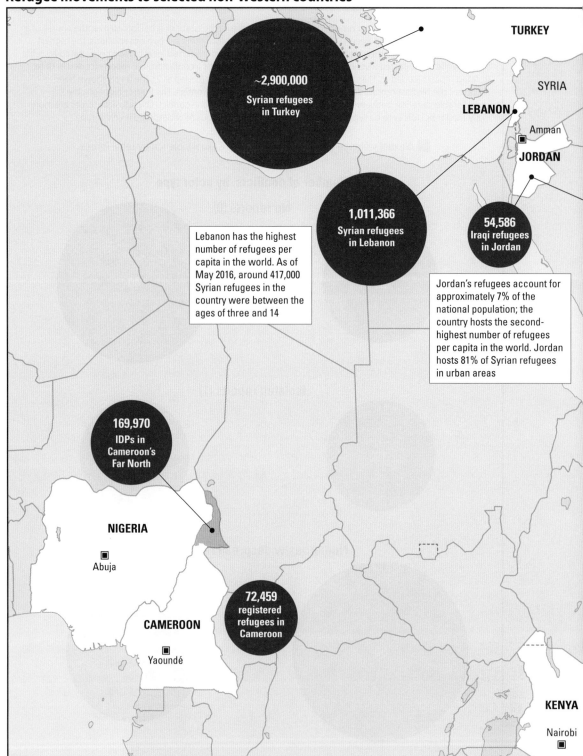

Sources: *Economist*; European Commission; Human Rights Watch; Relief Web; UNHCR

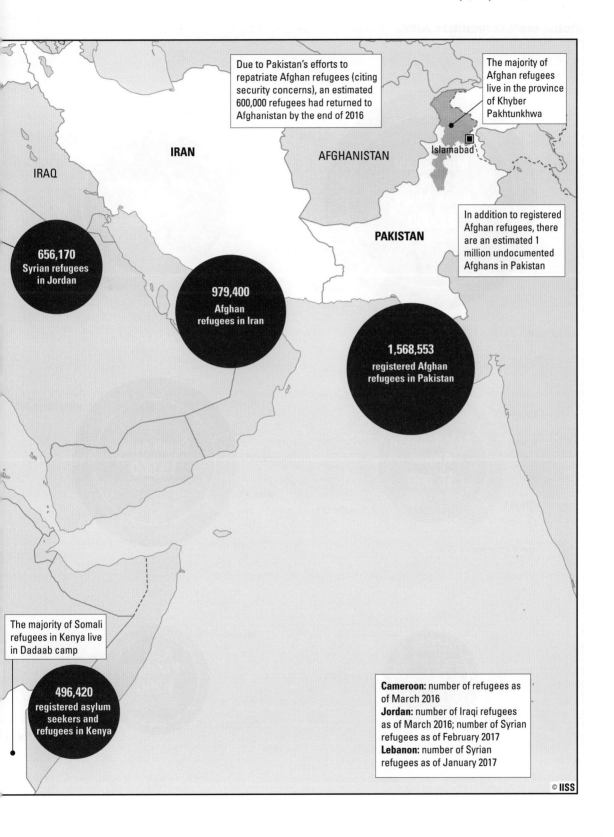

Global conflict fatalities 2016

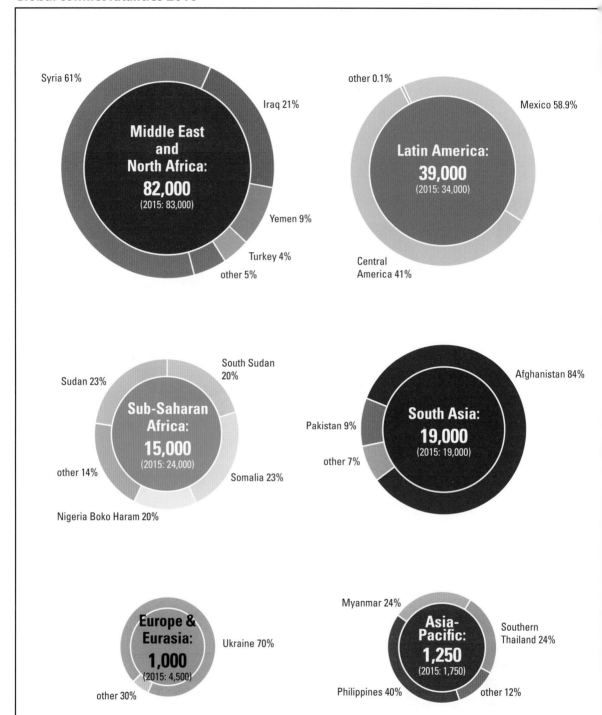

Source: IISS Armed Conflict Database

Global conflict-intensity data

Breakdown of conflicts by intensity

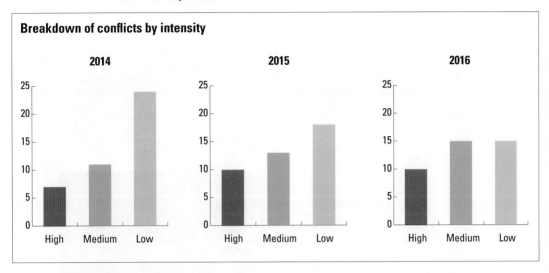

Percentage of fatalities by intensity

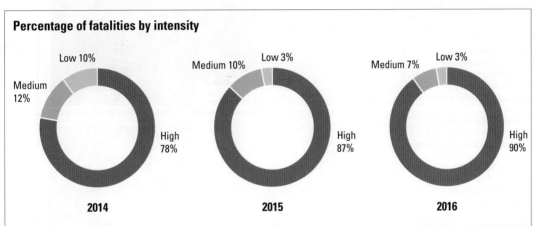

Conflict-intensity watchlist 2016

This watchlist covers the key conflicts that became notably more or less intense in 2016, compared to previous years; the changes they underwent might affect their intensity ratings in 2017.

Notable deterioration
India–Pakistan (Kashmir)
Myanmar
Philippines (ASG)
Nigeria (Delta Region)
Turkey

Notable improvement
Nigeria (Boko Haram)

© IISS

Myanmar's newest insurgency

Rohingya Muslim militants carried out an assault on border police in northern Rakhine State in October 2016, initially killing nine officers. The attack elicited a heavy-handed security crackdown that the United Nations says is likely to have killed hundreds of people and displaced at least 92,000 others, 69,000 of them to Bangladesh. At least 30 security-forces personnel have also been killed in the crackdown. Media organisations, impartial observers and aid agencies have had severely restricted access to the conflict area. The international community has warned of an attempt at ethnic cleansing, crimes against humanity and even genocide; Myanmar has denied that any crimes are taking place. A combination of historically rooted grievances and prejudices have led to repeated instances of mob violence against Rohingya, causing instability in Rakhine State and across the country. Rohingya have no rights to vote, own land, marry or travel, and many Myanmar citizens regard them as not 'Burmese'. The previous cycle of violence in Rakhine State began in 2012, when riots killed nearly 300 people – both Rohingya and ethnic Rakhine – and displaced about 150,000 others. Further riots in Meiktila, in central Myanmar, in 2013 killed another 44 people and destroyed thousands of homes.

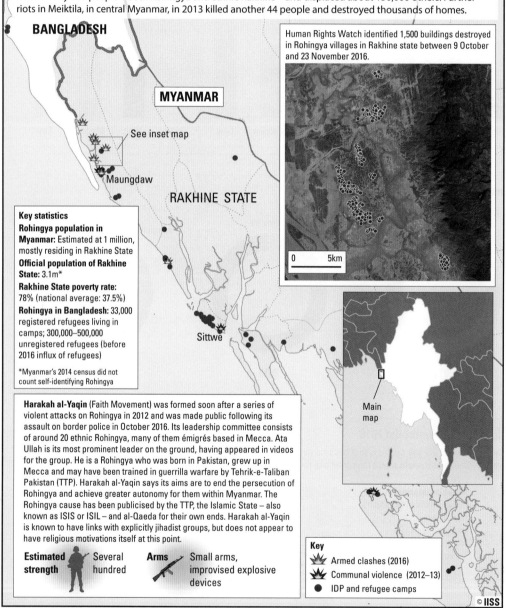

Human Rights Watch identified 1,500 buildings destroyed in Rohingya villages in Rakhine state between 9 October and 23 November 2016.

Key statistics
Rohingya population in Myanmar: Estimated at 1 million, mostly residing in Rakhine State
Official population of Rakhine State: 3.1m*
Rakhine State poverty rate: 78% (national average: 37.5%)
Rohingya in Bangladesh: 33,000 registered refugees living in camps; 300,000–500,000 unregistered refugees (before 2016 influx of refugees)

*Myanmar's 2014 census did not count self-identifying Rohingya

Harakah al-Yaqin (Faith Movement) was formed soon after a series of violent attacks on Rohingya in 2012 and was made public following its assault on border police in October 2016. Its leadership committee consists of around 20 ethnic Rohingya, many of them émigrés based in Mecca. Ata Ullah is its most prominent leader on the ground, having appeared in videos for the group. He is a Rohingya who was born in Pakistan, grew up in Mecca and may have been trained in guerrilla warfare by Tehrik-e-Taliban Pakistan (TTP). Harakah al-Yaqin says its aims are to end the persecution of Rohingya and achieve greater autonomy for them within Myanmar. The Rohingya cause has been publicised by the TTP, the Islamic State – also known as ISIS or ISIL – and al-Qaeda for their own ends. Harakah al-Yaqin is known to have links with explicitly jihadist groups, but does not appear to have religious motivations itself at this point.

Estimated strength: Several hundred
Arms: Small arms, improvised explosive devices

Key
- Armed clashes (2016)
- Communal violence (2012–13)
- IDP and refugee camps

Sources: IISS Armed Conflict Database; Human Rights Watch; UN Office for the Coordination of Humanitarian Affairs; International Crisis Group; *Dhaka Tribune*; UN Development Programme; Al-Jazeera; International Organization for Migration; Amnesty International; Google Earth/DigitalGlobe

Chapter Three

Middle East

Egypt

Despite reports of a decrease in violence in the Sinai Peninsula, Egypt failed to quell the jihadist insurgency in the region in 2016. There, the security forces often faced deadly attacks by Wilayat Sina (Sinai Province), an affiliate of the Islamic State, also known as ISIS or ISIL, that dominated the insurgency throughout the year. Elsewhere in Egypt, radicalised cells of Muslim Brotherhood members and anti-coup operatives became more focused in their terrorist campaign – as was underlined by the formation of Hassm (Determination) and Liwa al-Thawra (Brigades of the Revolution) in the second half of 2016. Although most Egyptians remained largely unsympathetic to Islamist activity, they grew increasingly disenchanted with President Abdel-Fattah Al-Sisi. Police brutality, forced disappearances and arbitrary detention stoked heavy criticism of the security forces, and made international headlines with the death of Italian doctoral student Giulio Regeni in January. Sisi's popularity was severely damaged by his attempt to cede control of two islands in the Red Sea, Tiran and Sanafir, to Saudi Arabia. The agreement led to significant tension between Cairo and Riyadh, which had

Key statistics	2015	2016
Conflict intensity:	Medium	Medium
Fatalities:	3,000	1,750
New IDPs:		
New refugees:	2,500	

Middle East

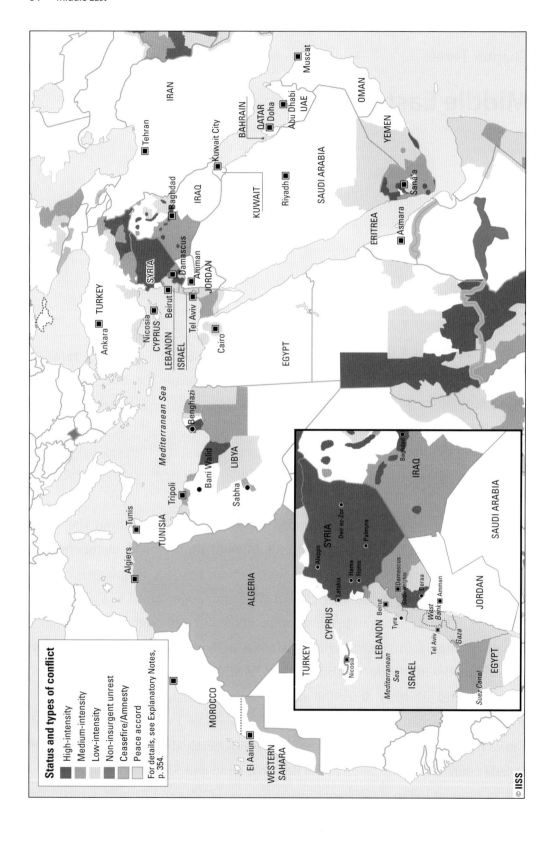

been the Sisi regime's most important backer since it came to power in 2013. Forced to contend with severe economic problems, Sisi also faced significant challenges in attempting to stabilise the country.

Wilayat Sina: enduring threats

The seeming decline in violence in the Sinai Peninsula was attributed to a decrease in attacks on international and other high-profile targets (such as the Russian MetroJet plane that was downed in October 2015, killing more than 200 people). Furthermore, the Egyptian Air Force killed the leader of Wilayat Sina, Abu Duaa al-Ansari, in August, along with hundreds of other Wilayat Sina operatives throughout the year.

Despite experiencing these setbacks, the insurgents maintained a high frequency of attacks on military and police personnel in 2016. They were particularly active between September and December, launching 20–30 such operations each month in the period, compared to 10–20 in the last four months of 2015. Although the more recent attacks were less spectacular, they were not always small-scale. In November 2016, an attack on al-Gaz army checkpoint in Sabeel village, south of the city of Arish, killed 12 soldiers. This was the third consecutive month in which at least ten soldiers had been killed in a single attack on a military checkpoint. The trend reflected the new boldness of the insurgents, who earlier in the year had begun to kill policemen in their homes near Arish.

Wilayat Sina maintained its monopoly on jihadist activity in North Sinai, with no other group claiming responsibility for an attack in the region in 2016. In addition to targeting military and police forces, Wilayat Sina executed civilians accused of collaborating with the army and launched attacks on Sufi clerics and Christian communities – moves that many perceived as being inspired by ISIS. Wilayat Sina also appeared to adopt other features of its parent organisation, particularly in its rhetoric. Although it was unable to hold territory as ISIS did in Iraq and Syria, Wilayat Sina remained highly mobile and proved its resilience through its capacity to repeatedly regroup. As observed by analysts, Wilayat

Sina's strategy focused on sustained engagement akin to attrition and attacks on soldiers' morale, which had the primary aim of eroding their will to fight.

Meanwhile, ISIS claimed responsibility for attacks in Egypt outside the Sinai Peninsula. The fact that the group took credit for such operations directly – rather than through Wilayat Sina – suggested that either it engaged in only minimal cooperation with its affiliate or that it was trying to establish a distinct presence in Egypt outside Sinai. Thus, ISIS claimed to have been behind attacks on civilian gatherings such as the 11 December suicide bombing at Saint Mark's Coptic Orthodox Cathedral in Abbassia, in Cairo, which killed 27 civilians. The group conducted larger-scale assaults in Egypt in 2016 than it had the previous year. However, such attacks did not necessarily suggest a qualitative increase in the group's capability to strike in Cairo as much as its capacity to skilfully exploit 'lone wolves' there.

Non-Salafist and revolutionary Islamism

Hassm and Liwa al-Thawra were among several non-Salafist and revolutionary Islamist groups to emerge in the second half of 2016. Former supporters of the Muslim Brotherhood, or individuals who were radicalised after the dissolution and outlawing of the organisation, established these groups by building on revolutionary entities such as the Popular Resistance Movement – adopting their rhetoric and goals, and thereby keeping their mission alive. The new militant groups predominantly relied on small arms and rudimentary explosives to carry out their operations. From July onwards, Hassm claimed responsibility for several high-profile attacks, including assassination attempts – only some of which succeeded. In August, the group tried to assassinate Ali Gomaa, Egypt's former Grand Mufti (the country's highest-ranking Islamic cleric). An outspoken critic of Islamist groups, Gomaa was viewed as having a close relationship with the military establishment. Liwa al-Thawra announced its formation in August, following an attack that killed three police officers in Menoufia province, in northern Egypt. The main adversary of these new groups was the military government, particularly Sisi.

The rise of Hassm and Liwa al-Thawra suggested a more concerted and focused effort by Islamists who had not accepted the outcome of Egypt's 2013 coup. They aimed to escalate the low-level insurgency outside Sinai, which appeared to be disintegrating in the first half of 2016. The Popular Resistance Movement also took responsibility for some of the attacks claimed by Hassm and Liwa al-Thawra, suggesting that it was trying to act as an umbrella group for this strand of violence in Egypt. However, it remained unclear whether the Muslim Brotherhood leadership was involved in these groups. It was likely that these groups, comprised of disenchanted young people, were working almost independently. Although the groups had similar objectives and praised one another's activities, the insurgency appeared to lack internal cohesion.

The goals and narratives of the new militant organisations differed from those of Wilayat Sina and ISIS. Rather than attempting to create an ISIS-style 'caliphate', they sought to replace the military government with one sympathetic to Islamists – and perhaps even led by former president Muhammad Morsi. The new groups mostly targeted the security forces rather than civilians. For example, both Hassm and Liwa al-Thawra were quick to release statements distancing themselves from the attack on the Coptic Church. As had been the case with Brotherhood-affiliated groups that branded themselves as anti-coup resistance movements, they emphasised their refusal to target civilians. Hassm was often described as an organisation formed to exact revenge for a massacre at Rabaa al-Adawiya in August 2013.

The Egyptian government continued to view the Brotherhood as the main threat, killing or arresting its supporters. Sisi rejected any possibility of dialogue or reconciliation with the organisation, stating that the 'conflict' is not between the state and Brotherhood but between the Brotherhood and the Egyptian people. This hardline approach meant that the insurgency would likely continue to mutate and re-emerge, as the grievances fuelling it were left unresolved.

Struggles of stabilisation

The Sisi government's main objective was to restore calm and a semblance of normality in the country, following revolutionary upheaval. One step towards such stabilisation was the meeting of parliament on 10 January, the first in four years. This process was also fundamentally flawed: an investigative report published in mid-March revealed that the government actively intervened in the election to engineer a parliament amenable to its interests.

However, the state remained highly insecure and appeared to be affected by internal discord. The extent of the state's insecurity was demonstrated on 25 January, the fifth anniversary of the revolution, when it deployed tens of thousands of security personnel to Tahrir Square, the iconic site of the protests in 2011. The security forces massively outnumbered the few hundred protesters who congregated there, suggesting a high level of paranoia. The political climate remained hostile to any dissent: on 24 January, the president had told the population, 'don't listen to anyone but me'. His comments came amid the government's continuing crackdown on activists and closure of online groups, which led to the arrest of their members for alleged incitement against the state.

The anniversary of the revolution coincided with Regeni's disappearance. His body, found nine days later, showed signs of torture: a post-mortem examination revealed that he had been interrogated for up to seven days at 10–14-hour intervals, in line with methods used by the interior ministry. The government denied that the security forces were involved in his murder and tried to deflect responsibility for the crime. It initially stated that Regeni had died in a car accident, before blaming drugs, espionage, sexual misadventure and criminal activity. It also suggested that the Brotherhood had killed him, with Sisi blaming 'enemies' who sought to destabilise Egyptian relations with Italy. The government said in March that a criminal gang had killed Regeni, and that the police had killed all four members of the group in a raid on their home. The story was largely dismissed as a cover-up and an attempt to end the bad press that the case was attracting.

Regeni's murder underlined the impunity of the Egyptian security forces, and led to a resurgence of public anger about police brutality and the practice of enforced disappearances. According to a report by Amnesty International, there were between three and four forced disappearances in Egypt every day. The authorities shut down the Nadeem Centre for Rehabilitation and Victims of Violence, which helps victims of torture, as part of a large-scale crackdown on civil-society organisations. Activists and researchers faced travel bans and criminal investigations into the sources of their funding. The Egyptian parliament overwhelmingly endorsed a repressive law on non-governmental organisations, which effectively restricted their operations to developmental and social work.

Regeni's death cast doubt on the government's ability to control the police force, which reported to an interior ministry that since 2013 had had carte blanche in dealing with political opponents, citing purported security concerns. There was no investigation into, or even acknowledgement of, the majority of arbitrary detentions and forced disappearances. This approach mirrored Egypt's handling of the MetroJet bombing, indicating the inner workings of a state that shunned threatening information and resorted to conspiracy theories. Italian investigators highlighted this lack of transparency, frequently expressing their frustration with a lack of progress in the Regeni case.

Spiralling economic challenges compounded these stabilisation problems. Egypt faced a severe shortage in foreign-currency reserves that caused a boom in the black market for US dollars. The country's central bank adopted a more flexible exchange rate and devalued the Egyptian pound by almost 13% in mid-March, before taking the long-awaited measure of allowing the pound to float in early November. The latter move led the International Monetary Fund to approve a US$12-billion loan to Egypt. Although it made the pound more competitive abroad, the measure risked aggravating domestic discontent by causing shortages in basic commodities such as sugar.

Shifting alliances

In April, thousands of Egyptians took to the streets to protest against a Saudi–Egyptian maritime agreement that ceded control of Tiran and Sanafir to Riyadh. Most Egyptians viewed the deal as a sale of national territory, and even national dignity, as it was signed alongside other agreements during Saudi King Salman bin Abdulaziz Al Saud's official visit to Cairo. Egypt and Saudi Arabia had disputed ownership of the islands since 1982, when Israel returned them to Egypt. Hundreds of people were arrested in the protests, while escalating tension between the authorities and the journalists' syndicate prompted the police to raid its offices. The protesters were subsequently emboldened by evident discord within the establishment, after a top court overruled the agreement.

Since Sisi came to power, Saudi Arabia had provided Egypt with an estimated US$25bn in aid – support that was arguably crucial to avoiding economic collapse. The Saudi–Egyptian relationship deteriorated after Cairo voted in favour of a Russian-backed UN Security Council draft resolution on Syria in October. This led Saudi Arabia's national oil company, Aramco, to halt shipments to Egypt, terminating a US$23bn deal involving monthly deliveries of 700,000 tonnes of petroleum. Egypt discussed an agreement with Iraq for one million barrels of oil per month. Despite this tension, Saudi Arabia still transferred US$2bn to Egypt's central bank. Although Cairo suggested that it would change some of its policies on Egypt, there was unlikely to be a complete rupture between the sides, as this would risk further destabilising the Middle East. Saudi Arabia was aware that Egypt's descent into deeper chaos would create another significant threat to the region.

Sisi was increasingly vocal in his support for one of Saudi Arabia's most significant opponents, Syrian President Bashar al-Assad. There were reports that Sisi had deployed Egyptian troops in Syria alongside their Russian counterparts, as part of a perceived tilt towards Moscow that involved strengthened military and diplomatic ties. There were also suggestions that this shift might bring Egypt closer to Saudi Arabia's main rival, Iran. However, Egypt's prioritisation of its relationship with Israel – above all other states – limited such a realignment. Nonetheless, Cairo's adoption of a 'third way' risked alienating allies on both sides.

Long after the 2011 revolution, and its undoing via the 2013 coup, Egypt continued to suffer from political and economic instability. The jihadist threat from Wilayat Sina had not diminished, despite the military's heavy-handed campaign against the group in North Sinai. Furthermore, the emergence of organisations such as Hassm and Liwa al-Thawra demonstrated that although the Sisi government had largely defeated the Muslim Brotherhood, remnants of the Brotherhood would continue to pose a threat in the near future. The forceful strategy of, and lack of accountability among, state forces arguably exacerbated many of Egypt's security problems. Yet major international actors acknowledged that the destabilisation of the country would have severe regional repercussions – a fact that Egyptian leaders played to their advantage.

Iraq

The Iraqi security forces and their allies made significant territorial gains against the Islamic State, also known as ISIS or ISIL, in 2016. They won a series of substantive, symbolically important victories in areas ranging from Anbar province to Ninewa province, where the operation to retake Mosul finally began. However, the widespread destruction and internal displacement evident in many cities, towns and villages previously controlled by ISIS generated concern that some recaptured territory was no longer habitable. The militia structures within the security forces became increasingly embedded in Iraqi politics and society, compounding these stabilisation problems. As an array of actors sought to advance their interests, Iraqi politics remained in turmoil throughout the year, with Prime Minister Haider al-Abadi struggling to implement reforms, and Baghdad and the Kurdistan Regional Government (KRG) failing to clarify the uncertain terms of their relationship.

Key statistics	2015	2016
Conflict intensity:	High	High
Fatalities:	13,000	17,000
New IDPs:	1,100,000	250,000
New refugees:	44,000	

Erosion of ISIS-controlled territory

By the end of 2016, ISIS had lost control of nearly one-quarter of the 30,200 square miles of territory it controlled across Iraq and Syria at the start of the year. Furthermore, ISIS made no territorial gains in Iraq. Given its ideological emphasis on building a state, the group suffered severe damage to its credibility as a result of these territorial losses, as well as the financial, administrative and other challenges that accompanied them. The retreat highlighted the group's failures in governance and state-building, which appeared to have dented its legitimacy and popularity. Internal ISIS documents indicated that it was experiencing shortages in personnel. While there were no verified accounts of these shortfalls, an estimate published in July 2016 stated that the group had around 12,000 members – a marked decline from the 31,500 identified by the US Department of Defense in 2014. This trend related to both the decline of ISIS and the introduction of stricter measures to prevent potential recruits from reaching Iraq and Syria. Although military victories contributed to the group's apparent loss of momentum and morale, they were unlikely to destroy it entirely. Many observers of the conflict in Iraq warned that ISIS might morph into a more traditional insurgent group. If ISIS abandoned its efforts at state-building and governance, it might instead deploy the full extent of the resources and manpower it had amassed in launching indiscriminate attacks against civilians. In that capacity, it would continue to inflict considerable damage.

The areas of Iraq recaptured from ISIS were in Anbar, Ninewa and Salahuddin provinces, where the Iraqi security forces, backed by airstrikes from the US-led coalition and an array of pro-government militias, launched a broad range of operations. In Anbar, the Iraqi security forces recaptured Falluja and Hit, and prepared to retake the towns of Rawa, Ana and Qaim, in the west of the province. The coalition pushed ISIS out of Qayyarah, in Ninewa, and Shirqat district, in Salahuddin province, in August and September respectively. Using Qayyarah Air Base as a staging ground, the Iraqi security forces and their allies launched in mid-October the battle to recapture Mosul, widely viewed as the stronghold of ISIS in Iraq.

The group also held on to pockets of territory elsewhere in the country, including parts of Anbar and all of Hawija, a city in Kirkuk province home to an estimated 500,000 people. Yet the long-delayed advance on Mosul was seen as the most important operation against the group in Iraq. As more than 1.5 million people lived in Mosul, occupying the city gave the group its strongest claim to being a 'caliphate', as well as access to a major revenue stream largely based on smuggling and 'taxes' on residents. Part of the US-led coalition's strategy involved targeting the economic infrastructure and capabilities of ISIS, especially financial and oil facilities, as a means of reducing the group's revenue (estimated to be around US$1 billion annually). Achieving victory in Mosul, Iraq's second-largest city, was also symbolically important, given that Iraqi security forces stationed there disintegrated before the ISIS advance in June 2014. Therefore, the operation to retake the city was depicted as a test of the effort to rebuild the Iraqi security forces.

Iraq closely cooperated with the US in planning the operation. An estimated 30,000 Iraqi soldiers and militiamen from the Kurdish Peshmerga and the Popular Mobilisation Units (PMU) participated in the battle. Despite making some major territorial gains around Mosul, retaking the towns of Bashiqa and Hammam al-Alil, the operation made slow progress overall. In a bid to boost coalition morale, the Iraqi government publicised numerous military victories on the outskirts of Mosul. Nonetheless, by the end of 2016, much of the territory seized by the security forces had been lightly defended at most.

Aftermath of military operations: destruction and displacement

Humanitarian agencies and international organisations issued a series of warnings about the consequences of violence and displacement in Mosul, with some believing that the operation to retake the city threatened to create the largest and most complicated humanitarian crisis in the world. The governor of Ninewa province, Nawfal Hamadi, said the operation was likely to result in the displacement of 1m people, adding to the 3.4m internally displaced persons (IDPs) already in Iraq. In an attempt to prevent an exodus from the city and a resulting humanitarian disaster, the Iraqi security forces urged residents of Mosul to remain in the city. These warnings were inspired by the experience of previous operations in urban environments such as Falluja and Ramadi, cities that had been badly damaged in the fighting. A UN team that visited Ramadi in early 2016 described the destruction of the city as 'staggering' and worse than that anywhere else in Iraq. The battle for the city resulted in the destruction of 5,700 buildings, including 64 bridges and much of the city's electricity infrastructure, as well as its main hospital and train station.

Large-scale destruction forms part of the strategy of warfare adopted by ISIS. The group's fighters often laid improvised explosive devices (IEDs) in urban infrastructure to slow down their enemies. Moreover, ISIS operatives frequently set fire to oilfields with the aim of deterring the advance of the Iraqi security forces – as occurred at 19 of these facilities in Qayyarah in August, prior to the group's expulsion from the district. The move created severe health risks for people living nearby and caused extensive environmental damage. An official from the state-owned North Oil Company noted that three months after the capture of Qayyarah, the Iraqi government was only in control of one of the oilfields, while more than a dozen were still ablaze. The group also set fire to the Mishraq sulphur plant, between Qayyarah and Mosul, in October, creating a 5-kilometre-wide toxic cloud that prompted the evacuation of 200 families.

The same pattern of destruction was repeated even where it had no obvious military purpose – aside from damaging Iraq's economic base and strategic industries. For example, ISIS reportedly destroyed at least 1m acres of arable land in Iraq, systematically looting agricultural equipment, which it used to make IEDs, and sabotaging storage facilities and harvests. Before escaping a major oil refinery in Baiji, the group rendered the facility unusable. The refinery had the capacity to produce 310,000 barrels of oil per day, enough to serve more than one-third of Iraq's energy requirements. Repairing the refinery would cost the cash-strapped Iraqi government billions of dollars. Almost all of Baiji's 200,000 residents had fled the city, leaving behind only militias involved in illicit activity. These militias reportedly looted the remaining functional parts of the refinery, before selling them on the black market. Indeed, several non-state armed groups other than ISIS also damaged infrastructure, and were responsible for attacks on civilians' homes.

These events indicated the long-term stabilisation and rehabilitation challenges Iraq faced – problems that Baghdad, registering a US$20bn deficit in its 2017 budget, lacked the financial capabilities to address. Although the World Bank allocated US$250m to Iraq for the reconstruction of cities and towns that the Iraqi security forces seized from ISIS, this was only enough to repair a fraction of the damage. The reconstruction effort was thought to require as much as US$10bn in Ramadi alone, and up to US$60bn in the country as a whole.

Popular Mobilisation Units and military arrangements

Members of the PMU continued to play a controversial role in military operations and other security arrangements, using the fight against ISIS to boost their power. The umbrella organisation comprised dozens of militias; most of them were Shia, but some were Sunni or Christian. Many of these Shia militias had ties to Iran and, as such, were often seen as following Tehran's agenda or otherwise pursuing sectarian or political goals separate from defeating ISIS. The presence of Qassem Suleimani, commander of the Islamic Revolutionary Guard Corps' (IRGC) Quds Force, on battlefields in Iraq demonstrated the close relationship between Iran and these militias. The militias argued that although Suleimani was not on the front-line of the advance on Mosul, the Quds Force was no different from the 5,000 US troops deployed to Iraq to advise and train the security forces. Abadi sought to mediate such tensions between the US and the PMU, addressing US objections by forbidding the militias from entering Mosul.

The PMU seized in November a military air base on the outskirts of Tal Afar, west of Mosul. Mostly inhabited by Turkmen, the city was strategically important as it provided the PMU with easy access to Syria, where some Shia militias had fought on behalf of President Bashar al-Assad. The PMU's involvement in Tal Afar damaged the relationship between Baghdad and Ankara, which had deployed Turkish forces to Iraq earlier in the year, after coordinating with the KRG.

Members of the PMU perpetrated several revenge attacks against Sunnis in 2016, as seen in January in Diyala – a province in which 56 militias operated. The attacks took place in public or inside the homes of civilians, who were warned to leave Diyala city or face execution. The militias assaulted Sunni mosques and other property using incendiary bombs; two journalists attempting to cover the events were shot dead outside the city. Similarly, Shia militias stationed on the outskirts of Falluja were implicated in an array of human-rights violations involving Sunnis in the city. In some cases, these militias oversaw the security-screening process for civilians fleeing Falluja, who reported that the groups detained them en masse, before separating men and boys from women and girls. There were reports that the detainees were subjected to systematic torture and looting. A committee set up by the governor of Anbar linked the PMU to the disappearance of around 600 people, and the deaths of 49 others who had turned themselves in to the authorities.

Furthermore, the militias have attempted to govern territory, often competing directly with Baghdad by undercutting the state's authority. Members of the PMU described the deployment of Iraqi forces in the Shia-dominated city of Basra as a provocation, intensifying this rhetoric after an altercation between the military and the Kata'ib Sayyid al-Shuhadaa militia at a checkpoint in January. And the militias continued to interfere in Baghdad's security structures, claiming that the government had failed to provide security. Following four large-scale bombings in Baghdad, influential Shia cleric Moqtada al-Sadr deployed his supporters to key positions in the city.

Meanwhile, Abadi made several attempts to reform the PMU, announcing plans to integrate the militias into the armed forces as a military formation. The Iraqi parliament passed in November a law that turned the PMU into a separate and legal military corps, a move designed to prevent members of the militias from participating in politics or swearing allegiance to a party. The decree also stipulated that PMU groups could comprise Iraqi volunteers from any ethnic or religious group. Abadi presented the decree as a means for disciplining the PMU, but it was widely interpreted as enhancing the militias' status by preserving the ambiguous, ad hoc arrangements behind their formation. Although the law purported to bring the PMU into the legal architecture of Iraq's military, critics of the measure argued that it legitimised the militias' lack of accountability, akin to that of the IRGC. The law built on Abadi's February decree formalising the PMU as the replacement for a national guard. The proposal for establishing a national guard had been seen as a way to provide more autonomy to local groups, particularly Sunnis who had been marginalised by existing security arrangements. Thus, the upgrade in the status of the PMU threatened relations with the Sunni population, and risked undermining efforts at national reconciliation.

Politics in turmoil
The PMU law followed nearly a year of political dysfunction that often brought Iraq to a standstill. Abadi's reform efforts were blocked in February, after he attempted to appoint several technocrats and academics to his cabinet, breaking from the tradition of nominating ministers according to their party, ethnic or sectarian identity. Abadi had made attempts at reform in 2015, with the objective of overturning the system of patronage institutionalised by the post-2003 system. The bid for a technocratic government was also an attempt to boost the private sector and rebuild the Iraqi economy, which had shrunk due to the fall in the global price of oil. However, Abadi barely gained parliamentary approval for six new cabinet members – whose appointments were then overturned by the Supreme Court. His announcement that he would remove the three vice-presidential positions was also overturned by the court.

Parliament was virtually shut down for three months on 30 April, after protesters stormed Baghdad's fortified Green Zone and occupied the parliament and cabinet building. Demonstrators occupied the building again on 20 May. Off limits to most Iraqis for 13 years, the Green Zone and parliament had become symbols of a decision-making system that many viewed as inept and corrupt. The protesters, most of them supporters of Sadr, overran the barriers of the Green Zone, chanting slogans, waving flags and taking photographs while parliamentarians fled or hid in the basement (some of them were assaulted as they attempted to flee). Sadr subsequently ordered the protesters withdraw from the Green Zone, using the political crisis to increase his stature and boost his role as a central actor on the Iraqi political scene. Although he claimed to be acting in support of Abadi, Sadr's populist vision for reform ultimately undermined both the prime minister and the cohesion of the state, particularly due to his reliance on unofficial mechanisms and approaches.

Abadi was also under threat from former prime minister Nuri al-Maliki, who exploited sectarian polarisation to challenge his successor. Maliki used the Reform Front, an opposition party in the Council of Representatives, to launch attacks against Abadi's government. The organisation called for votes of no-confidence in several Iraqi ministers accused of corruption, exploiting internal rivalries within the Sunni political bloc. In this way, the Reform Front ousted defence minister Khaled al-Obeidi in August and finance minister Hoshyar Zebari in September. By the end of the year, their posts remained vacant – as did that of interior minister – despite Abadi's rhetorical focus on economic improvements, technocratic governance and security operations against ISIS.

Maliki and Sadr were also sharply divided. Sadr's anti-corruption campaign had targeted Maliki from the start, but the tension between the two leaders escalated in December, when Sadrist protesters took control of a building in Dhi Qar in which Maliki was scheduled to speak. The demonstrators accused the former prime minister of corruption and held him responsible for the massacres committed by ISIS during his tenure. Maliki threatened a military response, describing the Sadrists' actions as lawlessness and criticising Abadi's weakness for being unable to stop them. With both Sadr and Maliki controlling militias in Baghdad, these political crises had the potential to escalate into violence.

Fragmentation in Kurdistan

The political turmoil in Baghdad momentarily brought Abadi and KRG President Masoud Barzani closer together. The impeachment of Zebari, who is from Barzani's Kurdish Democratic Party (KDP), highlighted the risks to Abadi and Barzani. While Abadi was vulnerable to attacks from parliamentarians, particularly Maliki's supporters, Barzani recognised that Zebari's ouster might create space for rival Kurdish groups to assert themselves through closer coordination with Baghdad. As a consequence, Abadi and Barzani agreed in September to resolve a long-running economic dispute by splitting the profit from the Kirkuk oilfields. This had been a contentious issue: before Baghdad shut down production at the oilfields in March, the KRG had been receiving all the income generated by the facilities.

The KDP's Kurdish rivals viewed rapprochement with Baghdad as the only viable solution to the economic crisis, which had prevented the authorities in the largely autonomous north from paying many public-sector employees for months. The crisis prompted many Kurdish parties to vote for a budget that gave the Iraqi government control of all Kurdish oil exports, allowing Baghdad to pay government salaries in the north, as well as the expenses of the Peshmerga. Thus, Kurdish politicians such as members of the Gorran party increasingly attempted to counter Barzani's influence in the north by turning to Baghdad.

The new budget reflected the progressive fragmentation of Kurdish parties; by the end of the year, the Kurdish federal bloc no longer operated in a unified manner. In November, Barzani offered to stand down as president and called for the formation of a new government in the KRG, but this was largely interpreted as a political manoeuvre rather than a serious effort at reconciliation. Barzani's presidential term expired in August 2015, precipitating a crisis between parties demanding his resignation and the KDP, which argued that the unusual circumstances facing Iraq required him to remain in position.

Barzani and the KDP subsequently sought to boost Kurdish unity by appealing to nationalist sentiment. In what the party billed as a 'victory speech' for the Kurds at Bashiqa in November, Barzani declared that ISIS had been driven from almost all 'Kurdish areas' of Iraq, emphasising that the KRG would continue to control the territory occupied by the Peshmerga at the start of the operation to retake Mosul. This statement directly contradicted Abadi's demand that the Peshmerga withdraw from these territories.

Tension over the areas grew following reports that the KRG had unlawfully destroyed Arab homes – and even entire villages – there without having a legitimate military purpose for doing so. According to Human Rights Watch and Amnesty International, the KRG evicted many Arabs in Kirkuk and destroyed their homes, detaining some of them in camps on suspicion of having conspired with ISIS.

The conflict in Iraq killed at least 6,878 civilians in 2016, according to the United Nations Assistance Mission in Iraq. The UN described this figure as an 'absolute minimum' because it did not include unverified fatalities in conflict zones and other areas in which it was unable to operate due to security restrictions, such as Anbar province in May, July, August and December.

Although ISIS had lost a considerable amount of territory and the Mosul operation had begun, by the end of 2016 it was still unclear how a potential defeat of the group would be defined, and how such a defeat would affect Iraq. The country continued to struggle with stabilising territories seized from ISIS, which witnessed large-scale displacement and destruction that threatened to undermine the economy for years to come. The prospect of a return to normal life remained distant in most of these areas, especially as many of their residents had been displaced.

The incorporation of paramilitary groups into the Iraqi security forces also undermined the process of rebuilding the state, as well as the military, by creating a basis in law for the sectarian organisations in the PMU. Abadi proved unable to exert his influence throughout the year, failing not only to reign in the militias but also to implement key governance reforms. Relations between Baghdad and Erbil were also in flux, largely due to divisions within the Kurdish bloc and an economic crisis that has significantly undermined the KDP's popularity.

Israel–Palestine

After experiencing a turbulent end to 2015, the conflict in Israel–Palestine saw a de-escalation in violence the following year, with a marked decline in fatalities and injuries. Yet the overall security and political situation remained fragile, against a backdrop of increasing Israeli settlement activity and demolition of Palestinian homes and infrastructure in the West Bank, along with sporadic 'lone-wolf' attacks by Palestinians. The prospect of peace remained dim, as the Palestinian

Key statistics	2015	2016
Conflict intensity:	Medium	Low
Fatalities:	200	125
New IDPs:	800	1,600
New refugees:	1,750	

and Israeli leaderships both consolidated their hold on power and remained uncompromising in the face of mounting international pressure to preserve the viability of a two-state solution.

Violence in the West Bank

The West Bank continued to be the epicentre of the conflict, as the site of more than 90% of recorded fatalities and injuries. According to the UN Office for the Coordination of Humanitarian Affairs, Israeli forces killed around 100 Palestinians, while Palestinians killed 13 Israelis (most of them settlers) in 2016 – compared to roughly 160 and 23 respectively the previous year. Moreover, the annual number of Palestinians injured in the conflict fell by more than 75% between 2015 and 2016.

The southern West Bank city of Hebron remained a hotspot of violence due to regular clashes between Israelis and Palestinians there. According to reports, at least one in three Palestinian attackers were residents of the city, and at least 64 Palestinians and eight Israeli settlers had been killed in the West Bank since the rise in violence in late 2015. The divided city, which served as a microcosm of the broader conflict, was regularly stifled by curfews, travel restrictions and heavy-handed security operations. Two incidents in 2016 typified the nature of violence there. The first of these occurred on 30 June, when a 19-year-old Palestinian from the nearby village of Bani Naim stabbed to death a 13-year-old Israeli girl in her bedroom in the Israeli settlement of Kiryat Arba. The incident formed part of the growing violence stemming from young Palestinians' resentment of Israeli settlers, who often live close by but enjoy better access to services and goods than is available in impoverished Palestinian villages. Some settlers' participation in 'price-tag' attacks on Palestinian farms and other property – acts of vengeance that were rarely prosecuted by the Israeli authorities – only fuelled this sense of injustice. The second incident was the killing on 24 March of a subdued Palestinian assailant in Hebron by Israeli combat medic Elor Azaria. The incident stoked Palestinian anger, and lent credence to the claim that the Israeli security forces routinely engaged in overly forceful – even extrajudicial – measures, often with impunity. Human-rights organisations consistently criticised Israel for its excessive use of live ammunition as a first response.

Although improvements in the efficiency of Israeli security raids and detention operations stemmed the violence, increased Palestinian involvement in security operations (such as arrests and weapons seizures) had a significant impact in de-escalating clashes, particularly those in Nablus and Hebron. Cooperation between the Israeli and Palestinian security forces remained key to containing violence in the West Bank. Israel's army central command said Palestinian security forces were responsible for 40% of all arrests made in May – an increase of 30 percentage points over three months. Another contributing factor was arguably Palestinian fatigue and frustration with the inefficacy of attacks. According to surveys conducted by the Palestinian Center for Policy and Survey Research, the proportion of Palestinians who supported knife attacks fell from 67% to 51% in the first half of the year.

Palestinian organisations in the West Bank and Gaza appeared to believe that the costs of conducting attacks continued to outweigh those of maintaining the status quo. As such, Gaza-based militant group Hamas refrained from engaging in another all-out confrontation with Israel. Indeed, in August, an uncharacteristically assertive series of Israeli airstrikes on around 50 Hamas targets in northern Gaza – conducted in response to the launch of a rocket from the territory into the Israeli city of Sderot – drew only a perfunctory warning of a military response. The new dynamic was also apparent in April, when Hamas was quick to distance itself from a bus bombing in Jerusalem that injured 20 Israelis. A type of operation commonly associated with Hamas during the second intifada, bus bombings had theretofore been conspicuously absent from the conflict since 2005. Hamas primarily focused on rebuilding Gaza and clamping down on rival militant organisations (including Salafist jihadist groups) that sought to challenge its leadership and reignite the armed struggle against Israel. Since 2014, an Israeli-imposed blockade had caused a severe deterioration in living conditions in the enclave, as significant water and electricity shortages combined with high levels of youth unemployment. Nickolay Mladenov, the United Nations' Middle East envoy, stated in November that people in Gaza had 'lost hope', making the area 'more dangerous and more explosive'.

Despite the overall decline in fatalities, lone-wolf attacks across the West Bank, East Jerusalem and Israel provided a reminder that the conflict could still escalate at any moment. This was especially true of assaults carried out in Israel proper. There was a slight increase in the number of Israelis killed in Israel in 2016. On 8 June, two Palestinians opened fire on a supermarket in Tel Aviv, killing at least four Israelis. This was the second major attack in the city during the year. The first came on 1 January, when a Palestinian citizen of Israel shot to death two people and wounded seven others on Dizengoff Street. Following the incident, Israeli Prime Minister Benjamin Netanyahu stated that 'Islamist incitement' had motivated the attack – in line with the Israeli government's typical language on the drivers of the unrest. For the Israeli government, political and religious incitement is the primary cause of the conflict. Several world leaders had consistently called on Palestinian President Mahmoud Abbas to condemn the attacks, and to curb incitement. Yet the Palestinian government considers the unrest to be a natural response to the occupation, the expansion of Israeli settlements and heavy-handed security measures – all of which undermine Palestinian aspirations to statehood.

Consolidation of power

Against this backdrop, Abbas and Netanyahu strengthened their hold on power. Netanyahu's top priority, following his election to a fourth term in 2015, was to increase his one-seat majority in the Knesset. In May, Avigdor Lieberman joined the ruling coalition as defence minister, marking a consolidation of the most right-wing government in Israel's history. The move allowed Netanyahu to placate his hardline support base, distance himself from defence minister Moshe Ya'alon (who resigned after his post was offered to Lieberman) and set back opposition leader Isaac Herzog. Ya'alon had strongly backed Gadi Eisenkot, chief of the Israel Defense Forces, in condemning the handling of the Azaria case. Herzog faced setbacks within his own coalition after he entered negotiations in May to join the Netanyahu coalition. Although he withdrew after learning of parallel talks between Lieberman and Netanyahu, key figures within the Zionist Union called for Herzog to resign. In January, the coalition had departed from its left-leaning position by stating that the two-state solution was unrealistic in the near future, and instead advocated a more assertive security response to the unrest. The resulting political turmoil created opportunities for centrist leaders such as Yair Lapid, whose approval rating increased throughout the year.

President Abbas agreed in September to hold the seventh Fatah convention, aiming to consolidate his position and stave off challenges from political rivals such as exiled former Fatah leader Mohammed Dahlan. In November, Abbas was unanimously re-elected as chairman of Fatah and head of the party's 23-member Central Committee. In the months leading up to the convention, Abbas had prohibited Dahlan and his supporters from participating in the event. The move escalated tension on the ground, leading to a series of clashes between pro-Abbas and pro-Dahlan militants in the West Bank settlements of Ramallah, Nablus and Jenin. Abbas remained wary of Dahlan despite the fact that his rival publicly stated in October that he would not vie for the presidency, instead endorsing popular imprisoned leader Marwan Barghouti. The endorsement was seen as a way for Dahlan to gain legitimacy in the long run by portraying himself as a potential prime minister. With Barghouti likely to remain in jail, Dahlan was positioned to become the de facto head of the Palestinian Authority if Barghouti was appointed as president. The exiled leader is understood to be largely backed by the United Arab Emirates, Egypt and Jordan, which desire reconciliation among Palestinian factions, starting with Fatah. Yet Abbas announced on 12 December that Dahlan would be stripped of his parliamentary immunity and charged with embezzlement – a move likely to create discord between the Palestinian Authority and these Arab states. Moreover, there seemed to be little prospect of a meaningful reconciliation between Hamas and Fatah, despite a meeting intended to resolve their differences between Abbas and Khaled Meshal, chief of Hamas's political bureau, in Qatar in October.

International frustration and the two-state solution

Preoccupied with strengthening their domestic power bases, Palestinian and Israeli leaders failed to adequately address a more pressing concern: preserving the viability of the two-state solution. Advances in Israeli settlement construction coupled with an aggressive campaign of demolitions in the West Bank appeared to spur attacks by Palestinians and fuel anxiety in the international community. According to official figures, Israel demolished twice as many Palestinian buildings and structures in the West Bank in 2016 than in the previous year, internally displacing at least 1,601 Palestinians. In May, Israel-based settlement watchdog Peace Now reported a 40% increase in Israeli settlements in the West Bank in the first half of the year. These developments cast doubt on the Israeli government's long-term commitment to peace.

In a bid to preserve work towards a two-state solution, the Quartet on the Middle East released in July a report underlining the need to avoid 'entrenching a one-state reality'. The Quartet urged Israel to stop settlement expansion in the West Bank and East Jerusalem, and called on the Palestinian leadership to proactively prevent the incitement of violence. Driven by worrying political and security trends on the ground, several countries led unsuccessful diplomatic efforts to revive the peace talks, which had stalled in 2014 after US-led negotiations broke down. More strikingly, France held a preliminary meeting in June to lay the foundations of the 2017 Paris peace conference, which sought to force Israeli and Palestinian leaders into a constructive debate on outstanding obstacles to peace within a fixed time frame. Egypt and Russia also sought to open separate channels for peace discussions. However, the intransigence of both Netanyahu and Abbas frustrated all international efforts. The Israeli leader firmly rebuffed any external involvement, consistently stating that Israel would only engage in direct negotiations with his Palestinian counterpart without preconditions. Abbas remained content to pursue peace through international brokers and institutions, refusing to meet Netanyahu face to face as long as Israel persisted with settlement construction in the West Bank and East Jerusalem.

In December, with the prospect of peace fading, the UN Security Council voted 14–0 in favour of Resolution 2334, which condemned Israeli settlements in the West Bank as illegal and an obstacle to peace. The resolution also criticised the Palestinian authorities' complacency in curbing incitement to violence. Although it was non-binding, the resolution reflected an international consensus that there could be no viable solution to the conflict while Palestinian violence, Israeli settlement expansion and the political impasse continued. The resolution was particularly significant in that it passed due to the United States' decision to abstain from the vote. The abstention broke with US policy since 1979, and prompted Israeli officials to accuse Washington of abandoning Israel. In the immediate aftermath of the vote, Netanyahu announced that Israel would have 'limited' diplomatic relations with 12 of the 14 UN Security Council members and halted millions of dollars in funding to UN agencies. Samantha Power, US ambassador to the UN, explained that facts on the ground suggested Israel's policies were no longer in line with the commitments on settlement activity it had made to the US. Since 2011, Israel had built over 30,000 settlement units. Despite the Obama administration's efforts to justify the abstention as being in line with half a century of bipartisan policy on Israeli settlements, the move met with strong opposition from within the newly Republican-controlled US Congress and from president-elect Donald Trump.

The resolution was the culmination of what had been a rocky period in US–Israeli relations, characterised by the deteriorating relationship between Netanyahu and US President Barack Obama. Keen observers of US–Israeli politics had anticipated the resolution as a means for the outgoing president to leave a mark on what had otherwise been an ineffective policy on the peace process. Some interpreted it as a way to hedge against the uncertainty surrounding Trump's approach to the issue. However, others regarded the abstention as too little too late from a president at the end of his tenure, one that had signed a record US$38-billion US–Israeli military aid package a few months earlier.

The Palestinian leadership viewed the resolution as a victory, having worked throughout the year to bring it before the UN Security Council. There was a possibility that the measure would allow the Palestinian leadership to bring a war-crimes case against Israel at the International Criminal Court. Yet Abbas had to carefully weigh the costs of pursuing such a confrontational effort.

The incoming Trump administration looked set to have a significant influence on the decisions of Palestinian and Israeli leaders. Although Trump had stated that he would seek to achieve the 'ultimate deal' between Israelis and Palestinians, his advisers also confirmed that he had an interest in moving the US Embassy in Israel from Tel Aviv to Jerusalem. This gesture appeared almost certain to increase tension and escalate the conflict. Key US and international figures – including some Arab leaders, whose domestic standing could be damaged by failure to comment on the issue – cautioned Trump against going through with the move. Nonetheless, Trump's decision to nominate far-right, pro-settlement lawyer David Friedman as his ambassador to Israel suggested that the peace process could face significant new challenges.

Lebanon–Hizbullah–Syria

Despite suffering from political paralysis and regional insecurity, Lebanon was relatively stable in 2016. In late October, after two and a half years without a president, the Lebanese parliament elected Hizbullah ally

Key statistics	2015	2016
Conflict intensity:	Low	Low
Fatalities:	250	20
New IDPs:		
New refugees:	450	

Michel Aoun to the post as part of a deal in which Saad Hariri, leader of the Future Movement and a supporter of the Gulf Arab states, became prime minister. Jihadist groups posed a threat to Lebanon throughout the year, carrying out several attacks on civilian targets and engaging in continuous fighting in the northeast of the country. Many analysts and politicians accused Hizbullah of conducting an assault on BLOM Bank headquarters, in the Verdun neighbourhood of Beirut, in June. The group also became increasingly involved in the Syrian war, as demonstrated by the key role it played in the siege and eventual takeover of eastern Aleppo by pro-government forces. Hizbullah's participation in the conflict, as well as the war in Yemen, increased tension between Lebanon and Gulf Arab countries. And, in March, Saudi Arabia suspended military aid to the Lebanese armed forces worth US$4 billion. Nonetheless, President Aoun's promise to visit Saudi Arabia on his first official trip abroad, along with his invitation to Gulf delegations to visit Lebanon shortly after his election, suggested that these relationships might improve.

Growing public discontent

Aoun and Hariri began their new roles on 31 October, following months of negotiations and multiple initiatives to end the political deadlock. The breakthrough came with Hariri's decision to back Aoun's candidacy after consultations with Lebanese political parties, foreign powers and Aoun himself. Christian leader and Hariri ally Samir Geagea had already thrown his weight behind Aoun in early 2016, creating an unprecedented consensus in his parliamentary bloc. Yet Hariri received criticism from his subordinates and other March 14 alliance leaders for supporting the presidential bid of a close ally of Hizbullah. For many, Hariri's support of Aoun shook Lebanon's traditional political alliances.

Hizbullah became a kingmaker in the election process thanks to its ambiguous stance in the presidential negotiations. While it publicly supported Aoun, it did not ask Sleiman Frangieh – another close ally of the group, whom Hariri supported early on in the process – to withdraw his candidacy. However,

Hizbullah only fully supported Aoun after ensuring that its key allies would go along with the choice and negotiating key governmental portfolios for itself, as well as the Amal Movement and the Marada Movement. Formed in mid-December, the new government had 30 cabinet members covering eight state ministries to accommodate all power brokers. The Future Movement and the Free Patriotic Movement gained control of the interior, defence, energy and foreign-affairs ministries, while the Amal Movement and Hizbullah were appointed to run the finance and industry ministries. The new cabinet's agenda reflected Hizbullah's influence, with Prime Minister Hariri committing to introduce a new electoral law in 2017. Hizbullah believed that the existent electoral system did not truly represent the demographic distribution of Lebanese communities. Although many others agreed with this point, they feared that a proportional electoral law would facilitate a Shia majority in parliament.

Lebanon successfully held local elections in May, although there were reports of voting irregularities throughout the country. Low voter turnout – at 20% in Beirut, the country's capital and most populated city – spoke to the population's disaffection with the country's political system and elite. Major political parties won municipal seats in their traditional strongholds, but were only able to do so by forming alliances. For instance, a coalition of traditional Muslim and Christian parties won in Beirut against Beirut Madinati, a campaign supported by civil-society activists and progressive actors exasperated with the country's corrupt politicians. Despite losing, the campaign energised factions seeking political change. There was another surprise result in Tripoli, a traditional Sunni stronghold of Saad Hariri and Najib Mikati. Former justice minister Ashraf Rifi, a one-time ally of Hariri, defeated the Hariri–Mikati alliance, winning 18 of the 24 seats in the city council. The outcome of the elections undermined the credibility of Hariri and Mikati by revealing the public discontent in their constituencies. Although Hizbullah and the Amal Movement won in eastern and southern Lebanon, their lead narrowed in comparison to previous elections. Indeed, Hizbullah's involvement in Syria and the decline in public services in these areas discouraged many citizens from voting.

In early 2016, paralysed by a dysfunctional parliament and the vacuum in the presidency, Lebanon struggled to find a sustainable solution to the waste-disposal crisis that had engulfed the country since July 2015. Separate efforts to export the rubbish accumulating in Lebanon's cities fell through in January and February. In March, the government agreed to reopen the Naameh landfill for two months, along with two other temporary landfills close to Beirut. Although rubbish was disposed of from 19 March onwards, the opening of temporary landfills led to significant tensions between the government, civil-society groups and local communities. The groups stressed the public-health and environmental risks of the approach, while local communities blocked access to temporary landfills nearby, which they feared would become permanent.

Persistent terrorist threats

Lebanon remained vulnerable to operations by jihadist groups throughout the year. In late June, the Islamic State, also known as ISIS or ISIL, carried out two suicide bombings in the predominantly Christian village of Qaa, killing nine people (five civilians and the four bombers) and injuring more than 15 others. Although these incidents led the Lebanese security forces to strengthen their deployments along the border, they were forced to continue battling ISIS and Jabhat al-Nusra in Arsal on a daily basis. Security officials also stated that in May, they had foiled several attempted ISIS attacks in areas of Hamra and central Beirut popular with tourists.

The Lebanese government implemented the Hizballah International Financing Prevention Act in summer 2016, following the US Congress's enactment of the measure the previous December. Lebanese banks closed the accounts of many individuals, businesses and charities affiliated with the group, hitting it hard. Tensions between the Lebanese central bank, other banks and Hizbullah peaked in June with the attack on the BLOM Bank headquarters. Although Hizbullah did not claim responsibility for the operation, many observers argued that the group had carried it out in response to the bank's application of US sanctions.

Hizbullah increased its involvement in Syria by sending additional fighters – including one of the bodyguards of its leader, Hassan Nasrallah – to Aleppo in November, accelerating pro-government forces' takeover of the city. The war continued to kill many of the group's fighters, including Mustafa Badreddine, a top commander and founding member who had ties to Nasrallah. He died in unclear circumstances near Damascus in May, sparking rumours that he had been assassinated by Israel. However, the group did not hesitate to display its military strength, conducting a parade that included US-made armoured personnel carriers and other equipment in the Syrian city of Qusayr in mid-November. An Israeli airstrike reportedly hit a suspected Hizbullah weapons convoy on the Beirut–Damascus highway later that month.

The members of the Gulf Cooperation Council designated Hizbullah as a terrorist organisation in March due to what they saw as its destructive role in Lebanon, Syria and Yemen, as well as the Lebanese government's failure to support Saudi-sponsored UN resolutions against Iran. The Arab League followed suit with the designation. In response, Nasrallah pledged to continue the resistance and repeatedly criticised the leaders of Gulf Arab states. Saudi Arabia also suspended military aid to the Lebanese military worth US$4bn (US$3bn of which was earmarked for an arms delivery) and mobilised other Gulf countries against the group. This was due to Beirut's failure to condemn an attack on the Saudi Embassy in Tehran, as well as Nasrallah's criticism of the execution of Saudi Shia cleric Nimr al-Nimr by Riyadh. The United Arab Emirates, Qatar and Kuwait threatened to deport Lebanese citizens, a measure that threatened to have a severe effect on a Lebanese economy reliant on remittances. Thus, President Aoun announced that he would travel to Saudi Arabia on his first official foreign visit with the aim of improving Lebanon's troubled relationship with Riyadh and its allies.

Increasingly vulnerable refugees

Palestinian and Syrian refugees in Lebanon were forced to live in deteriorating conditions, amid mounting public opposition to the Syrians' naturalisation or integration into Lebanese society. Meanwhile, rising conflict between various militias in Ain al-Hilweh Palestinian refugee camp, in the southern city of Sidon, spurred the Lebanese security forces to announce that they would build a controversial wall around the facility.

Syrian refugees remained vulnerable due to laws restricting both their stay and their movements in Lebanon. Acting on a 2015 law, the Lebanese authorities limited Syrian refugees' entry to the country and made it extremely difficult for them to obtain new or extended residence permits. In January, Lebanon forcibly returned at least 250 Syrian refugees to Syria after their arrival at Beirut airport, following Ankara's introduction of new visa regulations that prevented them from travelling on to Turkey. The move placed them in 'mortal danger', according to Amnesty International. Human-rights groups reported that the government's discriminatory policies resulted in the exploitation and abuse of Syrians, particularly women and girls, by human traffickers. The United Nations reported that at least 70% of Syrian refugees in Lebanon lived below the poverty line, despite several years of humanitarian initiatives designed to improve their living conditions.

The government demanded that Syrian refugees either return to their country or be resettled elsewhere, citing the tremendous costs of hosting them and the instability they created in Lebanon. While Tammam Salam and Gebran Bassil, then prime minister and foreign minister respectively, asked the UN and Western nations to resettle more of the Syrian nationals living in Lebanon, labour minister Sejaan Azzi proposed a plan to return them to Syria over the next two years.

Across Lebanon, the security forces carried out a series of raids on Syrian refugee camps suspected of harbouring Jabhat al-Nusra and ISIS fighters. Following the attacks on Qaa in June, they detained more than 200 Syrian nationals who lacked residence papers in Baalbek, Bekaa Valley and Sidon. Governors in eastern and southern Lebanon also targeted Syrian refugees, imposing daily curfews on them.

The security forces' plan to build a wall around Ain al-Hilweh was suspended following protests from Palestinian refugees, who denounced the move as racist and oppressive. The 70,000 or so residents of the camp already faced discriminatory treatment, particularly as the security forces maintained numerous checkpoints, metal fences and watchtowers there. Yet despite these security measures, there were frequent outbreaks of violence in the camp, leading many Lebanese citizens to suspect that such facilities had become breeding grounds for armed groups that used the refugees as cover.

Libya

Despite establishing the UN-brokered Government of National Accord (GNA) in early 2016, Libya made little progress towards unity in the year. The country's political and security environment remained highly unstable as its eastern and western coalitions continued to vie for control of territory, infrastructure, resources and institutions. In these contests, the sides sought and received backing from a range of international actors, who contributed to a stalemate in the civil war. The conflict threatened to create a humanitarian disaster, with an estimated 1.3 million Libyans still in need of humanitarian aid, the country harbouring as many as 1m migrants attempting to travel to Europe, ongoing decline in the oil-dependent economy and a security sector dominated by largely unaccountable local militias.

Key statistics	2015	2016
Conflict intensity:	High	High
Fatalities:	2,000	2,000
New IDPs:		
New refugees:	2,500	

Struggle to establish the Government of National Accord

Following the rushed finalisation of the UN-brokered Libyan Political Agreement – despite opposition to the deal from the presidents of both the Tripoli-based General National Congress (GNC) and the Tobruk-based House of Representatives (HoR) – the UN Security Council quickly passed Resolution 2259, recognising the GNA as Libya's legitimate government. Faiez al-Serraj, prime minister-elect and head of the newly formed Presidency Council, was subsequently tasked with forming a government within 30 days.

Under the Libyan Political Agreement, the democratically elected HoR would serve as the country's sole parliament, while former members of the GNC would be integrated into the political system as part of advisory body the High Council of State. The deal initially gave the Presidency Council until 17 January to submit a government to the HoR, which would then vote on establishing the GNA in law. But while the Libyan populace broadly supported an agreement of some kind, the parliaments disagreed on important details. One controversial aspect of the deal was the future role of General Khalifa Haftar, who was despised by the militias of western Libya for his anti-Islamist agenda and ties to the former regime, but had provided essential support to the HoR through his leadership of the self-proclaimed Libyan National Army (LNA). Article 8 of the agreement stipulated that the powers of all senior military, civil and security posts should be transferred to the Presidency Council, which would then appoint a new army commander. The HoR insisted that they would only approve the deal after this article was removed, allowing Haftar to remain in place as commander. However, such a move was unacceptable to the groups that supported the GNC, including the Misratan Islamist militias that made up the bulk of the western coalition.

The backers of the UN process had taken the gamble of pushing through the deal on a unity government without resolving these disputes primarily because they believed that there would not be another opportunity to reach an accord. At the time, the heads of the two parliaments had begun parallel talks; UN Special Envoy Bernardino León had resigned after being accused of having a conflict of interest in Libya; the Islamic State, also known as ISIS or ISIL, was increasing its presence in the country; the European Union was under pressure to deal with the flow of migrants from North Africa; and several Western nations had signalled that they would begin counter-terrorism operations in Libya. The parties to the conflict feared that the peace talks would collapse at this crucial moment, empowering those who favoured a military solution to the war. Unfortunately, those pushing for the deal underestimated the significance of the disputes between the sides. As a consequence of these disagreements, the HoR voted to reject the proposed GNA cabinet.

Over the course of the year, the major fault line of the conflict shifted to divide those who backed the GNA from those who opposed it. Initially, the GNC also opposed the GNA, threatening to arrest members of the Presidency Council if they set foot in Tripoli. But, in March, Serraj and the council called the GNC's bluff by travelling to Tripoli in a Libyan frigate and setting up camp in the city's Abu Sitta naval base. This was an effective move, as the Central Bank of Libya and the National Oil Corporation immediately recognised them, as did a substantial number of western municipalities. The GNC government, led by Khalifa Ghwell, reportedly stepped down a few days later. Nonetheless, it still took the GNA a month to gain control over several ministries, while Ghwell later denied that he had resigned.

The GNA's control of the capital and government institutions remained uncertain: the momentum built by its arrival in Tripoli dissipated as it became bogged down in the fragile politics of western Libya. Tripoli was controlled by rival militias that supported either the GNA or Islamist rule and frequently clashed with one another, often for control of vital infrastructure. Serraj struggled to run the government. The Tripoli Revolutionaries' Brigade and the Special Deterrence Force, or RADA – led by Haitham Tajouri and Abdul Rauf Kara respectively – supported the GNA. Around 30 much smaller Misratan militias, including Salah Badi's Steadfastness Front, backed Ghwell. All these groups controlled different districts and key infrastructure while primarily following independent agendas – which sometimes resulted in battles between pro-GNA groups. Due to the GNA's limited control of the militias, there was a possibility that they would use their control of key infrastructure to extort payments from the Ministry of Defence. Indeed, armed groups to the south of Tripoli had threatened to cut off the water supply. Inside the city, forces supporting Ghwell seized several ministries, declaring that the GNA had failed and calling for talks among Libyan parties without international interference. These efforts followed a failed coup attempt by Ghwell, in which the newly established Presidential Guard defected from the GNA to Ghwell's reinstated GNC.

The religious and ideological affiliations of the militant groups further complicated matters. For instance, many members of the Special Deterrence Force reportedly adhered to the Salafist ideology of Saudi cleric Rabi bin Hadi al-Madkhali, who promoted obedience to the authorities and forbade political activism or armed uprisings. This ideology accorded with the largely neutral stance of the group, which claimed to focus on fighting crime. However, Madkhali had also called on his supporters to confront groups affiliated with the Muslim Brotherhood, ISIS, al-Qaeda and organisations that had the support of the Tripoli-based Grand Mufti of Libya, Sadiq al-Ghariani. Ghariani, in turn, accused Madkhalists of supporting Haftar, while denouncing the GNA and the Libyan Political Agreement as having been imposed by outside forces. These dynamics also undermined the power of the GNA, which continued to depend on the support of western militias and had to navigate a churning sea of rivalry and interests as it attempted to govern.

End of ISIS rule in Sirte

The Presidency Council and its international backers set out to expel ISIS from Sirte, viewing the effort as a means to establish the GNA's credibility and legitimacy. The group likely assumed that Sirte would be a productive recruitment ground as the birthplace of former Libyan leader Muammar Gadhafi, the home of his Qadhadfa tribe and the final stronghold of Gadhafi loyalists in the revolution. However, the reality proved different: ISIS was only able to maintain control of the city by brutally repressing a local uprising. Another aspect of Sirte's appeal for the group was its strategically important location at the western edge of the 'oil crescent', thought to hold 80–90% of Libya's oil reserves. After ISIS took over Bouri and Mabruk oilfields as the forces of the eastern and western coalitions fought each other, other oilfields in the region were shut down out of fear that they would also be attacked. The group's strategic objectives in taking control of these oilfields remained unclear, as it would have been almost impossible for it to export or otherwise sell the oil they produced. The attacks may have been carried out to sabotage attempts to reach a power-sharing arrangement between east and west, or perhaps even to elicit a military response from European or American militaries that would assist local recruitment drives. In line with this second possibility, it was reported in January that up to 1,000 British Special Air Service troops would be deployed to secure Libya's oilfields, and in March that these forces were working alongside Jordanian military personnel in the country.

The operation to drive ISIS out of Sirte – named *Al-Bunyan al-Marsous* (BAM) in reference to the Koranic *Surah al-Saff* – was launched in May and quickly achieved results. Initially, the GNA and its international backers had planned to restructure the Misratan forces into an army for the GNA with Italian and British support, importing relatively advanced equipment – by partially lifting the arms embargo on Libya – and marching on Sirte with an army that was a government institution. However, after ISIS captured the strategically important crossroads of Abu Grein, halfway between Misrata and Sirte, the Misratans overtook the GNA plan, struck back and within weeks had entered Sirte. This led to months of urban warfare in which battles moved forward block by block, dislodging ISIS militants from fortified buildings connected by a network of tunnels. Although it was estimated early on that ISIS had as many as 6,000 fighters in Sirte, BAM commanders put this number at 2,000. There were reports that a sizeable number of the jihadists had fled before BAM forces entered the city; as the conflict there wore on, as many as 400 ISIS fighters launched attacks in Sirte from the surrounding region.

The conditions in the city favoured the dug-in defenders, making progress slow and very costly for the poorly trained and supplied BAM forces. The heavy losses started to create rifts in the GNA–BAM coalition. As the GNA had no forces of its own to contribute to the offensive, it had relied on the Misratan armed groups, supported by some smaller militias such as the Madkhalist

604th Infantry Battalion. Misratan commanders accused the GNA of providing insufficient support, claiming to lack equipment such as helmets, flak jackets and night-vision goggles to deal with ISIS snipers. The sale of these items to Libya was prohibited under the arms embargo, which the GNA had unsuccessfully tried to lift. Although the GNA secured an exemption that allowed it to request the supply, sale and transfer of arms for troops under its control that were fighting jihadist groups (subject to approval by the UN Security Council), little support was provided through this mechanism. At the request of the GNA, the United States, which had been reluctant to become involved in the conflict due to its presidential elections, supported the assault on Sirte with an intensive bombing campaign. Africa Command reported flying 495 sorties from the USS *Wasp* and, later, the USS *San Antonio* off the Libyan coast in only four months – operations coordinated by special forces on the ground. Italian forces also established a field hospital for the Misratan fighters, which was protected by Italian paratroopers. Despite receiving this international support, BAM forces only cleared Sirte in early December.

Haftar's strengthened position

While BAM forces assaulted the city from the west, the Petroleum Facilities Guard (PFG) moved towards Sirte from the east, capturing the towns of Bin Jawad, Nawfaliyah and Harawa. The PFG had formed in an earlier phase of the Libyan conflict; initially comprising forces that controlled oil infrastructure throughout the country, it later broke apart into autonomous militias. In the east, the largest remaining part of the PFG controlled the main oil facilities at Es Sider, Ras Lanuf and Zueitina. Led by Ibrahim Jathran, the PFG had declared the Libyan government corrupt and attempted to export oil. After failing in this effort, Jathran used his control of the facilities to extort wages for his men from the National Oil Corporation. Having alternately backed the GNC and the HoR, he struck an agreement with the GNA during the Sirte offensive, in late July.

The agreement between the PFG and the GNA likely provoked Haftar, who could not allow the western coalition and the GNA to gain control of the oil crescent. Taking advantage of their preoccupation with the battle against ISIS, Haftar's LNA seized control of the Es Sider, Ras Lanuf and Zueitina terminals in August. While an earlier attempt to capture the Zueitina terminal had been repelled by the PFG, this later effort reportedly involved little significant fighting. Shifting tribal alliances played an important role in the takeover (Jathran's forces had been living off local civilians for years): many PFG fighters reportedly fled their posts while others defected to the LNA. Haftar's use of Chadian and Sudanese mercenaries in the assault on the terminals, out of an apparent reluctance to use LNA troops, indicated that his position was less secure than often assumed. Nonetheless, his control of the oil crescent granted him significant influence on the GNA.

Haftar left the agreement between the PFG and the National Oil Corporation in place, giving the latter control over export infrastructure and thereby channelling funds into the Central Bank of Libya. The government came under pressure to resume oil exports to prevent bankruptcy, with a budget deficit of 61% of GDP in 2016 and foreign reserves that had dropped from US$107.5 billion in 2013 to US$43bn three years later. The central bank had previously been reluctant to provide the GNA with anything more than emergency funding without the approval of the HoR. But the GNA agreed on temporary financial arrangements for 2017 with the bank, while avoiding the term 'budget'. Nonetheless, the arrangement required the approval of Haftar, who had allowed oil production and exports to continue – perhaps because the funds they generated went to government services throughout Libya, and he had an interest in being perceived as reliable by the Libyan people, as well as by international actors.

Haftar's status among foreign powers seemed to grow; Moscow allowed him to visit the aircraft carrier *Admiral Kuznetsov* and to send wounded LNA soldiers to Russia for treatment. Russian troops also held joint exercises with the LNA. Although he was primarily backed by the United Arab Emirates and Egypt, Haftar hoped to use increased Russian support to outmanoeuvre the GNA, lift the arms embargo and sustain operations by his air force (which provided his main advantage over forces in the west). Like the GNA, Haftar had tried to project an image of himself as an enemy of ISIS. Having launched *Operation Dignity* in 2014 to tie together eastern forces in the fight against extremism, Haftar had been battling for control of Benghazi with the Shura Council of Benghazi Revolutionaries, which was formed in response to the operation and the territorial gains made by ISIS. Consisting of Ansar al-Sharia and the Benghazi Defence Brigades (affiliates of al-Qaeda and the Misratan brigades respectively), the council had clashed – but had also been accused of cooperating – with ISIS.

The LNA continued to focus on defeating jihadist and Islamist groups. The revelation that French forces had supported Haftar in this effort damaged the relationship between foreign powers and the GNA, causing many Libyans to question whether the international community was abandoning its push for a unity government. Reports that special forces from the United Kingdom, Italy, Jordan and the UAE had supported Haftar indicated that these states also saw the LNA as the only force capable of stabilising eastern Libya.

Increasing risk of confrontation between east and west

With the defeat of ISIS in Sirte, the areas controlled by forces in the east and west once again bordered each other across Libya. Haftar's strengthened position in the east and growing international support allowed him to increasingly turn his attention westward. It seemed likely that he would capture Benghazi in 2017, having begun an offensive on the city's Ganfouda district in November 2016 – after which Derna would follow. In the west, the Misrata Military Council announced that its brigades would join the Central Military Zone of a Libyan army, nominally putting them under the command of a national institution. The fight against ISIS had turned the Misratan forces into a more coherent army (similar to the LNA), reflecting what appeared to be a trend towards greater cohesion within the eastern and western coalitions. However, the announcement by the military council was seen as an attempt to gain international support, as the arms embargo would only be lifted after a Libyan army under the control of the GNA was in place.

In a possible sign of growing conflict between east and west, the LNA conducted airstrikes on Al-Jufrah Air Base, and the Misratan 3rd Force fought the

12th Infantry Brigade for control of Gwairat al-Mal checkpoint, as 2016 came to a close. Moreover, GNA Minister of Defence Mahdi al-Barghathi allegedly supported a coalition of former Benghazi militias and PFG forces in a failed effort to recapture facilities in the oil crescent in December. Although the GNA denied that Barghathi, a rival and former subordinate of Haftar, had been involved in the operation, this may have indicated that Serraj was unable to control even members of his government.

It seemed unlikely that the GNA would survive without the international support that loosely bound it to the Misratan forces. However, this co-dependency made cooperation between the GNA and Haftar all but impossible. Yet both sides' reliance on foreign powers also presented an opportunity to stabilise Libya, particularly by ending battles between Misratan forces and the LNA for control of the oil crescent, Al-Jufrah Air Base and Fezzan.

Mali (The Sahel)

Key statistics	2015	2016
Conflict intensity:	Medium	Medium
Fatalities:	350	600
New IDPs:		
New refugees:	13,000	

Despite achieving political and military successes in 2015, the parties to the conflict in Mali were unable to achieve similar progress the following year. There was only a minor fall in the number of Malian refugees – from 140,700 in 2015 to 139,700 in 2016. In contrast, the number of internally displaced persons in Mali decreased by almost half – from 61,900 to 36,690 in the same period. While this humanitarian success indicated that some regions of Mali had become less volatile, there were few major steps towards implementing the Algiers Accord, designed to promote stability in the country. Key stages of the peace process – including that to establish interim authorities in northern Mali – experienced multiple delays, while violence and allegations of corruption marred local elections held on 20 November. The political situation remained relatively unchanged; yet the perceived stagnation exacerbated existing inadequacies within the peace process, contributing to rising tensions in Bamako and throughout the countryside. Attacks by jihadist groups were the primary cause of conflict-related fatalities in 2016. Amid persistent inertia in development, political representation and good governance, the frustrations of both civilians and armed opposition groups grew, threatening the progress that had been made and undermining the legitimacy of Malian institutions. As signatories to the Algiers Accord failed to adhere to the official ceasefire and extremist organisations became bolder in their attacks, it became increasingly likely that the conflict would escalate dramatically. While Mali had many complex interrelated problems, they could be roughly divided into four main categories: security concerns; an inert peace process; humanitarian issues; and the decline of governmental authority.

Security vacuum

The worsening security situation was perhaps the most obvious challenge facing Mali. The UN Multidimensional Integrated Stabilization Mission in Mali (MINUSMA) remained the most dangerous peacekeeping mission in the world, with 35 of its personnel killed in 2016. Violent actors continued to target foreign personnel, especially in ungoverned territory in northern Mali.

As the Algiers Accord was still technically in force, the greatest threat to international forces was not the official armed opposition but fringe extremist groups conducting terrorist attacks. These latter groups, particularly Ansar al-Dine, grew bolder and their operations crept further south throughout 2016. In the latter half of the year, Islamist groups launched several major well-coordinated attacks against army camps in southern and central Mali. Over the course of three such assaults, 24 people were killed and roughly 47 were injured. The swiftness with which the militants defeated the Malian army was especially worrying, given that there had been extensive investment in the capacity of the national security forces. Additionally, 114 prisoners (many of them with connections to extremist groups) were freed during attacks on two prisons north of Bamako, in central Mali. The attacks occurred relatively close to the capital, and well within the Malian army's purported zone of control.

The Malian army and international forces achieved some minor victories in their counter-terrorism efforts, separately arresting three senior leaders of Ansar al-Dine between July and November. Moreover, Malian special forces killed the leader of al-Qaeda in the Islamic Maghreb (AQIM) in July. Yet the successes of the counter-terrorism operations, widely supported by the international community, were not enough to meaningfully damage the organisations. Indeed, Islamist groups claimed responsibility for twice as many attacks in 2016 as in the previous year. Furthermore, there were many small-scale attacks (both lethal and non-lethal) on individuals that were not officially claimed by extremist groups. Nevertheless, witnesses identified the attackers as 'terrorists' who carried jihadist flags and professed to have religious motivations. Although these assaults could be categorised as banditry, their association with religious extremism created greater unease among civilians while enhancing the perceived power and reach of extremist groups. As such, the incidents had much the same effect as other extremist attacks.

Nonetheless, for most Malians, banditry was the greatest threat to their safety. Between December 2015 and March 2016, bandits were responsible for 45% of all violent incidents in Timbuktu, and 25% in the Mopti and Gao regions. Lacking sufficient government or international protection from these criminals, citizens tended to look to armed groups to ensure the safety of their families and possessions. Thus, the weakness of security infrastructure, especially in the north, aided the groups' recruitment efforts.

Slow dissolution of the peace agreement

The Algiers Accord faced several problems that could pose a serious threat to its survival. One of these was the apparent failure of all actors to implement the agreement, undermining public faith in it. This was particularly true of the multiple delays in the disarmament, demobilisation and reintegration process, which drew criticism from across the political spectrum. Meanwhile, the Malian government failed to fulfil its promise to establish interim authorities in northern regions in 2016. Although opposition groups criticised the government for failing to honour the peace agreement, attempts to establish the authorities in July were met with widespread protests in the streets of Timbuktu and Gao. There, demonstrators refused to recognise the legitimacy of the authorities, repeating the claim that they had been unfairly appointed by the ruling powers in Bamako and did not adequately represent the north.

The increasing fragmentation of organisations that had signed the Algiers Accord – including the Coordination of Movements for the Azawad (CMA) and the pro-government Platform – further complicated the implementation process. The resulting splinter groups often sought to rejoin the Algiers Accord, thereby gaining preferential access to benefits that may have been diluted if they had remained part of a larger cohort (including access to the disarmament, demobilisation and reintegration process) and a proportionately louder voice at the negotiating table. This factionalisation created confusion among both armed groups and the Malian population; reduced the effectiveness of peace negotiations; risked further violent incidents; and motivated other actors to establish new sub-groups. Above all, it undermined the main purpose of the Algiers Accord: to stabilise Mali. Meanwhile, the broken ceasefire in Kidal proved that the Malian government and international forces could do little to ensure that the peace negotiations made progress. In July and August, Platform affiliate the Self-Defence Group of Imrad Tuareg and Allies clashed multiple times with CMA fighters in Kidal and surrounding regions, killing or injuring more than 100 people. Although a series of summits held in Niamey and Bamako in subsequent weeks eased the tension between the groups, they continued to engage in minor skirmishes throughout the remainder of 2016.

By the end of the year, the Algiers Accord remained in place. Yet it was under threat from a general loss of faith in the peace process, disruptions of the negotiations (which could prompt the creation of more armed groups) and frequent ceasefire violations. While it was unlikely to collapse in the near future, the accord risked succumbing to slow disintegration if the parties to the conflict were unable to make meaningful progress.

Persistent humanitarian problems

Human security remained a critical issue in Mali throughout 2016. A difficult lean season and significant flooding throughout central Mali over the summer months exacerbated major shortfalls in development, particularly in food security. Ethnic violence and widespread banditry further threatened human security, as neighbouring ethnic groups engaged in small-scale clashes throughout the region. Many incidents of banditry were likely linked to intercommunal tensions, although this was often difficult to verify.

Humanitarian non-governmental organisations (NGOs) had diminishing access to civilians in need of aid, limiting their capacity to address development crises. The closure of Kidal airport in April, and multiple suicide attacks against Gao and Timbuktu airports late in the year, restricted access to northern communities. Roads connecting northern and southern Mali continued to be dangerous and poorly maintained, especially during the rainy season. As extremist groups and bandits repeatedly targeted aid workers, many NGOs decided that operating in the north had become too dangerous. Indeed, by the end of November, such organisations had been affected by 97 security incidents in 2016, 86.6% of which involved robbery. Due to these security concerns, NGO activities were largely concentrated in southern Mali. While these geographic restrictions still allowed for the provision of aid to most of the population, they served to widen existing inequalities in service and resource delivery to impoverished and alienated civilians in the north. Furthermore, future development efforts would be impaired by the fact that many humanitarian agencies pulled out of Mali entirely, reflecting their lack of faith in the peace process.

Mali's ongoing refugee crisis increased the strain on development aid. Roughly 135,000 civilians lived as refugees in neighbouring states in 2016, and their prolonged presence put greater pressure on institutions and systems that may have been unable to bear the burden. Around 37,000 Malians were internally displaced. The UN High Commissioner for Refugees (UNHCR) reported that September 2016 saw the largest monthly exodus of Malians from the north into neighbouring countries since the height of the conflict in 2013.

An estimated US$2.66 billion would be required to implement emergency-response measures that could support the 15 million most destitute people in the Sahel. The UN Office for the Coordination of Humanitarian Affairs (OCHA) estimated that US$293.1m would be required for Mali alone. Yet there was a drop in international funding for peace and reconciliation efforts between 2015 and 2016, as the UNHCR's budget fell from US$68m to US$49m. This decline partly resulted from the fact that after the Algiers Accord was signed, many donors redirected funds away from further stabilisation projects.

Declining governance capacity and legitimacy

A continued lack of governance across Mali, particularly in the north, compounded these humanitarian challenges. It was estimated that by the end of 2016, only 33% of elected officials actually lived and governed in the north, and no government officials remained in the Kidal region. This had significant negative effects on the provision of desperately needed services, as well as the promotion of good governance, in the north. For example, the 20 November elections did not take place in at least 40 northern communities due to security concerns.

Weak governance in the north also allowed smugglers to exploit the porous borders between Sahel countries, maintaining a steady flow of people, arms, drugs and other black-market material across the region. As such, Malian instability continued to be both a product and a cause of greater regional instability, as was reflected in the movement of fighters from Algeria to extremist groups in northern Mali. Furthermore, attacks allegedly committed by Malian terrorist groups destabilised border towns and refugee camps in neighbouring Mauritania and Niger.

Regional and international actors continued to address gaps in governance where possible: MINUSMA extended its official mandate to June 2017, and increased its total number of personnel to 15,209. However, high-level engagement by international actors such as the United Nations and the European Union continued to focus primarily on bolstering military capacity – which, while crucial in the short term, threatened to be inadequate in the long run without the implementation of a comprehensive security and development transition.

Another disquieting trend emerged in 2016: civilians increasingly questioned the legitimacy of the government and Malian institutions more broadly. This was particularly worrying given that the peace process depended on the existence of a unified government backed by the people, as this would provide the strongest defence against separatist movements and extremist groups.

Civilians and government officials often questioned the leadership of President Ibrahim Boubacar Keita, while several members of the ruling party's parliamentary coalition publicly abandoned him, citing corruption and ineffective governance. Discontent with the transparency of institutions and the president's management of the peace process also gave rise to protests in Bamako, as well as in cities such as Gao and Timbuktu. The government's heavy-handed responses to these demonstrations – which sometimes involved blocking access to social media – only fuelled further accusations that the leadership was undemocratic and corrupt.

The dissatisfaction with Keita came to a head during the 20 November local elections, the first since he had taken office, in which Malians appointed around 12,000 councillors across the country. After being announced in 2014 and subsequently delayed several times, the vote was marred by violent incidents in the northern communities that had previously been deemed secure enough to participate in elections. These incidents included protests, the burning of election material by extremist groups and the abduction of a candidate. Furthermore, there was widespread anger at the fact that many people, including residents of the north and recently returned refugees and internally displaced persons, were unable to participate in the elections. Members of the opposition contested the results of the election, and publicly accused the ruling party of rigging the vote in parts of Bamako. However, although pervasive discontent with the results of the election damaged the legitimacy of the government, it simultaneously showed that there remained widespread support for Malian democratic processes.

While Mali remained relatively stable in 2016, the events of the year demonstrated that the country would likely continue to face serious threats amid delays in the peace process and meaningful reform. The security situation markedly worsened in 2016, as extremist groups such as Ansar al-Dine attained greater influence by conducting coordinated attacks against Mali's army, as well as foreign personnel, throughout the country. Despite minor victories by counter-terrorism forces, the growing power of these groups was cause for concern in the international community. The stagnation of the peace process also fuelled public discontent, as did the failings of efforts to establish interim authorities and hold local elections. Persistent humanitarian problems plagued the people of Mali, who experienced no great improvements in their food security or access to resources. The withdrawal of many aid organisations from the country, prompted by the dangers facing their employees, appeared to create further challenges for vulnerable populations. Meanwhile, poor governance throughout the country continued to contribute to a growing legitimacy crisis for both President Keita and Malian democratic institutions. Although the Algiers Accord was unlikely to completely fall apart, consistent inertia in the implementation process underlined significant problems in Malian society and its approach to the peace process.

Syria

The conflict in Syria turned dramatically in favour of the regime of Bashar al-Assad in 2016. The political process to resolve the war was derailed early in the year as, with the direct military support of Russia, the regime led an escalation in the fighting. With the regime in the ascendant, there was little prospect that the talks would resume at any stage later in the year. Russia's military intervention paid dividends: after years as a divided city, Aleppo fell outright to the regime and, through cruel but effective use of siege tactics, pro-Assad forces made significant gains in the suburbs of Damascus. The opposition took a heavy blow with

Key statistics	2015	2016
Conflict intensity:	High	High
Fatalities:	55,000	50,000
New IDPs:	1,300,000	900,000
New refugees:	1,300,000	

the loss of Aleppo and – confronted with the combined efforts of the regime, Russia, Iran and foreign Shia militias – faced the prospect of further significant losses in the months ahead. The regime's campaigns took a significant toll on civilians, especially those living in besieged areas. Russia imposed a number of ceasefires, but the regime repeatedly blocked requests from UN and other aid agencies for humanitarian access to these areas.

The military threat posed by the Islamic State, also known as ISIS or ISIL, diminished throughout 2016. Backed by the United States and other Western powers, the Syrian Democratic Forces (SDF) alliance captured large swathes of territory from the group, while airstrikes by international forces reduced its capacity to fund operations through oil production. Ankara began direct military involvement in Syria, sending mechanised infantry and special forces into the north of the country, along with thousands of Turkish-backed rebels. Although they primarily fought against ISIS, these forces also became embroiled in clashes with the Kurdish-led SDF, as Turkey sought to contain Kurdish advances in northern Syria.

Frustrated diplomacy

In late 2015, the international and regional powers of the International Syria Support Group agreed to a programme of talks between the regime of President Assad and the opposition, prompting representatives of Arab opposition factions to gather in Riyadh to form a delegation for these talks, the High Negotiations Committee (HNC). In early 2016, after several delays, the parties met in Geneva. Despite remonstrations by the Kurdish-led Syrian Democratic Council, Staffan de Mistura, UN special envoy for Syria, invited the regime and the HNC to participate as the two main parties in the talks, with Kurdish politicians invited to attend alongside representatives of women's groups and other civil-society organisations. However, as the talks began on 31 January, regime and Russian forces escalated their attacks on opposition-held territory in Aleppo province, infuriating both the opposition delegation and international backers of the peace process. The talks were suspended and only resumed after Russia and the United States, acting on behalf of the regime and the opposition respectively, agreed on a nationwide temporary ceasefire that would begin on 27 February.

In the second round of talks, like the first, the UN delegation held separate meetings with each party in turn. As the talks were conducted with the express purpose of discussing a political transition, the HNC produced detailed proposals for the formation of a transitional government. However, the regime delegation, led by Bashar al-Jaafari, refused to address the issue. As a result, the discussions were suspended again. They restarted in mid-April, but once more the regime delegation refused to discuss a political transition, despite conciliatory gestures and concessions from the HNC. Frustrated by the intransigence of the regime, armed opposition groups demanded that the HNC take a harder stance. De Mistura made optimistic attempts to restart the talks, but the opposition refused to engage with the regime while its forces escalated the fighting on the ground and continued to prevent humanitarian access to the many opposition areas under siege. Eventually, on 22 April, the opposition delegation ended the talks by leaving Geneva.

As discussions between the regime and the opposition began to falter in February, US Secretary of State John Kerry and Russian Foreign Minister Sergei Lavrov met on the sidelines of the Munich Security Conference to discuss the prospect of a ceasefire. This set a pattern that held for several months. As the military situation deteriorated, the US determinedly pursued the support of Russia to de-escalate the fighting. Between July and September, diplomatic efforts almost exclusively took the form of bilateral talks – but these came to an end in October. Frustrated by Russia's repeated failure to live up to its commitments to secure the regime's cooperation in reducing the fighting, Kerry ended the talks.

Western states increasingly turned to the UN Security Council as a forum for condemning events in Syria, although this brought them into direct confrontation with Russia, which vetoed two resolutions on the matter. The first of these resolutions – jointly submitted by France and Spain, and vetoed on 8 October – called on the US and Russia 'to ensure the immediate implementation of the cessation of hostilities' nationwide. Then, on 5 December, both Russia and China voted against a resolution demanding a seven-day truce in Aleppo. Believing that the Security Council was failing to perform its role of maintaining peace and security, UN Secretary-General Ban Ki-moon called on the UN General Assembly to use the terms of a 1950 resolution to bypass the council and consider measures to exercise the responsibility to protect civilians in Syria. On 9 December, the General Assembly voted overwhelmingly in favour of a resolution demanding an immediate end to all attacks on civilians and civilian infrastructure. On 21 December, it voted to adopt a resolution establishing a mechanism to prepare the case for the prosecution of war crimes in Syria 'in national, regional or international courts or tribunals that have or may in the future have jurisdiction over these crimes, in accordance with international law'.

The main players in Syria's stalled diplomacy were no longer Western states but increasingly Russia, Turkey and Iran, with Egypt and China also showing a growing interest in their relations with the regime. The relationship between Russia and Turkey had deteriorated markedly in November 2015, when Turkish fighter jets downed a Russian warplane, but was normalised the following June after Turkish President Recep Tayyip Erdogan apologised for the incident. It improved further with Erdogan's visit to Moscow in August and a reciprocal visit by Russian President Vladimir Putin to Istanbul in October. As Turkey stepped up its role in the Syrian conflict throughout the second half of 2016, the dialogue between the two countries became closer. With the Obama administration soon to leave the White House, the US was overtaken by Turkey as the main interlocutor with Russia; in late December, Ankara and Moscow led the discussion on a renewed ceasefire.

With the backing of Russia, the regime – and Assad personally – looked increasingly secure. In April, Syria's ruling Ba'ath party won parliamentary elections (held only in government-controlled areas, and under less than free and fair conditions) with 80% of the vote. A subsequent cabinet reshuffle saw Emad Khamis replace Wael al-Halqi as prime minister, but the changeover resulted in no material policy changes. Western powers dismissed the election as a sham.

Russian reinforcement

When Russia began its direct military support to the Assad regime in late 2015, the military balance on the ground quickly shifted in the regime's favour. Throughout 2016, the regime translated this advantage into significant territorial gains, as was first seen in northern Latakia province. In January, regime forces, backed by an intense campaign of Russian airstrikes and with the support of Russian military advisers on the ground, secured control of all the major towns and cities in the province, capturing Khan al-Assal, Salma and Rabia. In a counter-offensive by Jaysh al-Fateh, al-Qaeda affiliate Jabhat al-Nusra and various Free Syrian Army affiliates in June and July, rebels briefly captured Kansaba. But the regime retook the town in August, leaving it in control of all of Latakia province except for a few villages in its northeast.

The value of Russian support also became clear in March, when regime forces launched a campaign to drive ISIS out of the city of Palmyra, in Homs province. Russian warplanes launched 150 airstrikes against the group's positions on the first day of the offensive alone. The regime regained control of the whole city in 20 days, as ISIS forces withdrew to nearby Sukhnah. Although most of Palmyra's residents had fled the city, the victory was symbolically important as it re-established regime control of one of Syria's most important historical sites. Regime forces then pushed further into Homs province and captured Qaryatayn. The nearby Shaer gas field and other smaller fields repeatedly changed hands between the sides from April onwards, but Sukhnah remained under the control of ISIS. The group eventually launched a counter-offensive from the town in December, recapturing Palmyra.

Following the regime's initial success in Palmyra, Putin announced that Russian forces had achieved their objectives and that a large part of those forces would be withdrawn from Syria. However, although Russia immediately flew 16 of its 35 warplanes out of Syria, the promised withdrawals did not materialise; instead, Moscow stepped up its commitment. In October, Russia announced it would expand and upgrade its naval facility in Tartus; deployed an S-300 air-defence system to the base; and dispatched to the Syrian coast a fleet of warships, including the navy's flagship aircraft carrier *Admiral Kuznetsov*, the battlecruiser *Pyotr Veliky*, the destroyer *Vice-Admiral Kulakov* and multiple anti-submarine ships. The Russian Federal Security Service also reportedly undertook a campaign of cyber attacks against the websites of Syrian opposition groups, activists and monitors.

Fall of Aleppo city

Regime and Russian forces focused their efforts on Aleppo throughout 2016, targeting the city and surrounding areas with the most intense aerial bombardment of the war, which paused only for occasional, short-lived ceasefires.

At the start of 2016, the opposition controlled most of eastern Aleppo city, which – otherwise encircled by regime-held territory – was supplied via a route running to the northwest via Castello Road. Therefore, the regime's ground campaign initially focused on cutting off this remaining supply route to opposition-held territory. In February, regime forces north of Aleppo advanced west, linking up with the regime-held towns of Nubl and Zahra, which rebel forces had besieged since July 2012. In late June, regime fighters (comprising Syrian Arab Army forces, pro-regime Syrian militias and Iranian-funded foreign militias such as Hizbullah) began a large-scale offensive on Castello Road. Within a month, they had captured positions overlooking the road, using constant bombardment to make it impassable. Shortly thereafter, they captured the districts of Bani Zayd, Khaldiyeh and much of Layramoun, placing eastern Aleppo city under siege.

Faced with the looming threat of losing Aleppo outright, rebel forces operating in two major alliances, Jaysh al-Fateh and Fateh Halab (Conquest of Aleppo), launched a joint offensive in the southwest of the city, establishing a new supply route to eastern Aleppo through the districts of Ramouseh and Hamdaniyeh. However, rebel forces only maintained control of this route between 6 August and 5 September before regime forces again took over the south and southwest of the city, re-establishing their siege of opposition-held districts. In October, rebel forces made one last attempt to reach eastern Aleppo city through the districts of Minyan and Dahiyat al-Assad, but regime forces thwarted the attack.

Regime forces then pressed into eastern Aleppo from all directions. In October and early November, they advanced south into northeastern Aleppo, taking Handarat and Owaija. Attacking from the east in an attempt to divide territory held by the opposition in the northeast of the city from areas in the south, the regime captured Masaken Hanano, which had been the first district of Aleppo to fall to the opposition. This made the northeastern districts impossible to defend, forcing rebel forces to withdraw from the area. By 28 November, they had lost control of Halak, Bustan al-Basha, Haydariyeh, Sheikh Khader and Sakhour. However, the southern parts of the city proved no easier to defend, as district after district fell to the regime. By 6 December, the rebels had lost around 70% of the territory they once controlled in the city, ceding Jabal Badro, Myassar, Karm al-Jabal, Qadi Askar, Shaar and the Old City to the regime. Periodic halts in the fighting allowed thousands of civilians to flee the city, although the United Nations reported that rebels were attempting to prevent some of them from doing so. Subsequently, Sheikh Lutfi, Marjeh, Bab al-Nayrab, Maadi and Saliheen districts fell to government forces, followed by Sheikh Sayed and Fardous.

On 13 December, Jaysh al-Fateh and Fateh Halab agreed to a ceasefire mediated by Russia and Turkey. In the following days, around 3,000 opposition fighters left Aleppo and the regime assumed control of all of the city except for Sheikh Maqsoud, which remained in the hands of Kurdish groups.

Kurdish expansion

The Kurdish Democratic Union Party (PYD) and its armed wings also fared well in 2016. The group expanded the areas under its control in Hasakah province, as well as – through the SDF alliance, which includes secular Arab groups and various ethnic and tribal militias – Aleppo and Raqqa provinces.

In February, the PYD's main military wing, the People's Protection Units (YPG), led the SDF into direct confrontation with Islamist rebels in northern Aleppo province. With the Islamists under pressure elsewhere in the province and suffering heavy bombardment by Russian forces, the SDF opportunistically seized rebel-held Tel Rifaat, Sheikh Issa, Kafr Naya and Kafr Naseh. The Islamists attempted to recover lost ground in April, but suffered significant losses.

Kurdish forces also clashed with the regime. The first significant confrontation between the two parties took place in Qamishli in April, when an altercation at a regime checkpoint sparked three days of serious fighting. However, the largest confrontation between the two sides took place in August when small-scale clashes broke out in Hasakah city between the PYD's internal security force, Asayish, and the pro-regime National Defence Force (NDF) militia. The regime responded with its first airstrikes against Kurdish forces, in Hasakah. The fighting was briefly suspended after the US scrambled fighter jets to intercept regime planes and protect both the SDF and its own special forces in Hasakah, while the city's Kurdish governor and regime authorities attempted to agree on a ceasefire. However, the violence resumed when the YPG and Asayish conducted an offensive to oust the NDF from Hasakah. By 23 August, Kurdish forces had established control over more than 90% of the city. A subsequent ceasefire forced the regime to withdraw all military units from the city, leaving behind only police forces to protect the government quarter.

Despite these sporadic clashes with the regime and the opposition, Kurdish forces were primarily focused on the war with ISIS. The SDF expanded its offensive against the jihadist group from Hasakah and Raqqa provinces to Aleppo province, bringing it into conflict with Turkey and Turkish-backed rebels. In Hasakah province, the SDF launched a major offensive to capture Shaddadi and, despite an effective counterattack by ISIS forces, had established complete control of the town by 26 February. The SDF's first offensive in Raqqa province was less successful: it sought in early May to advance on Raqqa city with the support of US airstrikes, but was only able to capture a small strip of villages just south of Ayn Issa. The SDF then turned its attention to the ISIS-held city of Manbij and, on 31 May, launched an offensive across the Euphrates River into new territory in Aleppo province. The group made rapid progress, encircling the city in the following eight days and forcing ISIS to withdraw its forces from nearby Marea and Azaz to defend the city. However, the SDF's progress slowed thereafter, as it made gradual, hard-fought advances into Manbij. It was not until 12 August that the group captured the city outright. Unable to defend the areas of the Manbij still under its control, ISIS fighters withdrew from the city to Jarabulus in a large convoy of vehicles, apparently using thousands of civilians as human shields to ward off US airstrikes.

After its success in Manbij, the SDF looked to begin a fresh offensive on Raqqa city, and to push further east into Aleppo province towards ISIS-held Jarabulus and Al-Bab. Both efforts caused the US-backed SDF to enter into direct conflict with Turkey and Turkish-backed rebel groups in northern Syria. As a consequence, the SDF's advances towards Jarabulus and Al-Bab met with airstrikes launched by Ankara and offensives by Turkish-backed rebels. Washington and Ankara openly disagreed over the former's support for the SDF, contributing towards a general deterioration of relations between the governments. The US continued to support the SDF and confirmed that group would play a leading role in the offensive to isolate ISIS-held Raqqa, although it added that Turkish-backed Arab forces might be allowed to conduct the planned advances into the city. However, the Kurdish authorities exacerbated tension between Washington and Ankara by announcing that they would accept no role for Turkey in capturing Raqqa. The Kurds also stated that the

SDF's political wing, the multi-ethnic and secular Syrian Democratic Council, would establish a civilian administration in the city as part of the wider federal scheme it envisaged for Syria.

Earlier in the year, representatives of Kurdish and other ethnic-minority groups, as well as some secular multi-ethnic groups, convened a conference to declare the creation of the Democratic Federal System of Rojava–Northern Syria, establishing a new federalist government in territories under SDF control. Although rival rebel groups condemned the announcement – as did many countries, the regime and other Syrian political and civil-society actors – it seemed increasingly probable that the conflict would result in the adoption of a federalist system. Foreign powers were reportedly discussing this option, and in November de Mistura explicitly proposed such a model.

However, relations between Kurdish political factions deteriorated in 2016, as the PYD clamped down on its rivals. In August, Asayish exiled Ibrahim Biro, leader of the Kurdish National Council and its constituent Yakiti party, and detained several of the council's senior members, forcing the closure of its office in Qamishli. The crackdown continued in October as Asayish arrested Biro's deputy in the Yakiti party and detained the leader of the Kurdish Reform Movement, sparking protests by activists and supporters of these groups.

ISIS diminished

By the end of 2016, ISIS had lost a series of major towns in Syria and faced US-backed offensives against its two most important cities – Raqqa in Syria and Mosul in Iraq – as well as attacks by both Turkish-backed forces and the SDF on its remaining territory in Aleppo province.

The US-led international coalition's *Operation Inherent Resolve* (and *Operation Tidal Wave II*), as well as Russian airstrikes, debilitated ISIS. The group's revenues fell due to aerial assaults on its oil infrastructure and its loss of control over oil and gas fields, as well as strengthened security on the Turkish border. The US Department of Defense reported in October that ISIS revenues from oil had fallen by 60% compared with 2014, while satellite imagery suggested that the group's oil production fell even further in the remainder of the year. As early as January 2016, ISIS was forced to halve the salaries of all its fighters and, by the end of the year, it was not paying them for weeks at a time. The pay reductions and growing barriers to crossing the Turkish border prevented ISIS from maintaining its flow of foreign fighters. Meanwhile, US airstrikes killed several senior ISIS figures in Syria, including Abu Ali al-Anbari, the group's deputy leader and finance minister; military commander Abu Omar al-Shishani; official spokesperson Abu Muhammad al-Adnani; and information minister Abu Mohammed al-Furqan. Airstrikes in Syria also killed Salah Gourmat and Sammy Djedou, who were involved in plotting the November 2015 Paris attacks, as well as Walid Hamam, who was behind a thwarted plot to conduct a terrorist attack in Belgium in January 2015.

Planning for an offensive against Raqqa, the de facto ISIS capital in Syria, began when the states participating in *Operation Inherent Resolve* met in Brussels in February. Among them were Saudi Arabia, Bahrain and the United Arab Emirates, who had already volunteered ground forces to fight alongside rebels in a US-led advance on the city. However, the prospect of foreign-military incursion into Syria prompted the regime to announce its own impending offensive, making the involvement of international ground forces untenable. Instead, the SDF launched in November a campaign to isolate Raqqa with the support of coalition airstrikes, military advisers and a small number of special-forces personnel. In the first phase of the offensive, the SDF advanced south from Al-Rai, directly north of Raqqa, to capture Tel as-Saman and a cluster of villages, before taking nearby Hadaj one month later. The second phase of the operation, which began in late December, primarily involved Arab units of the SDF advancing along the Euphrates River towards Tabqa and eventually capturing the village of Jabar, which had served as the main local weapons-storage and supply centre of ISIS.

ISIS also continued to lose territory in Aleppo province. Following the SDF's capture of Manbij, Turkish forces directly intervened in the conflict, crossing intro Syria from Karkamis with 1,500 massed rebel fighters to capture the ISIS stronghold of Jarabulus before advancing SDF fighters could reach the town. In anticipation of the attack, ISIS fled to Al-Bab, leaving the Turkish-backed rebels to swiftly capture Jarabulus and several villages, including Dabiq (the location, according to ISIS doctrine, of the final battle before the apocalypse). Both the Turkish-backed rebels and SDF then pressed towards Al-Bab. Under pressure, ISIS withdrew from its westernmost territories near Halisa, allowing regime forces, the SDF and the Kafr Saghir Martyrs Brigade to take control of them. Although Turkish and Turkish-backed forces entered the outskirts of Al-Bab while the Turkish Air Force bombarded the town from the air, ISIS held onto it. This led to the deployment of an additional 1,000 Turkish and 2,000 rebel fighters across the Turkish border to Al-Bab on 27 December.

Rebels north and south
Turkey's intervention in support of moderate Free Syrian Army-affiliated rebel groups in northern Syria served to tackle ISIS and check the advances of Kurdish groups near the Turkish border. For most of the year, moderate rebels gradually ceded territory to the regime, the SDF and ISIS. These moderate groups were left to defend the territory in northern Aleppo province while the major Islamist factions – with whom they frequently clashed – redeployed to Aleppo city. These moderate rebels required Turkish support to hold territory. Turkey's cross-border shelling of Kurdish forces in Afrin served to deter them from further advances towards rebel-held Azaz, and Turkish artillery support was vital to the rebels' defence of territory around Marea and Azaz from an ISIS offensive in May. With increased support from Turkey, the moderate rebels began to have real success in August, first capturing Al-Rai and then advancing into ISIS-held territory.

Islamists largely dominated the rebel movement, as they controlled more territory and had far more fighters than their rivals in the opposition. In northern Syria, the two most significant rebel alliances were the moderate Fateh Halab, based in Aleppo city, and the hardline Jaysh al-Fateh, which was led by Ahrar al-Sham and worked closely with Jabhat al-Nusra.

Throughout 2016, Jabhat al-Nusra repeatedly sought to ingratiate itself with these groups. Its leader, Abu Muhammad al-Golani, proposed in January a merger of Islamist factions fighting in the north, but refused to disassociate his group from al-Qaeda. By maintaining its loyalty to al-Qaeda, Jabhat al-Nusra was excluded from all ceasefires and often targeted by US and Russian airstrikes. In July, the global leader of al-Qaeda granted its Syrian affiliate permission to break formal ties with the wider organisation; the same day, Golani announced that the group would be re-established as Jabhat Fateh al-Sham, and would have no affiliation with al-Qaeda. However, many states dismissed the change as a mere rebranding and quickly added the renamed entity to sanctions lists. The US continued its airstrikes against the group's leadership, killing Abu Omar Saraqeb, Abu Faraj al-Masri, Haydar Kirkan and Abu Afghan al-Masri.

However, these alliances often failed to prevent infighting between their constituent rebel groups, making their front-lines vulnerable to the regime's attacks. There were particularly frequent clashes between two Salafist groups, Jund al-Aqsa and Harakat Ahrar al-Sham al-Islamiyya, which fought each other in March, September and October. Harakat Nour al-Din al-Zenki also clashed with its supposed ally the Fastaqim Union.

In southern Syria, the most active opposition groups were Islamist factions in Rif Dimashq, while moderate rebels in Daraa province largely remained locked in a stalemate with the regime. However, as in northern Syria, infighting between Islamist groups weakened the opposition's defence against a dogged regime campaign. Between April and November, Jaysh al-Islam and Faylaq ar-Rahman came into conflict in Eastern Ghouta. Jaysh al-Islam had previously dominated the area due to its size and the forceful leadership of Zahran Alloush, but it felt threatened by Faylaq ar-Rahman's merger with the Ajnad al-Sham Islamic Union, particularly following Alloush's death in December 2015. Supported by Jabhat al-Nusra, Faylaq ar-Rahman clashed repeatedly with Jaysh al-Islam, despite civilian protests against the fighting. Regime forces took advantage of the disarray to make regular advances in Eastern Ghouta, capturing the towns of Deir al-Asafir and Zabdeen in April and May. The regime then seized Maydaa, Bahariyah, Hawsh al-Farah, Hawsh Nasri, Tel Sawwan and Tel Kurdi as the rebel groups continued to clash with each other between June and October. On 20 November, Jaysh al-Islam and Faylaq ar-Rahman finally reached an agreement, prompting them to redeploy their forces to the front-line with the regime and halt its advance.

However, the Islamist opposition also lost large swathes of territory to the regime in Rif Dimashq, in Western Ghouta. Regime forces made advances across the region, subjecting a series of towns to devastating sieges. Daraya had been under siege for years, but in June the regime stepped up its bombardment of the town and began forays into its outskirts. Suffering heavy casualties and lacking support from rebels elsewhere, Daraya succumbed to the siege in August, as rebels and civilians there agreed to move to the opposition stronghold of Idlib province. The regime achieved the same feat in October in the nearby towns of Muadimiyat al-Sham, Hameh and Qudsaya, and in November forced rebel fighters out of Khan al-Shih and Al-Tal, although the latter continued to be governed by unarmed opposition groups.

Rebels in the south also had to defend themselves from the Yarmouk Martyrs Brigade and the Islamic Muthanna Movement, purported ISIS affiliates that in May merged to form Jaysh Khalid ibn al-Walid.

War crimes, sieges and starvation

The regime's strategy of subjecting city after city to siege warfare took a heavy toll on civilians. However, it shifted its approach in 2016. Having previously pressured civilians in besieged areas to surrender by cutting off their access to food and medical supplies, the regime instead began to attack these areas with intense bombardment from aircraft and artillery. It alternated between these attacks and offers of negotiation, eventually forcing the opposition authorities to capitulate. This tactic was used to devastating effect in Rif Dimashq. Elsewhere in the country, the regime continued to subject opposition-held areas to long-term sieges and starvation. Independent monitor Siege Watch estimated that by the end of October, around 1.3 million people were trapped in at least 39 besieged areas.

The regime consistently refused to grant humanitarian access to these areas for organisations such as the Red Cross, the Syrian Arab Red Crescent and the UN. Jan Egeland and Stephen O'Brien, the UN's humanitarian adviser for Syria and under-secretary-general for humanitarian affairs respectively, repeatedly called on the UN Security Council to intervene, to no avail. By mid-November, people living in the besieged eastern region of Aleppo city had run out of UN humanitarian supplies. Both the regime and Russia refused to approve a UN plan for delivering food and medicine to the city.

In August, the UN Commission of Inquiry on Syria condemned the sieges as a breach of international law, having reported earlier in the year that the Assad regime had conducted a campaign of 'extermination' in its detention centres. The UN high commissioner for human rights called for the Security Council to refer the alleged crimes to the International Criminal Court, but Russia prevented such a resolution with the threat of a veto. In response, the UN General Assembly voted to establish a mechanism that would gather evidence and build the case for any future prosecution of war crimes in Syria.

However, Russia's intervention in Syria seemed to ensure that Western states would not use military means to punish the regime. This emboldened Assad's forces to openly target civilian infrastructure, particularly medical facilities. According to the Syrian American Medical Society, there were at least 250 attacks on Syrian healthcare facilities in 2016; data produced by Médecins Sans Frontières showed the regime to be responsible for more than 90% of these assaults, including those on maternity and children's hospitals. The destruction of infrastructure forced civilians and combatants to travel significant distances to receive medical treatment, and reduced the chances of survival for those who were injured. Airstrikes also hit a wide variety of other civilian facilities. Bread ovens, which helped supply food to thousands of Aleppo's residents, were often destroyed over the course of the year. The regime also repeatedly targeted schools, killing almost 40 people in one such attack in Haas, in Idlib province, in October.

The regime and Russia made increasing use of incendiary weapons and cluster munitions in these airstrikes. The Syrian Civil Defence, a volunteer search-and-rescue organisation, reported that 130 attacks between 2 June and 8 December involved the use of incendiary weapons. The use of toxic gas (most likely chlorine) also continued in 2016, as seen in attacks in Aleppo, Idlib, Hama and Rif Dimashq provinces. There was little prospect of holding the Assad regime to account for such assaults, despite the fact that the Joint Investigative Mechanism (JIM) of the UN and the Organisation for the Prohibition of Chemical Weapons found evidence to attribute responsibility for three chlorine-gas attacks to the regime. One of these attacks occurred in Talamenes on 21 April 2014, and the others in Sarmin and Qamenas on 16 March 2015. The JIM also found ISIS responsible for a sulphur-mustard-gas attack in Marea on 21 August 2015. Western states sought to punish the regime through the UN Security Council, only to be blocked from doing so by Russia. Nonetheless, the Security Council extended the mandate of the JIM, allowing it to continue investigating alleged chemical-weapons attacks in Syria for another year.

Widespread displacement

As the fighting continued throughout 2016, tens of thousands of people were displaced from their homes. In the first four months of the year, the intense clashes between the regime, rebels, the SDF and ISIS in northern Syria displaced around 120,000 people, who fled towards the Turkish border, placing an enormous strain on camps and humanitarian resources. By the end of May, 180,000 people were left in temporary and informal shelters as Turkey refused to open the border to refugees. The subsequent SDF offensive on Manbij later in the year displaced tens of thousands of others.

Meanwhile, in Daraa province, battles between rebels and ISIS-affiliated groups in the west of the province and clashes between the regime and rebels in Sheikh Miskeen caused thousands of people to escape towards the Jordanian border. Thousands of them were stranded on the border as the careful checks undertaken by Jordan's authorities reduced the flow of refugees into the country to 50–100 per day. Jordan closed the border completely on 21 June, when a suicide bomber attacked a border checkpoint near the Rukban camp for displaced persons, killing several Jordanian soldiers. By November, the number of refugees in the camp had increased to around 85,000 and aid agencies were unable to deliver humanitarian supplies to people stranded there. Jordan sent convoys of water trucks to the camp, but between June and late November, when the UN regained access to it, the facility received only two major deliveries of aid – by crane, in August and October.

Despite the ongoing displacement, the flow of refugees to Europe slowed. A deal struck between Brussels and Ankara in March allowed the European Union to return to Turkey refugees (from Syria and other countries) who had entered Greece without official authorisation. In exchange, the EU would take an equal number of Syrian refugees through official channels. Although only around 1,000 migrants (of whom a small minority were Syrian) were returned to Turkey under this agreement, the hardening of the EU's rhetoric and response to the migrant crisis appeared to contribute to a dramatic fall in the number of people attempting the trip: only around 80,000 Syrian refugees entered Europe via Mediterranean routes in 2016, according to figures from the International Organization for Migration.

With at least 4.8m Syrians still stranded in Turkey, Lebanon, Jordan and Egypt, European states increased their contributions to the regional refugee response. Several countries also paid growing attention to Syrian refugees' access to employment. In 2016 the governments of Turkey and Jordan began to issue work permits for Syrian refugees – giving them a chance to enter legal, paid employment – while donors increased their development assistance for refugees from around US$812m in 2015 to approximately US$2.7 billion in 2016 under the UN's Regional Response Plan. Although these measures provided hundreds of thousands of refugees with a chance to make small but significant improvements to their lives, the number of refugees in the region continued to place Syria's neighbours under severe strain.

Turkey (PKK)

Key statistics	2015	2016
Conflict intensity:	Medium	Medium
Fatalities:	2,000	3,000
New IDPs:		350,000
New refugees:	1,500	

The conflict between the Turkish government and the Kurdistan Workers' Party (PKK) escalated dramatically in 2016. Both parties pursued security-driven agendas that pushed violence to a level unseen since the 1990s – a dark period in the history of conflict – and resulted in the destruction of the Diyarbakir, Sirnak and Hakkari city centres. To re-establish its presence and control in these cities, the government built larger police stations and more checkpoints, increased its use of airstrikes and attacks with unmanned aerial vehicles (UAVs), and recruited additional village guards.

The fighting was initially concentrated in urban areas in southeastern Turkey, where the PKK imposed its rule. But the focus of the conflict eventually shifted to rural areas in the southeast, while the Kurdistan Freedom Falcons (TAK) – a semi-autonomous group placed on the US Department of State's list of terrorist organisations in 2008 – carried out deadly targeted attacks on the security forces and civilians in city centres in western Turkey.

Meanwhile, there was a narrowing of the political space available to elected Kurdish and pro-Kurdish officials from the People's Democratic Party (HDP) and the Democratic Regions Party (DBP): the government arrested ten Kurdish lawmakers and more than 60 Kurdish mayors and co-mayors for allegedly supporting the PKK. A parliamentary vote to rescind Kurdish lawmakers' immunity from prosecution in May helped clear the way for these arrests, as did the government's decision to declare a state of emergency after an attempted coup on 15 July.

The Turkish government launched its long-anticipated intervention in Syria in August, after the Kurdish People's Protection Units (YPG), the military wing of the Democratic Union Party, made territorial gains west of the Euphrates

River, and the Islamic State, also known as ISIS or ISIL, increased its attacks on civilians in Turkey. The Turkish armed forces also launched airstrikes against the YPG and the PKK in northern Iraq, adding another regional dimension to the conflict and underlining the Turkish government's view of the groups as sister organisations that cannot be separated.

The government initiated a US$39-billion programme to construct housing complexes, factories and other buildings in 23 cities damaged by security operations – with a view to facilitating the return of almost half a million internally displaced persons and creating livelihoods for civilians affected by the operations.

Rise in urban warfare

Violence between the parties to the conflict rose sharply in 2016, with most casualties occurring in trench battles in districts such as Nusaybin, Sur, Silopi, Cizre, Semdinli, Yuksekova and Idil in the first quarter of the year. Supported by the Patriotic Revolutionary Youth Movement, the PKK took over neighbourhoods in these districts by digging trenches and installing barricades. This effectively contested the Turkish government's control of the areas and demonstrated that the security forces had lost full control of the towns during peace negotiations between late 2012 and mid-2016. The Patriotic Revolutionary Youth Movement denied following direct orders from the PKK's leadership, but continued to receive backing from the group despite the dire effect of its actions on local civilians.

This support for youth militias conducting asymmetric urban warfare formed part of a shift in the PKK's tactics, as did its increasing use of improvised explosive devices in densely populated areas. The Turkish government waged full-scale urban warfare to restore public order and take full control of these neighbourhoods, declaring months-long curfews and establishing special security zones, while deploying special forces to battle the youth militias for several months. In the first half of 2016, the Turkish security forces gradually re-established their control of Kurdish neighbourhoods, maintaining daily curfews as part of strict security policies. The fighting displaced at least half a million people and killed at least 1,500 others, including civilians, members of the security forces and PKK fighters. The destruction of entire neighbourhoods in Sur, Sirnak and Nusaybin districts prevented civilians from returning home. Some local Kurds expressed discontent with the PKK for having allowed youth militias to dig trenches and provoke a heavy government response.

The PKK's eventual loss of control in urban centres had two main effects. Firstly, the PKK and the Turkish security forces renewed their intense fighting in rural and mountainous regions. The Turkish military started to use armed UAVs in rural areas for the first time, allowing it to strike PKK targets and limit the group's movements in the mountains. Meanwhile, the government recruited hundreds of village guards to collect local intelligence and support Turkish forces. The move drew criticism from human-rights groups and Kurdish politicians, as most of these guards were ethnic Kurds whose work with the security forces allegedly changed the dynamics of their community by setting its members against one another.

Secondly, the TAK responded by expanding the conflict beyond the southeast, conducting attacks in western city centres. The most important difference between the PKK and the TAK is that only the latter group carries out operations in urban areas that often kill or injure civilians (despite its claims that it does not target them). Indeed, the TAK killed at least 110 people, more than half of whom were civilians, in six major attacks. Two car bombs detonated in Ankara's military quarter and Kizilay quarter, in February and March respectively, killed 28 military personnel and 38 civilians. Another car bomb targeting off-duty policemen after a football game in Istanbul's Besiktas district claimed the lives of 44 people, 36 of whom were policemen. Combined with ISIS attacks on civilians in Istanbul and Gaziantep provinces, the assaults created a climate of fear that damaged Turkey's tourism sector and prompted more heavy-handed security measures.

Shrinking political space

The escalating violence put the HDP in a difficult position, as it relied on a loose alliance of leftist and Kurdish parties. While some members of the HDP called on the party to denounce the PKK's violent acts, others wanted it to take a tougher line against the increasingly authoritarian Justice and Development Party (AKP) government. However, the HDP failed to denounce the PKK's attacks and carve out political space in which it could operate independently. Against this backdrop, the government moved to revoke the immunity from prosecution of HDP deputies, due to their alleged support for the PKK and terrorist propaganda. The effort succeeded: a new nationalist alliance between the government, the Nationalist Action Party (MHP) and the Republican People's Party (CHP) passed in May a parliamentary motion to strip HDP deputies of their immunity, threatening to leave more than 6m voters unrepresented in the national legislature.

Following Ahmet Davutoglu's resignation as prime minister and chairman of the AKP that month, President Recep Tayyip Erdogan acquired the power to directly execute his policies on the Kurdish issue. Although he was seen as a loyal ally of Erdogan, Davutoglu had a more conciliatory approach to the conflict than the president, backing the Dolmabahce accord that outlined a framework for the peace process. Erdogan resorted to publicly opposing Davutoglu's position after the rift between them grew to obstruct his bid to replace Turkey's parliamentary system with a presidential one. Davutoglu's replacement, Binali Yildirim, was widely regarded as a close friend of Erdogan. Having assumed office, Yildirim took a tough line against not only the PKK but also the HDP, criticising the party's reluctance to denounce PKK attacks and support the adoption of a presidential system.

The failed coup attempt on 15 July facilitated major structural changes in the Turkish military and other security institutions, introducing additional civilian oversight mechanisms while turning the ceremonial position of president as commander-in-chief into one with real power, even in peacetime. The government curbed the power of the chief of the general staff, and increased its control of military agencies through the introduction of ministry reporting lines and the requirement that the chief of the general staff report directly to the president. Moreover, the government decreed on 31 July that more civilians would join the supreme military council, which plays a key a role in promoting senior officers. Likewise, the reporting lines of the air, land and naval commands were changed from the chief of the general staff to the minister of defence. The interior ministry assumed direct control of the gendarmerie command and coastguard command (in addition to that of the police). The government also announced that it would establish a new coordination body to communicate between intelligence agencies, the police and military institutions. While all these developments were likely to have long-term implications for the fight against the PKK, the most immediate effect of the attempted coup was to prompt Erdogan to purge the security services of thousands of personnel, in a move that damaged not only their capacity but also their morale and reputation.

The failed coup attempt equipped the government with two critical counter-terrorism tools: the state of emergency and the power to rule the country through decrees. The National Security Council extended the state of emergency beyond its initial three-month period, allowing the government to take extraordinary measures and, in some cases, suspend basic human rights. The decrees allowed the government to bypass parliament, effectively abolishing the separation of powers. By issuing such decrees and revoking the immunity of lawmakers, the government extended its counter-terrorism policies to the persecution of elected Kurdish representatives.

In October and November, the government arrested around 60 Kurdish mayors and co-mayors before dismissing them and 12 other Kurdish leaders for allegedly supporting the PKK and separatists, as well as for promoting terrorist propaganda. Selahattin Demirtas and Figen Yüksekdag, co-leaders of the HDP, were also arrested, along with eight other lawmakers from the party. The government shut down most Kurdish news outlets by decree for the same reasons, muting views on the conflict that contradicted the official line. In the context of the broader crackdown on journalists, these moves effectively ensured that only the government's narrative on the conflict could be reported. Ankara's suspension in September of more than 10,000 members of a pro-Kurdish teachers' union further narrowed the political space.

Thus, the government demonstrated its firm belief that the conflict could be resolved only through military action, drawing comparisons with the 1983–2009 Sri Lankan civil war. As well as constituting a radical departure from government policy between 2009 and 2015, its approach reinforced the PKK's argument that 'self-defence' was the only viable option for Kurds seeking to resolve the conflict. However, neither party had the personnel and resources needed to establish its primacy in the conflict, with the Turkish security forces undergoing a major reconstruction after the attempted coup and the PKK having incurred significant losses in prolonged urban battles. The shifting dynamics of the conflict, particularly the growing militarisation of both sides, suggested that the violence would intensify amid a mutually destructive stalemate.

Internationalisation of the conflict

The YPG's territorial gains and establishment of political governance in northern Syria changed the calculations of Ankara and the PKK. Viewing the PKK and the YPG as sister organisations, the Turkish government believed that any attempt to create a Kurdish political structure in Syria would threaten Turkey's unity and security, and as such should be prevented with military action.

Turkey's fears peaked in spring 2016 as US-backed YPG forces moved west across the Euphrates River, seizing several towns around Tishrin Dam from ISIS. Although it agreed to the YPG joining the US-led anti-ISIS coalition, Ankara made clear that YPG fighters should withdraw from recaptured areas, leaving them to be governed by Arab tribes and other groups. These demands plagued Turkish–US relations for months, with Ankara repeatedly threatening to use armed force against Kurdish groups if they remained west of the Euphrates River.

Ankara eventually intervened in Syria in late August, prompted by ISIS attacks in Turkey, the YPG's presence in the town of Manbij and disagreements with Washington over the Kurdish group's role in the fight against ISIS. An ISIS assault on Istanbul airport on 29 June killed more than 40 civilians, while another at a wedding party in Gaziantep claimed the lives of approximately 60 civilians, including 34 children. Turkey's cross-border operation *Euphrates*

Shield involved Free Syrian Army units that initially drove ISIS out of the towns of Jarabulus, Al-Rai and Dabiq. Nonetheless, it soon became clear that Turkish-backed forces would also create a zone of control that compelled YPG forces to retreat. There were repeated clashes between the Syrian Democratic Forces, an umbrella organisation in which the YPG is a primary actor, and Turkish-backed Syrian rebel groups from October onwards, sparked by the latter's advance on Al-Bab. Enjoying Western support due to its victories against ISIS, the YPG remained confident that it could counter Turkish forces in Syria in the short term. The YPG's success or failure in Syria could shape the course of the conflict in Turkey, making either the PKK or the Turkish government return to the negotiation table, or encouraging the PKK to seek an independent Kurdish state in the region.

The dispute with the YPG has also inhibited effective US–Turkish cooperation against ISIS. Relations between Washington and Ankara reached their nadir after the US announced that YPG fighters would play a key role in the offensive to recapture Raqqa, but Turkish forces would not. Turkey repeatedly asked the US to consider the YPG as an extension of the PKK, which Washington designates as a terrorist group. Yet such efforts were to no avail: the US government continued to make a distinction between the groups. Washington also refused to instruct the YPG to withdraw east of the Euphrates River as Turkey requested, although the group announced in November that it would do so as part of the Raqqa offensive. Turkey launched its intervention in Syria amid this tension with the US, having achieved rapprochement with Russia and participated in proxy negotiations with the Syrian government through Iran and Russia.

The PKK presence in Iraq also became a source of tension between Washington and Ankara, after the US objected to Turkish special forces training anti-ISIS Sunni militias in Bashiqa, against the wishes of the Iraqi government. The problem became worse as some Iraqi lawmakers suggested that their government should support the PKK in Iraq, and Turkey realised that the group was conducting a military build-up in the Sinjar mountains (where the PKK helped defend Yazidi communities against ISIS). Ankara feared that if the PKK maintained a base in Sinjar along with a headquarters in Qandil, Turkish forces would struggle to prevent the group's movements across the porous Turkey–Iraq border.

Deteriorating humanitarian conditions

Due to ongoing violence, curfews and security operations, civilians in southeastern Turkey faced dire humanitarian conditions. The fighting displaced around half a million people, while thousands of others experienced collective punishment through drawn-out curfews and the loss of their livelihoods. Moreover, attacks by the TAK and ISIS created a climate of fear and insecurity throughout the country.

Zeid Ra'ad al-Hussein, UN commissioner for human rights, reported in May that the Turkish security forces had deliberately shot civilians and destroyed homes and infrastructure in the southeast. Although Ankara denied the reports, the Turkish parliament passed in June legislation to protect members of the armed forces and civil servants from prosecution for alleged abuses during the counter-insurgency campaign. Under the new law, investigations into armed-forces personnel involved in operations in the southeast would require the permission of the prime minister's office, while those of civil servants would have to be approved by the local district governor.

As part of its US$39bn reconstruction programme in the southeast, the government planned to build 67,000 apartments and ten factories over ten years. It also pledged to pay the rent of internally displaced persons until the apartments could be built, allowing them to return home.

The year marked the beginning of a particularly violent period in the conflict between the PKK and the Turkish government, which had significant effects at the local, national and regional levels. The shift in battlegrounds from rural areas to urban centres in the first half of 2016 caused a sharp rise in casualties among civilians, the security forces and PKK fighters, as well as widespread damage to infrastructure. Although the focus of the fighting shifted back to rural areas in summer, after the security forces completed their assertive counter-insurgency campaigns in towns and cities, the arrest of elected Kurdish lawmakers and mayors left Turkey's Kurds with few opportunities to pursue their interests through the political system. The TAK's attacks in western Turkey suggested that the spiral of violence would claim many more civilian lives as terrorism became increasingly common, while the outcome of the YPG's actions in Syria looked set to shape the next phase of the conflict.

Yemen

By the end of 2016, after several rounds of failed UN-sponsored peace negotiations, the conflict in Yemen remained in a violent stalemate. With neither side close to a decisive victory, the country was effectively divided: the northern part controlled by the Houthis and their ally, former president Ali Abdullah Saleh, and the southern part by the government of President Abd Rabbo Mansour Hadi, which established the city of Aden as its temporary capital. This division became increasingly institutionalised, leading to the creation of parallel or split administrative processes in entities such as the central bank (theretofore one of the few remaining impartial, functional institutions). The Saudi-led coalition backing Hadi proceeded with an aerial campaign that killed many civilians, prompting its main backer, the United States, to start distancing itself from the campaign. According to the Office of the United Nations High Commissioner for Human Rights, airstrikes were responsible for two-thirds of civilian deaths in Yemen. Meanwhile, Houthi operatives escalated their attacks in Saudi Arabia, sustaining an insurgency on the country's southern border. To quell militant activity there, the coalition deployed in northern Yemen and advanced southwards into

Key statistics	2015	2016
Conflict intensity:	High	High
Fatalities:	7,500	7,000
New IDPs:	2,200,000	500,000
New refugees:	13,000	

Houthi-controlled territory. Concurrently, it cracked down on jihadist activity in southern Yemen, expelling al-Qaeda in the Arabian Peninsula (AQAP) from the city of Mukalla in Hadhramaut province. However, as long as the conflict continued, jihadist groups were likely to take advantage of the insecurity to launch large-scale attacks. Yemen also faced a dire humanitarian situation, with acute shortages of medical care and food.

Failed peace talks

Although the parties to the conflict participated in UN-sponsored negotiations in Kuwait in May–October 2016, their achievements were limited to agreeing on a temporary reduction in hostilities. The UN's special envoy to Yemen, Ismail Ould Cheikh Ahmed, pushed forward with discussions on the withdrawal of coalition forces, Houthi withdrawal from Sana'a, disarmament and other security and governance arrangements. However, the talks ultimately revealed that neither side was willing to make the concessions necessary to reach a negotiated settlement.

Furthermore, there were signs of a schism between the Saudi-led coalition and the Hadi government. Perhaps viewing the war as having become too expensive or risky, the coalition took substantive measures to reach a political solution in the first half of 2016, holding direct talks with Houthi leaders in Riyadh. Saudi Foreign Minister Adel al-Jubeir adopted a softer tone towards the Houthis, describing them as 'part of the social fabric of Yemen' and 'neighbours' of Saudi Arabia with whom Riyadh could hold talks.

Despite this discernible shift in Saudi Arabia's position and priorities, the Hadi delegation appeared determined to obstruct the talks. The main obstacle to finalising an agreement was the issue of a consensus government, particularly Hadi's role in the post-conflict political arrangement. While the Houthis have insisted that he should not retain his position, Hadi refused any settlement in which he was not the transitional leader. For example, the Hadi delegation rejected the last UN proposal of the year – presented to the parties in October – on the grounds that he would be left as only a 'figurehead' following the proposed Houthi withdrawal from Sana'a. Yet given the emphasis it had placed on Hadi's legitimacy, the coalition was reluctant to exclude him from any settlement.

Hadi's dismissal in April of his vice-president and prime minister, Khaled Bahah, further complicated the search for a consensus candidate. Bahah was considered to be a more conciliatory figure than Hadi, and his dismissal was widely interpreted as a sign of the Hadi government's growing unilateralism. Further aggravating the situation, Hadi appointed General Ali Mohsen al-Ahmar as the new vice-president. The move signalled the adoption of a harder, more militaristic position on the Houthis, as Ahmar had led six bloody wars against them between 2004 and 2010.

Division of Yemen and its institutions

Although the Houthis attempted to distance themselves from Iran at the beginning of the year – a gesture aimed at Saudi Arabia – they subsequently took unilateral measures to entrench their rule in Sana'a. In July, the Houthis' political arm, Ansar Allah, and Saleh's party, the General People's Congress (GPC), replaced the Houthi Supreme Revolutionary Committee with the Supreme Political Council – in essence a power-sharing arrangement between the two. The establishment of the council marked the official return of Saleh and the GPC to Yemeni politics, defying expectations that the Houthis would retreat from state institutions. Indeed, the move entrenched the partition of the country, with the warring parties strengthening their authority in the areas they controlled. For instance, there were separate institutional processes for visiting Sana'a and Aden, while Aden's security personnel, who are linked to southern secessionists, undertook a mass deportation of northerners that may have been designed to encourage the secession of the south. They continued to do so despite Hadi's orders to cease the deportations.

The Central Bank was based in Sana'a until September, when there was a breakdown of the sides' tacit agreement to preserve its neutrality. Until that point, the institution had continued to pay state salaries throughout the war, guaranteed the import of basic commodities and protected the value of Yemen's currency. However, that month, Hadi appointed a new governor of the bank and relocated it to Aden, in an apparent effort to reduce the flow of money to the Houthis and their allies by cutting off their access to state financing. The Hadi government particularly objected to the fact that the central bank still paid YER25 billion (around US$100 million) to the defence ministry, which the Houthis controlled and ran. Although the government pledged to continue paying the salaries of civil and military staff who had been hired before September 2014 (when the Houthis captured Sana'a), an estimated 1.2m of these employees – including soldiers, police officers and bureaucrats – said that they stopped receiving their wages after the move.

The changes to the central bank were an attempt to hijack the functional aspects of the state under Houthi–Saleh control and shift the conflict onto an economic front. But the practical outcome of the move was to create two parallel institutions. Houthi officials sought to revive the Sana'a branch of the central bank by depositing money from sources such as customs in Hodeida, treasury bonds and deposits in local banks, as well as donations solicited from civilians in northern Yemen. Houthi leader Abdul Malik al-Houthi made a broad appeal to support the central bank, describing this as a form of resistance. Although many citizens mocked and rebuffed the request, the Houthis later claimed to have received donations worth YER1bn (around US$4m).

Analysts expressed concern that decimating the Yemeni economy formed part of the coalition's strategy; Jamie McGoldrick, the UN's humanitarian coordinator for Yemen, described the economic dimension of the war as a 'tactic'. According to estimates by the Sana'a Chamber of Commerce and Industry, airstrikes had destroyed or damaged more than 200 businesses nationwide since the start of the war. There were signs that the coalition had targeted large institutions that continued to operate after the Houthi takeover of Sana'a. It conducted intense strikes against economically key infrastructure in Houthi territory such as bridges, ports and power stations, as well as industrial facilities such as factories producing cement, ceramics, agricultural equipment and a variety of foods.

Controversial aerial campaign

The coalition's often indiscriminate aerial attacks killed many civilians, attracting broad international criticism. In one particularly high-profile incident in October, more than 140 civilians died in a coalition airstrike on a funeral hall in Sana'a. Coming seven months after coalition aircraft killed 106 people in a crowded village market in Hajjah province, the attack prompted thousands of Yemenis to demonstrate in Sana'a. According to statistics published in August by the UN's high commissioner for human rights, the coalition was responsible for around 60% of the 3,800 civilian deaths in the conflict since March 2015. The UN repeatedly suggested that these attacks might amount to war crimes.

In contrast to some previous attacks that had killed many civilians, the airstrike on the funeral home did not prompt the coalition to deny responsibility for the operation or attribute it to the Houthis. Instead, Saudi Arabia launched an investigation into the incident, concluding that it was the product of intelligence and communications failures. American officials also attributed the attack to flawed targeting, adding that the US provided logistical and intelligence support to the coalition but did not designate targets. The US had previously insisted that attacks against civilians could be attributed to 'errors of capability or competence, not of malice', arguing that inexperienced Saudi pilots tended to fly at high altitudes to avoid the threat of fire from the ground.

As Riyadh's main weapons supplier, Washington came under increasing pressure to minimise the number of fatalities caused by the coalition's aerial campaign. As a consequence, the US stated that it would reassess the merits of this assistance. President Barack Obama decided in mid-December to limit US military support for Saudi activities in Yemen, leading to the cancellation of sales of air-dropped and precision-guided munitions to Riyadh.

These measures built on an earlier attempt by the US to distance – or, arguably, detach – itself from the coalition. Washington had suspended sales of cluster munitions to Riyadh in May, and reduced the number of US military personnel coordinating with the coalition from 45 to six in August. The same month, the

US relocated a planning team that coordinated with the coalition's aerial campaign from Saudi Arabia to Bahrain.

Washington was attempting to respond to public pressure while maintaining the foundations of the US–Saudi alliance and balancing its commitments to allies in the Gulf. The US continued refuelling coalition aircraft and maintained some weapons sales to Saudi Arabia. Moreover, Washington said that it would share additional intelligence with the coalition on the Saudi Arabia–Yemen border, which became a prime concern for Riyadh in 2016.

Escalation on the northern border

The Houthis and their supporters intensified their fatal cross-border attacks into Saudi Arabia, most of which targeted Asir, Jizan and Najran provinces. These operations involved ground incursions, long- and medium-range missiles, and short-range rockets and mortars. Houthi operatives killed or captured hundreds of Saudi soldiers, while also targeting civilians and civilian infrastructure, including mosques, homes and schools. During their raids into Saudi territory, Houthi operatives took over facilities in the area, ambushed military convoys and occupied depopulated towns. These raids resulted in hours-long battles with Saudi forces and tribesmen. Although the Houthis were always pushed back, the incursions allowed them to boast that they had taken over Saudi territory – propaganda victories that were symbolically important, particularly as Saudi Arabia and Yemen were engaged in a long-running dispute over ownership of parts of the southern border area.

The Houthis' numerous, occasionally successful attempts to infiltrate Saudi territory had a devastating impact on civilians living in border towns. The Saudi authorities evacuated civilians from border areas to create a buffer zone, while thousands of other residents relocated of their own volition. The airport in Najran was shut to all civilian traffic. Saudi Arabia typically retaliated for the cross-border incursions with airstrikes against military positions, with the goal of degrading Houthi capabilities. Although Saudi air-defence systems intercepted most of the missiles, the short-range rockets could not be intercepted;

according to one estimate by Saudi officials, Najran had been hit by more than 10,000 rocket-artillery rounds since the beginning of the war. To bolster Saudi Arabia's defences against these attacks, Bahrain and the United Arab Emirates deployed in January an additional 1,000 troops to the border. These reinforcements joined around 20,000 Saudi troops stationed on the 1,800-kilometre frontier.

The coalition was coordinating with the Popular Resistance, an umbrella organisation comprising groups fighting on the ground on behalf of Hadi, who had received support and weapons from the coalition. Popular Resistance fighters launched in July an operation to halt these attacks by retaking Houthi-controlled areas of Yemen near the border; by October, it had seized swathes of this territory in the north. These operations accorded with the goal of advancing on the Houthis in Sana'a – although, by the end of the year, Popular Resistance fighters had not yet entered the city. Advancing from the northern front-line, the Popular Resistance seized strategically important positions in Buqa and Saada. Yet the advance was risky, as the Popular Resistance could only enter Sana'a from the north by passing through territory in which the Houthis were dominant and still relatively popular.

On Yemen's southern coast, there were assaults from Houthi-controlled territory on vessels in the Bab al-Mandeb Strait. In October, a missile attack sank the Emirati HSV-2 *Swift* and spurred American destroyer the USS *Mason* to deploy countermeasures, prompting the US to send warships to the area and declare its commitment to upholding freedom of navigation in the strait. Another destroyer, the USS *Nitze*, subsequently launched *Tomahawk* cruise missiles at radar sites believed to be held by Houthi fighters. An official from US Central Command voiced the suspicion that the attacks on US vessels had been conducted with Iranian missiles. There was speculation that Tehran had increased its material support for Houthi fighters, as the coalition and the US Navy confiscated shipments of weapons – including missiles and small arms – believed to be en route to Yemen from Iran. Such shipments also allegedly flowed through Oman, although the country denied this.

Operations against jihadist groups

The Saudi-led coalition began to push back AQAP in February, reversing some of the significant territorial gains that the group had made by capitalising on the chaos of the war. By April, coalition fighters had expelled AQAP from Mukalla, in Hadhramaut province, a port city that AQAP had governed for more than a year. This victory was symbolically, as well as militarily, important: Mukalla had come to be viewed as AQAP's base in Yemen, where the group set up a kind of parallel state and informally governed more than 500,000 residents. The group was thought to have earned an estimated US$2m per day through oil smuggling, shipping and taxation in the city.

The campaign against AQAP reflected a shift in the coalition's strategy, which had previously focused on fighting the Houthis and Saleh. It suggested that the coalition had started viewing AQAP's activity as a direct threat to the stability of Aden. The coalition directly coordinated its military activities against AQAP with the US, which used unmanned aerial vehicles (UAVs) to kill hundreds of the group's operatives in 2016 – including 70 in March alone. The US also deployed a special-forces team to Yemen on 25 April, with the aim of aiding the coalition in the battle against AQAP. The team reportedly worked closely with Emirati troops, which had requested broader military support from the US – including in airpower, intelligence and logistics – to launch a new offensive against the group. Coastguard personnel trained by the UAE deployed in October to secure the Hadhramaut coastline. The coalition and allied militias also targeted AQAP positions in Abyan, Lahij and Aden provinces, but these campaigns failed to decisively reduce the group's operational capacity there.

Although the US and the UAE made substantive progress in the battle against AQAP, they did not entirely defeat the group, which sought to recover from its ouster from Mukalla. The organisation continued to promote itself as a protector of Yemen's Sunnis, placing great emphasis on popular support and describing its withdrawal from the city as a way to 'prevent the enemy from moving the battle into your homes, markets, roads, and mosques'.

The expulsion of AQAP from Mukalla, the only area it held in Hadhramaut province, paved the way for attacks by the Islamic State, also known as ISIS or ISIL. The latter group operated in Yemen by establishing an array of *wilayat* (provinces), while launching coordinated, large-scale attacks that killed dozens of people. These operations involved multiple suicide attacks using vehicle-borne improvised explosive devices. Yet as ISIS did not appear to have a fixed base in Yemen and was dispersed across the country – particularly in southern provinces – it remained a difficult target for coalition and allied counter-terrorism campaigns.

Growing humanitarian crisis

The conflict had a catastrophic effect on civilians, who faced acute shortages of food, water and medicine. Having maintained a naval blockade on Yemen, the coalition began to allow commercial vessels to offload supplies through a UN verification and inspection mechanism implemented in May, which significantly sped up the process. But major shortfalls of basic goods persisted due to widespread poverty, insufficient funding for humanitarian deliveries and insecure transportation.

Over the course of the war, Yemen's poverty rate had doubled to 62%. Shortfalls in funding compelled the UN World Food Programme to scale back its food-distribution activities in the country, despite the fact that the organisation classified more than 14m Yemenis as experiencing food insecurity at 'crisis' or 'emergency' levels. A similar number of people lacked access to safe drinking water or sanitation, relying on supplies of untreated water that contributed to outbreaks of cholera and other diseases. In 2016 there were more than 10,000 suspected cases of cholera, as well as 156 confirmed cases and 11 confirmed deaths from the illness, in Yemen.

The medical sector was close to collapse: 1,200 doctors and other health workers had reportedly fled the country since the conflict began, affecting 50% of hospitals and medical facilities; and at least 274 facilities had been damaged or destroyed. Following multiple attacks on facilities it ran, Médecins Sans

Frontières (MSF) withdrew in August from six hospitals in northern Yemen. The most recent attack, believed to have been conducted by the coalition, had killed 19 people in Abs Hospital in Hajjah province; it was the fourth assault on MSF facilities, and occurred despite the fact that the group has shared the locations of its hospitals with the combatants.

The conflict resulted in increasing child-mortality rates. This was partly because the number of children suffering from severe malnutrition had almost tripled to 462,000, while the number experiencing moderate malnutrition had risen to 1.7m. Save the Children estimated that there were 10,000 preventable deaths of Yemeni children every year. According to the organisation's figures, a Yemeni child died every ten minutes from conditions such as dysentery, malnutrition and respiratory-tract infection.

After almost two years of fighting, there was still no end in sight in the war. Both sides' inability to reach an agreement at the negotiating table demonstrated the depth of the power struggle in Yemen. Neither party was willing to make concessions in a power-sharing agreement, leading to growing political and institutional separation that risked another formal division of the country. Airstrikes by the Saudi-led coalition were widely criticised for their inaccurate targeting, spurring the US to attempt to disentangle itself from the conflict. Meanwhile, escalation on the battlefield in the north included frequent Houthi incursions into Saudi territory and Popular Resistance operations to advance into Houthi territory from the north. The coalition and the Popular Resistance also had some success in their operations against jihadist groups but despite driving AQAP out of Mukalla, such organisations remained active. As widespread insecurity exacerbated the parlous humanitarian situation in Yemen – causing extreme poverty and hunger, as well as a cholera epidemic – civilians bore the brunt of this intractable conflict.

Chapter Four

Sub-Saharan Africa

Central African Republic

Key statistics	2015	2016
Conflict intensity:	Medium	Medium
Fatalities:	500	600
New IDPs:	200,000	
New refugees:	95,000	

Central African Republic began 2016 on a surprisingly positive note, holding peaceful presidential and legislative elections in January and February. Despite isolated security incidents, logistical challenges and some allegations of irregularities, the vote proceeded in a calm, orderly manner. Following the nearly three-year political transition and suspension of electoral activity under President Catherine Samba-Panza, the elections were widely perceived as a major step towards stability. Faustin-Archange Touadéra, who was elected to replace Samba-Panza with more than 60% of the vote, began significant efforts to create a stable government, aiming to reconcile the parties to the conflict in the country and to address prevailing security issues. In mid-April, Touadéra appointed both political allies and former rivals to his first cabinet – although he left out supporters of Christian and Muslim militias.

While sporadic intercommunal clashes and the activities of roving non-state armed groups continued to affect rural areas of Central African Republic, as well as the cities of Bambari and Kaga-Bandoro, the country was generally

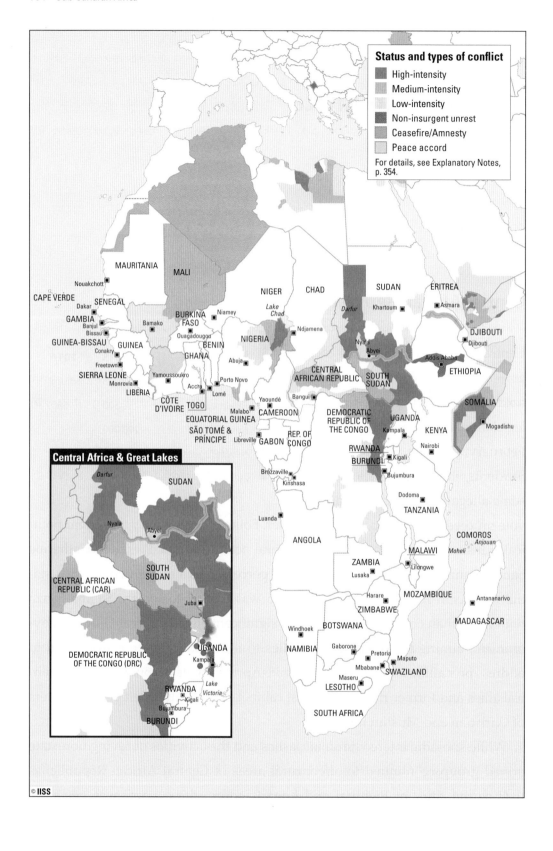

more peaceful in the first five months of 2016 than it had been for several years. However, there was a resurgence of instability and lawlessness across huge swathes of territory from June onwards, with conflict parameters such as the rates of recorded fatalities, violent incidents and internal displacement at higher levels than those in 2015. Sectarian attacks remained the predominant form of violence throughout 2016.

Stalled disarmament, demobilisation and reintegration

One dominant driver of the conflict – and challenge for the Touadéra administration – was the largely unaddressed need for the disarmament, demobilisation and reintegration (DDR) of the non-state armed actors proliferating throughout Central African Republic, ranging from fragmented Muslim Séléka militias to predominately Christian anti-balaka groups. President Touadéra demonstrated that he understood the importance of this process, starting a DDR initiative across the country as one of his first measures in office. But the effort immediately encountered strong resistance. Although several anti-balaka groups agreed to lay down their arms and join the peace process by May 2016, prominent ex-Séléka factions – including the Union for Peace in Central Africa (UPC) and the Popular Front for the Renaissance of Central African Republic (FPRC) – publicly opposed it, citing their lack of representation in government.

With security forces too weak to enforce stability, a lack of government presence in many parts of the country and access to only limited capabilities through the UN Multidimensional Integrated Stabilisation Mission in the Central African Republic (MINUSCA) outside of Bangui, the Touadéra administration confronted a resurgence of chaos and violence in the second half of 2016. As part of this, the eastern, northern and western periphery of the country once again became a focus of conflict activity.

At the turning point in the violence in June, armed clashes between Muslim and Christian militias erupted in several parts of Bangui, leaving dozens of people dead and forcing hundreds to flee their homes, particularly in the predominantly Muslim PK5 neighbourhood. The disorder continued for days:

Muslim 'self-defence' vigilante groups attacked and kidnapped policemen; a violent armed crowd assaulted and encircled a police station housing UN police officers in PK5; and MINUSCA forces clashed with violent protesters at several points across Bangui, causing the death of one peacekeeper. The same month, Peul militiamen previously associated with the Séléka movement started to attack several villages and cities in the remote Ouham-Pendé region, leading more than 6,000 people to register in refugee camps in neighbouring Cameroon and Chad. At the same time, northern areas around Kaga-Bandoro saw a flare-up of violence between armed Fulani herdsmen and ex-Séléka militias vying for territorial and economic control of the region. The fighting killed or injured dozens of rural civilians, while displacing hundreds more, on 21 June alone.

Meanwhile, the eastern part of the country experienced a significant rise in activity by the Lord's Resistance Army (LRA). The group was responsible for abductions, killings and looting in the eastern regions of Mbomou and Haut-Mbomou – and, more sporadically, further north, in Haute-Kotto. Reflecting the growing chaos, violence and brutality throughout the country, a new Fulani militia emerged and rapidly expanded in the west. Called Return, Reclamation, Rehabilitation, the group initially formed to protect tribesmen from Christian anti-balaka attacks but became aggressive and predatory in the second half of 2016. The United Nations and Human Rights Watch accused the militia of having killed at least 50 people and displaced 17,000 others in a series of attacks across the west in late November. Although it claimed to defend Muslims, the group seemed to target civilians indiscriminately, capitalising on the security vacuum to maximise its economic power and control of territory. Furthermore, its rapid expansion – which involved the recruitment of well-armed fighters and the adoption of an increasingly aggressive posture towards peacekeepers – quickly destabilised large areas of the country and had the potential to undermine MINUSCA's peacekeeping efforts.

The already poor humanitarian situation in Central African Republic deteriorated further due to a growing number of attacks on aid agencies. In May and June, unidentified gunmen looted Médecins Sans Frontières (MSF) aid convoys in the areas of Bossangoa and Sibut, leading to the temporary suspension of humanitarian operations in many northern and central parts of the country.

Peacekeeping crisis

Despite the relative calm in the first half of 2016, international peacekeeping efforts including MINUSCA and the French *Operation Sangaris* faced significant reputational and operational challenges in the period. A continuing stream of sexual-abuse cases involving MINUSCA's personnel severely damaged the legitimacy and operational capacity of its forces. Dozens of such cases became public, as did others involving personnel from *Operation Sangaris* and the European task force in Central African Republic. Under pressure to limit the reputational damage, the UN dispatched investigation teams and pushed for sanctions against implicated individuals and organisations. On 8 January, the UN asked the Democratic Republic of the Congo to withdraw its peacekeeping contingent from MINUSCA. A UN spokesperson said that the Congolese troops serving in MINUSCA were only partially meeting UN requirements in terms of equipment, vetting and preparedness. The reviews and evaluation of the contingent followed allegations of rape and sexual exploitation against Congolese troops stationed in Bambari.

The UN partially addressed this challenge by establishing stronger internal control and discipline measures, and there were no reports of new sexual-abuse cases in the first half of 2016. But operational issues remained. Firstly, French military engagement via *Operation Sangaris* – the spearhead of peacekeeping efforts in Central African Republic, and as such an important factor in ensuring the security of Bangui – gradually declined throughout the year and officially ended in October 2016, leaving more security responsibilities to the fragmented and sometimes incoherent MINUSCA force. At the same time, the escalating instability across the country, MINUSCA's limited operational capacity outside of Bangui and the obstacles to DDR initiatives called for a shift in approach and mandate.

Responding to the Touadéra government's inability to cope with widespread violence, the UN Security Council adopted Resolution 2301 (2016) in July, extending the already robust MINUSCA peacekeeping mandate until 15

November 2017 and shifting its focus to stabilisation, as well as the containment and disarmament of non-state armed actors. In line with this, MINUSCA contingents became increasingly active outside Bangui and confronted non-state armed groups in multiple reported incidents. For example, on 4 September in the northwestern city of Boguila, MINUSCA peacekeeping forces confronted, isolated and later expelled around 200 heavily armed fighters affiliated with the ex-Séléka Patriotic Movement for Central Africa (MPC). However, despite these efforts, reported operational inconsistencies and multiple failures to intervene to protect civilians generated substantial public discontent with MINUSCA. Beginning on 14 September, violent clashes between fighters associated with the MPC and local self-defence groups with links to anti-balaka militias continued for several days, killing dozens of people and displacing hundreds of families. Many houses were burned down and humanitarian facilities vandalised in the battles, which started in Ndomete and rapidly spilled over into neighbouring Kaga-Bandoro. The UN recorded 16 violent incidents against aid agencies in the area in September alone, resulting in the temporary withdrawal of humanitarian personnel and the postponement of urgently needed aid programmes. Reportedly, MINUSCA peacekeeping contingents stationed there merely watched the clashes from the sidelines.

The widespread escalation in violence and MINUSCA's operational shortcomings gradually shifted public perceptions of international peacekeeping efforts in Central African Republic towards at least partial hostility. In October, anti-MINUSCA protesters in Bangui demanded the withdrawal of the UN peacekeeping mission, following several clashes in which the locals felt that the peacekeepers had failed to protect them. At least four protesters died and 14 others were injured when the demonstrations turned violent, as armed men exchanged fire with peacekeepers at the UN base.

Escalation in late 2016
The last quarter of 2016 saw a massive rise in fighting across the country that echoed the intense violence of 2013–14. The MPC openly attacked camps for

internally displaced persons in Kaga-Bandoro; Muslim and Christian militias clashed in Bambari and Bangui; and there were heavy confrontations between rivals the UPC and the FPRC in Bria. Fighters from the FPRC specifically targeted Fulani people – the dominant ethnicity of the UPC – during door-to-door searches, killing more than 80 civilians and displacing 11,000 others.

As a consequence, the prospect that the militias would be disarmed and reintegrated seemed more distant than ever. Non-state armed groups and criminal gangs persisted across Central African Republic – and even appeared to flourish, as more actors joined the fragmented conflict. Allied militias such as the UPC and the FPRC began to target one another. Adding to the grim outlook for the country, amid the growing lawlessness and chaos there were isolated reports of armed groups once again engaging in ethnic cleansing.

Democratic Republic of the Congo

Political conflict stemming from the postponement of national elections dominated events in the Democratic Republic of the Congo in 2016, threatening to destabilise the wider region. President Joseph Kabila's strategy for remaining in power beyond the

Key statistics	2015	2016
Conflict intensity:	Medium	Medium
Fatalities:	1,500	1,500
New IDPs:	600,000	500,000
New refugees:	40,000	

constitutional two-term limit through a series of administrative delays, known as *glissement* (slippage), provoked staunch resistance from the opposition. The government's repression of protests led to human-rights violations across the country, the use of excessive force against demonstrators and the killing of dozens of civilians – most of them in the capital, Kinshasa. The crisis further delegitimised the regime, resulting in the deterioration of an already volatile security situation in the east of the country. As the government pressured armed groups through military operations, violence targeting civilians increased. Organisations such as the Rwandan Hutu-based Democratic Forces for the Liberation of Rwanda (FDLR) engaged in large-scale kidnapping for profit near Rutshuru, while the Allied Democratic Forces (ADF) continued to massacre civilians in the Beni area. There was intensifying ethnic violence between the Luba and Pygmy communities in Tanganyika province, and between Hunde, Nande and Hutu communities in North Kivu. Meanwhile, a new militia emerged in Kasai-Central.

Re-energised opposition

As the year wore on, it became increasingly clear that the government would be unable to conduct national elections as scheduled, having failed to make the necessary preparations in voter registration, budget planning and revision of the electoral calendar. The purportedly independent Congolese electoral commission announced in January that it lacked the funding to fulfil its duties. A possible cause of the shortfall came to light in October, when the grandson of Patrice Lumumba revealed that individuals linked with the regime had misappropriated money intended for the commission. However, instead of tackling these challenges, the ruling party filed a motion with the constitutional court to allow Kabila to stay in power beyond the end of his second term. On 11 May, the court ruled that the president and members of parliament could remain in office until their successors were elected. At the same time, the electoral commission seemed to prolong the delay. Corneille Nangaa, head of the commission, announced on 1 October that the elections might be postponed until December 2018. Nangaa estimated that the commission would only finish updating the voter registers at the end of July 2017, after which it would require an additional 504 days to organise the election. Opposition leader Vital Kamerhe described this as an exaggeration, arguing that it took only 180 days to set up the 2006 elections.

The opposition also criticised the *glissement* strategy more broadly, calling on Kabila to step down and for the elections to be held on time. Kabila's adversaries grew bolder following the return on 27 July of veteran opposition leader Etienne Tshisekedi, the so-called 'sphinx' of Congolese opposition politics and runner-up in the 2011 presidential elections who had spent the previous two years receiving medical treatment in Belgium. Four days later, Tshisekedi led the first large anti-Kabila rally. In the subsequent months, he made harsh statements against the government and asked Kabila to step down.

In response to the growing political resistance, the government used incentives and disincentives to increase its domestic and international appeal while disrupting the opposition. Kabila implemented confidence-building measures and made several concessions. For instance, he denied that he intended to illegitimately hang on to power and justified delays in the elections with references to organisational challenges and a lack of donor support. He repeatedly met with UN officials and ensured them of his government's intention to respect human rights. As a sign of good faith, the authorities released several prisoners, including political and human-rights activists Fred Bauma, Yves Makwambala and Christopher Ngoy. The president also travelled to the troubled east of the country, meeting with civil-society groups in Beni and young political activists from the Struggle for Change (LUCHA) movement, acknowledging their role and ensuring them of his support.

Flawed political dialogue

After long preparations, the political dialogue to resolve the election stalemate began on 1 September. International observers and the domestic opposition saw the process as the only mechanism for breaking the deadlock, and had been urging all interested parties to participate in it since February. However, there remained several flaws in the undertaking. Firstly, Kabila and, before him, President Mobuto Sese Seko had repeatedly used political dialogues for political manoeuvres and window dressing. Secondly, the government failed to refine the broad concepts behind the dialogue, leaving it without clear objectives or procedures. Thirdly, the regime organising the dialogue had (by delaying the elections) created the political tension that the dialogue sought to resolve. Therefore, large sections of the opposition and many civil-society groups refused to participate in the dialogue, perceiving it as a political ploy to buy time and distract from Kabila's attempt to stay in power. The only opposition party to participate in the dialogue was fringe group the Union for the Congolese Nation, led by Kamerhe, a former ally of Kabila. Finally, the African Union facilitator, former Togolese president Edem Kodjo, was widely perceived as biased towards Kabila. The opposition denounced Kodjo's involvement early on, persistently calling for him to be replaced.

Thus, it was unsurprising that the majority of the opposition rejected the dialogue and the instrument ultimately proved unable to develop a viable political solution. The process concluded on 18 October with an agreement to establish an interim government and hold presidential, legislative and provincial elections by April 2018. In the interim period, Kabila would remain president, while a prime minister would be recruited from among the section of the opposition that participated in the dialogue. Surprisingly, Samy Badibanga – not Kamerhe – was chosen as the new prime minister, and a new government was formed on 14 November.

The opposition rejected the agreement for failing to reflect the popular view that elections should be organised as soon as possible and Kabila should step down – or, at least, guarantee that he would not run for the presidency after the transition period. The Episcopal Conference of the Democratic Republic of the Congo shared this perspective, calling for a more inclusive dialogue to reach a consensus on the electoral process. The organisation subsequently started mediation efforts with members of the opposition and the president's party on 31 October. However, the effort seemed to have little prospect of success, especially following a violent crackdown on protesters on 20 December, the day after the original election date.

Nonetheless, the sides reached a surprise agreement on 31 December. The new deal stipulated that the elections would take place in 2017 and that Kabila would step down after an interim period. The interim government would comprise members of the opposition and Tshisekedi would lead a supervisory committee on the transition process. Although the agreement seemed promising, it also appeared challenging to implement even if it gained the support of all interested parties. For instance, there were significant obstacles to organising local, provincial and national elections in just one year.

Widespread repression

In parallel to its apparent efforts at conciliation, the government repressed political dissent. The Organization Stabilization Mission in the Democratic Republic of the Congo (MONUSCO) identified a narrowing of political space in the country due to violations of rights to freedom of speech and peaceful assembly. The government made use of the police and the national intelligence service to intimidate and arbitrarily arrest protesters and other political activists. In February, the government arrested 45 civil-society activists and opposition members – including Martin Fayulu, a member of parliament – in relation to *villes mortes* (dead cities) strikes. Although most of these people were released without charges, the courts sentenced six members of the LUCHA to two years in prison. Generally, demonstrations planned by opposition parties and/or civil-society organisations were violently repressed, dispersed by the security forces or banned by local authorities. For example, on 24 April, the police dispersed a rally organised by opposition party G7, arbitrarily arresting at least 35 supporters of its leader, Moise Katumbi. On 21 May, the government detained and subsequently released 27 people who had participated in a peaceful demonstration in solidarity with the victims of ADF attacks in Beni.

In April, the government opened a politically motivated legal case against Katumbi for his alleged recruitment of mercenaries. A court indicted Katumbi and the prosecutor's office issued an arrest warrant for him on 19 May. Although Katumbi was allowed to leave the country to receive medical treatment, the government appeared to have effectively disposed of the opposition presidential candidate with the highest approval rating. The judge in the case later admitted to having been pressured into condemning Katumbi by the national intelligence service, so as to prevent him from running for president.

Due to the increasing strain on the resources of the police, the government deployed the Republican Guard and the army against demonstrators. The security forces curtailed protests by blocking roads, public squares and the residences of key opposition figures such as Tshisekedi. They repeatedly used excessive force, employing tear gas and live ammunition against civilians who were demonstrating peacefully. Most dramatically, on 19 September, a large-scale demonstration in Kinshasa against a third Kabila term escalated into widespread violence. Although the authorities had authorised the demonstration, the security forces interfered with the event at key locations. In response, protesters attacked the police, allegedly killing at least four officers. During the escalating violence, government buildings, police stations, private property and the offices of eight political parties were set on fire. The security forces reacted with excessive force; in the subsequent two days of violence, at least 53 people died, 143 were injured and another 299 unlawfully arrested, according to a conservative estimate by the United Nations. The security forces reportedly killed 48 of the victims – 38 of whom died of gunshot wounds. The UN Joint Human Rights Office stated that the real number of deaths was likely to be even higher, as the government hindered access to some detention centres, hospitals and morgues.

Drawing severe domestic and international criticism, the violence led to a temporary halt in the political dialogue and prompted the government to ban further demonstrations – a measure that it stringently enforced. The authorities arrested Bruno Tshibala, deputy secretary-general of opposition party the Union for Democracy and Social Progress, at Kinshasa airport in October for his alleged role in organising the violent demonstrations. On 26 October, the government arrested several young LUCHA activists conducting a sit-in protest in front of MONUSCO's headquarters. It also detained Franck Diongo Shamba – president of the Lumumbist Progressive Movement, another opposition party – in December, before sentencing him to five years in prison for a scuffle with members of the Republican Guard. And at least 31 people died during anti-Kabila demonstrations held on 19–20 December.

The persecution of journalists also increased, drawing criticism from Congolese non-governmental organisation Journaliste en Danger on 2 November. Indeed, at least 87 journalists were attacked in the Democratic Republic of the Congo in 2016. Shortly after the 19–20 September demonstrations, the security forces obstructed, arrested and beat around a dozen journalists. On 14 November, Marcel Lubala became the sixteenth journalist to be killed in the preceding ten years.

The government also took measures against international reporters. From 5 November onwards, it blocked the signals of Radio France Internationale and the MONUSCO-supported Radio Okapi, introducing new regulations specifying that media outlets should be majority-owned by Congolese citizens. The government refused to restore the signal of Radio France Internationale, despite protests from France, the United States and the president of the Organisation Internationale de la Francophonie.

The government also pushed back against perceived international interference in its domestic affairs by expelling several international experts and researchers, such as Jason Stearns. Expelled in April, Stearns had published a report linking the armed forces to the Beni massacres. In July, the government expelled two Global Witness employees who were researching the practices of logging firms. In August, it declared Ida Sawyer, Human Rights Watch's senior researcher in the country, as persona non grata.

Heightened political instability reportedly led to an upsurge in localised conflicts driven by ethnic tension or competition for natural resources. Armed groups such as the ADF, the FDLR, the Front for Patriotic Resistance in Ituri, the Lord's Resistance Army and various Mai Mai factions continued to commit human-rights abuses in the east. By June, MONUSCO had recorded 1,153 allegations of such abuses in the year. However, the organisation estimated that armed groups were responsible for only 38% of alleged violations, attributing the remaining 62% to state actors.

Against this background, a new militia emerged in Kasai-Central. Named after its leader, Kamwina Nsapu, the group formed in response to alleged corruption among government officials. It became increasingly violent after the security forces killed Nsapu in August. On 23 September, the militia took revenge by attacking government buildings such as Kananga airport, reportedly causing the deaths of 49 more people, including 27 militiamen, 16 security personnel and six civilians.

Around the same time, there was a new round of ethnic violence between Luba and Pygmy communities in Tanganyika province, involving a series of revenge killings that left scores dead. For instance, on 17 October, fighting between the communities resulted in as many as 16 deaths. There were four violent clashes in the area in September alone. Yet the security and humanitarian needs of civilians living in the southeast were reportedly neglected as international donors and UN troops focused their efforts on Ituri, North Kivu and South Kivu provinces.

Limited military advances

On 11 January, President Kabila authorised resumed cooperation between the armed forces and MONUSCO. They signed a technical agreement on 28 January, and began joint operations against most armed groups on 23 February. These efforts, as well as the armed forces' independent operations, were relatively successful in inflicting losses on the groups, recapturing territory, destroying their bases and pressuring some of their members to join demobilisation programmes.

In North Kivu, *Operation Sukola* 1 focused on destroying the ADF. In addition to logistical support, MONUSCO provided the armed forces with unmanned aerial-surveillance systems, attack helicopters, artillery and medical-evacuation capability. In May, they destroyed five camps belonging to the ADF, while killing 24 of the group's fighters and capturing four others, in only seven days.

Operation Sukola 2 against the FDLR initially continued in North Kivu and South Kivu without MONUSCO support. In North Kivu, the armed forces conducted operations in Rutshuru, eastern Walikale and southern Lubero. From 23 May onwards, joint effort *Operation Nyamuragira* supported these operations.

In South Kivu, *Operation Sukola* 2 proceeded in Kabare, Kalehe, Mwenga and Shabunda.

The UN Group of Experts on the Democratic Republic of the Congo reported that the FDLR had been 'considerably weakened' in 2016. The Congolese authorities arrested several high-level militant leaders in the year. On 20 March, Ladislas Ntaganzwa, a senior FDLR commander, was transferred to the International Residual Mechanism for Criminal Tribunals. On 28 April, the Congolese authorities arrested Leopold Mujyambere, chief of staff for FDLR faction Combat Forces Abacunguzi, in Goma. The arrest led to an apparent split in the faction's leadership, as Laurent Ndagijimana, second vice-president of the FDLR, was accused of having given up Mujyambere. Ndagijimana and his followers subsequently created a new group, the National Council for Renewal and Democracy. The new faction comprised between one-third and half of FDLR troops, including 50 officers and the entire South Kivu operational sector. Low morale continued to affect FDLR fighters, seemingly spurring an increase in desertions. At least 15 FDLR commanders were captured or surrendered in 2016.

Although these advances reduced the operational capacity of all major armed groups, the military failed to decisively defeat any of them or improve the security situation for civilians. As had been the case in similar operations in the past, the security forces' pursuit of armed groups increased insecurity and abuses against civilians in the area of operations. Under pressure from the military, armed groups became more brutal in their treatment of local communities, adopting tactics such as robbery, pillage and, increasingly, kidnapping. For instance, the Front for Patriotic Resistance in Ituri conducted 41 attacks on civilians in April alone.

Despite facing dedicated operations by the armed forces and MONUSCO, the ADF continued its attacks against civilians in Beni. On 3–6 May, suspected members of the group assaulted two villages in the Eringeti area, killing 36 civilians and wounding 17 others, including children and pregnant women. The largest ADF attack of the year occurred on 13 August, when the group killed approximately 50 civilians in Rwangoma, a suburb of Beni town. The incident provoked demonstrations, some of them violent, against the government and UN peacekeepers in Beni, Bunia, Butembo and Goma. Although ADF rebels seemed to be carrying out the killings to impose costs on the government for its military campaign, the reported involvement of local strongmen and the security forces indicated that the violence had more complex causes. A report published by the Congo Research Group in March stated that the army had been involved in the massacres in Beni.

The Lord's Resistance Army became more active in 2016. On 5 June, suspected members of the group briefly kidnapped more than 100 civilians, whom they forced to carry pillaged goods. Meanwhile, the FDLR dramatically stepped up its kidnapping-for-ransom operations in Rutshuru, terrorising the local population and inhibiting the delivery of humanitarian assistance there.

Military operations against the FDLR also exacerbated communal violence. Communities targeted by the group resorted to establishing and/or supporting local militias for their protection. This resulted in the emergence of Nande militia the Union of Patriots for the Defence of the Innocent, known locally as Mai Mai Mazembe. The government's offensive against the FDLR also encouraged several other Mai Mai groups, such as Raia Mutomboki and Nduma Defence of Congo, to increase their attacks against presumed FDLR members, often brutally targeting Hutu communities viewed as supporting the latter group.

The displacement of Hutus due to military operations and the government's closure of camps for internally displaced persons (IDPs) exacerbated communal conflict, as the indigenous Nande, Hunde and Kobo communities perceived the Hutus as foreign invaders, particularly in southern Lubero Territory, in North Kivu. The most significant incident took place on 27 November, when around 50 Mai Mai Mazembe fighters attacked Luhanga IDP camp in the area, killing 30 people and injuring 21 others. The militias' persecution of Hutus continued a cycle of reprisal attacks involving them and the FDLR that had begun in late

2015. One such assault took place on 7 January, when FDLR militants killed 16 Nande civilians in Miriki village, in Lubero Territory.

Surge in displacement

The humanitarian situation in the eastern Democratic Republic of the Congo continued to deteriorate due to military operations, the activities of armed groups and communal violence – all of which contributed to an increase in the number of IDPs in the country from 1.5 million in March to nearly 2m by the end of the year. The UN secretary-general reported that by 30 November, there were approximately 439,000 refugees and asylum seekers in the country, including nearly 245,000 from Rwanda, 96,500 from the Central African Republic, 31,500 from Burundi and nearly 65,000 from South Sudan.

Although there were an estimated 7.5m people in need of humanitarian assistance in the Democratic Republic of the Congo, the international community's US$690m response plan had received only 56% of its requested funding by mid-November. The shortfall hampered humanitarian operations – particularly given that when the funding request had been made, aid agencies had not anticipated such a sharp increase in demand for their services.

Furthermore, insecurity in the east restricted humanitarian access to some areas. Several humanitarian organisations suspended their activities due to security threats in Ituri, North Kivu and South Kivu. Ambushes of humanitarian vehicles caused significant disruption to humanitarian operations and endangered the effective delivery of aid to those in need. The UN secretary-general reported that at least four humanitarian workers were killed in the year, while ten others were wounded and 29 abducted.

The government contributed to the dire humanitarian situation not only by exacerbating the instability through election delays, but also by closing several IDP camps in North Kivu. Conducted without prior consultation with aid agencies, the move further displaced some of the most vulnerable people in the country.

Ethiopia

Key statistics	2015	2016
Conflict intensity:	Low	Low
Fatalities:		500
New IDPs:		
New refugees:		6,000

The security situation in Ethiopia deteriorated in 2016, primarily due to a wave of protests by ethnic Oromos who demanded greater political freedoms, as well as increased access to governance and economic resources. Although they were sometimes peaceful, the demonstrations led to numerous violent confrontations with regional and federal security forces in various parts of Oromia (data on the resulting civilian casualties could not be independently verified). The Ethiopian government frequently blamed Eritrea and the Oromo Liberation Front for the attacks.

The instability extended beyond Oromia. Gambella, another of Ethiopia's nine federal states, continued to experience political volatility, insecurity and violence. Home to around 400,000 native Ethiopians and approximately 300,000 South Sudanese refugees, Gambella had experienced persistent ethnic and political violence that claimed thousands of lives in the preceding decade. The conflict in, and influx of weapons from, South Sudan further destabilised Gambella – with which the country shares a border. Ethiopia's federal security forces were reportedly forced to intervene in the fighting in Gambella after regional security personnel began to participate in the violence, siding with their respective ethnic groups. The federal government eventually disarmed the regional police and assumed control of security in the region.

In mid-June, the escalating tension and war of words between Ethiopia and Eritrea resulted in what was reportedly the largest clash between the countries since the end of their 1998–2000 war. Addis Ababa stated that the battle followed an Eritrean attack, while Asmara contended that aggression from Ethiopian forces had triggered the fighting. Both governments claimed victory. Ethiopia declared that it had caused significant casualties among Eritrean troops, while

Eritrea put forward a 'conservative' estimate that it had killed more than 200 Ethiopian soldiers. The clash highlighted the volatility of the border dispute between the sides and the risk of escalation into full-scale war, given their apparent inability to reach a settlement. Asmara blamed the instability on Addis Ababa for rejecting the 2000 Algiers Agreement and the Eritrea– Ethiopia Boundary Commission's ruling that the largest area of contested land, Badme, belonged to Eritrea.

Nigeria (Boko Haram)

On 30 January 2016, 30 Boko Haram fighters attacked the village of Dalori, near the Borno State capital, Maiduguri. They used firebombs to set alight huts containing women and children, while three female militants detonated suicide bombs after

Key statistics	2015	2016
Conflict intensity:	High	High
Fatalities:	11,000	3,000
New IDPs:	700,000	
New refugees:	75,000	

attempting to hide among civilians. As many as 86 people died in the assault, which ended only after soldiers from Maiduguri reached the village hours later. The attack took place around a month after Nigerian President Muhammadu Buhari claimed 'technically, we have won the war' against Boko Haram. Although the group's capacity to launch large-scale attacks in major cities had greatly diminished, the insecurity it created remained a major cause for concern, especially for civilians living on the fringes of large cities and in the Lake Chad region. Boko Haram was responsible for around 800 fatalities across Cameroon, Niger and Chad in 2016 – a considerable number, albeit around 2,200 fewer than in the previous year. As the security forces expelled the group from a growing number of areas, the scale of the humanitarian crisis in Nigeria became increasingly apparent. By the end of the year, experts estimated that along with three million registered internally displaced persons (IDPs), the country contained 5–7m unregistered IDPs, as well as at least 14.8m civilians whom the conflict had left facing poverty, food insecurity, health problems and dire living conditions.

In early 2016, under pressure from Buhari's election promise to defeat Boko Haram and the military's resulting deadline for completing the task, security and political officials prematurely declared that the group had been driven out of the northeast. In February, Borno Central Senator Baba Kaka Garbai refuted claims by Mohammed Ndume, a senator from Borno South, that the army had recaptured all local-government areas, telling journalists that Mobbar, Abadam and Kala-Balge were still '100% occupied by the insurgents'. On 7 February, Garbai said that the military was only in control of three areas – Maiduguri, Bayo and Kwaya Kusar – while Boko Haram still partially controlled the remaining 21 areas (covering around half of the state). Similarly, David Rodriguez, then commander of the US Africa Command (AFRICOM), told the US Senate Armed Services Committee that the group still held significant amounts of territory in northern Nigeria.

The attempt to hide the true nature of the conflict was corroborated by IDPs, who complained that the National Emergency Management Agency (NEMA) forced them to leave camps – despite insecurity in their communities – in an effort to demonstrate that the authorities had made significant progress against the insurgency. Throughout the year, security agencies failed to publicly acknowledge significant attacks, preferring to highlight their achievements instead. In November, Yaga Yarkawa, chairman of the Chibok local-government area, only informed the media that the territory was under siege after the insurgents had been launching attacks there for two weeks, loading vehicles with stolen food and setting homes alight. Such assaults demonstrated that despite significant military advances, the threat from the group was likely to persist for some time.

In a more positive step, the Nigerian authorities inaugurated the Borno State Islamic Preaching Board. Led by Chief Imam Zannah Ibrahim Ahmed, the board monitored Islamic clerics and the traditional Almajiri and Arabic schools that had gained a reputation as breeding grounds for the insurgency. Aside from setting preaching standards, the board identified any unusual or suspicious preaching by Islamic clerics, in recognition of the fact that Boko Haram would almost certainly attempt to spread its ideology once again.

Tension among Boko Haram leaders

The Islamic State, also known as ISIS or ISIL, proclaimed in the August issue of *Al-Naba* – a newspaper produced by the group – that its West African affiliate was under the leadership of Abu Musab al-Barnawi. The announcement appeared to prompt a public split between Barnawi and Abubakar Shekau, who had been widely regarded as leader of the ISIS-affiliated Boko Haram. Nigerian officials interpreted the dispute as evidence of Boko Haram's weakness. However, the development was alarming in the sense that the authorities would need increased intelligence capacity to monitor the two factions. Brigadier-General Donald Boulduc, commander of US special forces in Africa, reported to the US Senate that ISIS was providing operational and logistical support to its West African affiliate, confirming that troops had intercepted a large cache of weapons the group had sent from Libya to the Lake Chad region. The affiliate focused on weakening the capacity of the Nigerian armed forces through large-scale surprise attacks. In October, it claimed to have killed 20 soldiers during an ambush of an army position in Ghashghar, in northeast Borno State. In the days that followed, the army played down the severity of the incident despite multiple reports that a significant number of soldiers had been killed or injured. Local media outlets stated that as many as 83 soldiers went missing. On 26 October, Major-General Lucky Irabor said that a large number of the soldiers declared missing after the attack had returned; he also admitted that a few soldiers were yet to be found (without providing a specific number). Irabor stated that there would be an investigation into the attack to determine what had occurred and how the troops went missing.

Nigeria continued to rebuild diplomatic ties with the United States, following the repeal of the ban on US arms sales to Nigeria. Linda Thomas-Greenfield, US assistant secretary of state for Africa, announced that the US Army would resume its programme of combat training for Nigerian soldiers. The administration of Goodluck Jonathan, Buhari's predecessor, had terminated the scheme after Washington imposed the ban due to concerns about human-rights violations by the Nigerian security forces. Colonel Patrick Doyle, a US defence attaché to Nigeria, said that the US would continue to work with its West African partners to restore peace to the Lake Chad region, the site of a growing number of Boko Haram attacks in the year. In October, AFRICOM announced that it would establish a temporary expeditionary base in Agadez, in central Niger. Hosting unmanned aerial vehicles, the US$100m facility would bolster the counter-insurgency efforts of countries in the Lake Chad region.

Military gains by the government

Increased resources and an improved counter-terrorism strategy enabled the Nigerian military to excel in preventing suicide bombings, destroying insurgent camps and arresting several prominent members of Boko Haram. Thousands of the group's fighters surrendered following victories by the military – more than 800 of them in April alone. At the second Regional Security Summit, held in Abuja on 14 May, Multinational Joint Task Force commander Major-General Lamidi Adeosun stated that since January, 675 Boko Haram militants had been killed while 566 had been captured. In the same period, 32 camps had been destroyed and 4,690 hostages freed from the group. Yet Boko Haram continued to be adaptable, and to take advantage of the remaining pockets of insecurity in the region. After being forced out of most major cities, the insurgents moved further into the Lake Chad region, increasing their attacks on Cameroon, Niger and Chad. Despite greater coordination among its members, the multinational force still struggled to protect remote villages.

The military also had difficulty holding on to the gains that it had made in 2015; Boko Haram channelled its most significant resources into orchestrating a series of high-casualty attacks on army locations and towns it had previously held. In March, the insurgents attempted to retake the town of Gwoza, conducting simultaneous assaults on military formations at Bita, Pulka and Gamoru.

Similarly, in June, hundreds of Boko Haram militants launched a major attack on the town of Bosso, in the Diffa region of Niger, leading to the deaths of 35 Nigerien and Nigerian soldiers, as well as the injury of 70 others. The offensive displaced 50,000 people from the town, which had been a safe haven for IDPs and refugees. In October, the insurgents launched another large-scale attack on an army position in Ghashghar. Local media reported that 83 soldiers went missing during the attack, while at least 22 sustained serious injuries and an unconfirmed number were fatally wounded. The Islamic State's West Africa Province, a Boko Haram faction, took responsibility for both of the operations. The attacks revealed a serious error of judgement by Nigerian security officials who had dismissed the split within Boko Haram as propaganda from a weakened group. A more serious look at the politics within Boko Haram and analysis of different tactics used by its factions could have resulted in better intelligence gathering and heightened awareness among troops. In mid-November, the insurgents launched a series of attacks on Chibok local-government area. Although Nigerian security officials released no statement on the attacks, local leaders informed the media that militants were overrunning villages, forcing hundreds to flee. Yarkawa claimed that in two weeks Boko Haram had razed nine villages within 25 kilometres of Chibok town, some of them without encountering resistance from the security forces.

The previous month, the government had secured the release of 21 of the 276 girls kidnapped from a Chibok school more than two years earlier. The authorities had been negotiating with Boko Haram on the issue since July 2015, but the talks had broken down three times even after the government agreed to release some of the group's fighters from prison. In one instance, the discussions had collapsed after Boko Haram issued a new set of demands at the last minute; in another attempt, the talks failed after some of the group's negotiators were killed. On 22 September, media reports revealed that Boko Haram had asked human-rights activist and lawyer Aisha Wakil to represent it in new negotiations with the government over the release of the Chibok schoolgirls. Originally from Maiduguri, Wakil had been at the centre of peace negotiations between the government and Boko Haram since 2009 due to her contacts in the group. Following the release of the girls, presidential spokesman Garba Shehu announced that the deal to free them had been brokered by the Swiss government and the Red Cross. The press subsequently reported that the Swiss government had paid a large ransom to Boko Haram and that four of the group's commanders had been released in exchange. However, Minister of Information and Culture Lai Mohammed stated that he was not aware of any ransom payment, while Vice-President Yemi Osinbajo stated that no exchange of any kind had taken place. With many of the girls still being held prisoner, the Nigerian government stated that the negotiations would continue until all of them had been released.

Acute humanitarian challenges
Security improvements in Nigeria and the Lake Chad region enabled the government to realise the full extent of the humanitarian emergency there and to strengthen its emergency-response efforts. Aid agencies continued to experience shortfalls in funding, despite receiving millions of dollars from federal and local governments, international institutions such as the World Bank and the United Nations, and a host of other organisations. Around 2.2m people, more than half of them children, were thought to be trapped in areas controlled by Boko Haram and in need of humanitarian assistance.

In October, UN Assistant Secretary-General Toby Lanzer stated that the emergency created by Boko Haram would become the world's worst humanitarian crisis if Nigeria did not receive more humanitarian funding, adding that suffering in the northeast of the country and surrounding areas was the worst he had ever witnessed. In the Lake Chad area, attacks on farming areas led to food-security crises in Niger and Cameroon. Cameroon's government reported that the country had lost thousands of cattle, sheep and goats to terrorist attacks. Having lost resources in military defeats, members of Boko Haram resorted to stealing food and cattle, often disguising themselves as cattle ranchers when transporting cows to Nigerian markets.

The Nigerian government employed a mixture of long- and short-term policies to enable civilians to sustainably rebuild their lives. In April, Defence

Headquarters announced that it would establish a rehabilitation and reintegration camp for Boko Haram fighters who had surrendered. The facility played a part in the Operation Safe Corridor programme, which provided vocational training to former militants. The Nigerian Army also launched an investment effort, run by the army chief of staff, for the barracks community in Maiduguri, aiming to create employment opportunities for women and children. Under the scheme, ten women's cooperative groups would receive NGN1m (US$3,177) in soft loans to start their own businesses, and the army would distribute funds to different women each month. A ranch was also established as part of the initiative.

The Borno State government continued to regard education as its top priority. In March, Borno Governor Kashim Shettima pledged to fund free nursery- and primary-school education for 23,000 children who had been orphaned by the fighting. The chairman of Borno's State Emergency Management Agency said that children living in IDP camps who had no access to schools would also receive a free education. On 13 March, the Victims Support Fund launched a new educational-support programme for children affected by the insurgency in Adamawa, Borno, Yobe and Edo states at an IDP camp in Damaturu. The first phase of the project aimed to provide 21,291 schoolchildren – 10,000 in Borno, 7,000 in Adamawa, 3,000 in Yobe and 1,291 in Edo – with essential equipment such as books, writing materials, bags and sandals. The US Agency for International Development (USAID) also committed an additional US$4.1m to IDP camps in Nigeria, aiming to provide education to 60,000 children aged between six and 17, as a part of which the Borno State government would open 150 informal learning centres.

As part of their strategy to address human-rights concerns and increase community engagement, the defence services established in January a human-rights office at their headquarters in Abuja. The office would serve as a central base for training the Nigerian army on civilian relations, and for ensuring that all alleged human-rights violations were investigated. It also aimed to facilitate interaction with external human-rights organisations and strengthen the army's capacity in protecting human rights. Meanwhile, General Abayomi Olonisakin, Nigeria's chief of defence staff, inaugurated an independent human-rights committee tasked with investigating all accusations of human-rights abuses made against the Nigerian security services.

By the end of the year, the military had made significant progress in destroying insurgent camps and retaking territory, leading it and the government to declare victory against Boko Haram. However, relentless attacks by the insurgents – particularly the West African affiliate of ISIS – demonstrated that they still had significant manpower, and that the conflict was far from over. Boko Haram continued to conduct hit-and-run assaults, raids and suicide attacks in and around areas that had purportedly been cleared by the military, albeit at a significantly lower rate than in 2015. Militant cells remained in these territories, as remote terrain and a lack of personnel, materiel and surveillance capacity prevented the armed forces from holding and building there. The government's overeagerness to send IDPs back to their homes was also a serious cause for concern: many of the areas to which they were returned still contained pockets of insecurity and suffered from chronic underdevelopment.

Nigeria (Delta Region)

Oil and gas pipelines in the Niger Delta region came under attack an average of nine times per month in 2016, compared to just a few times in all of 2015. Although the Niger Delta Avengers (NDA) conducted most of the assaults, others were carried out

Key statistics	2015	2016
Conflict intensity:	Low	Low
Fatalities:	250	90
New IDPs:		
New refugees:		

by some of the many other new militant groups that emerged in the year. The attacks resulted in relatively few casualties but caused Nigeria's oil output to fall from 2.2 million barrels per day to 1.69mb/d, plunging the country's oil-dependent economy further into recession.

Nigerian President Muhammadu Buhari announced in early 2016 that contrary to a statement he had made upon assuming office, the Presidential Amnesty Programme (PAP) would continue for at least two more years. The decision followed a review of the programme by Buhari's Transition and Empowerment Task Force, which concluded that the scheme could not be shut down in December 2015 as planned due to the lack of an exit strategy. As the task force's report stated that it would take at least two years for the government to successfully terminate the programme, the government set a new deadline of December 2018. While it was unclear exactly how the government would find jobs for the 30,000 former militants in the programme, PAP coordinator Brigadier General (Retd) Paul Boroh said that the process would require the involvement of partners such as the Nigeria Content Development and Monitoring Board, the Ministry of Niger Delta Affairs, the Niger Delta Development Commission and the Ministry of Environment, as well as international oil companies. In reintegrating the participants in the PAP, the federal government also worked with international organisations such as the United Nations Development Programme, the European Union, the US Agency for International Development and the UK Department for International Development, as well as military and paramilitary organisations. By January 2016, around 17,000 of the 30,000 former militants had completed various educational and training courses.

Failed negotiations

In February, the NDA published a statement entitled 'Operation Red Economy', declaring war on the Nigerian government and international oil companies while threatening to bring the economy to a standstill. The group accused Buhari's government of nepotism, and of unfairly targeting allies of his predecessor in a nationwide crackdown on corruption and mismanagement. The NDA stated that the president would risk war if he failed to meet a series of demands within two weeks. It called for Buhari to immediately implement the report of the 2014 National Conference, which recommended, inter alia, the creation of 18 new states; the establishment of special intervention funds to support reconstruction and rehabilitation in areas affected by the insurgency and internal conflict; the reapportionment of funds in the Federation Account, with 42.5% going to the federal government, 35% to state governments and the remainder to local governments; and the adoption of a modified presidential system, a form of governance that effectively combined the presidential and parliamentary models. Moreover, the report recommended that the presidency rotate between the north and the south, as well as among Nigeria's six political zones, and that governorships rotate between the three senatorial districts in each state.

The NDA also demanded that the government issue an apology to former Bayelsa State governors Timipre Sylva and Diepreye Alamieyeseigha, as well as the people of the Niger Delta region; ensure that the ownership of oil blocks in the region was split 60/40 between indigenous and non-indigenous people; establish the Nigerian Maritime University; compel Minister of Transportation Rotimi Amaechi to apologise for his criticism of the Ijaw community; clean up Ogoniland and all parts of the region that had been polluted with oil, while providing compensation to communities affected by the spills; release pro-Biafra leader Nnamdi Kanu; guarantee the continuation of the PAP; and ensure that any corruption case involving a member of the ruling All Progressives Congress be brought to trial.

The group also repeatedly called for a referendum on the secession of the Niger Delta. Following the release of the statement, the NDA carried out a complex attack on Shell's Forcados oil-export terminal, one of the largest in the region. The assault forced the company to shut down the terminal for weeks, at the cost of 250,000b/d in production or around US$12m per day. The attack demonstrated a level of planning and technical ability unseen since the 2004–09 insurgency in the Niger Delta.

The key challenges for the government were in understanding the motivations, and identifying credible leaders, of the militant groups in the region. The

Movement for the Emancipation of the Niger Delta (MEND), a former militant group that had signed the 2009 peace deal, played a key part in the early stages of a new peace process. However, various active militant groups in the region stated that the organisation could not represent them in the dialogue. The NDA accused the community leaders behind MEND's revival of seeking involvement in the negotiations for personal gain. In June, the Minister of Petroleum Ibe Kachikwu led a delegation to the region to speak to representatives of various community groups and militant leaders, following which the government announced that the sides had agreed on a 30-day ceasefire. But the NDA denied any knowledge of such a deal, and attacks by militants continued. Lacking an effective representative in the community, Kachikwu and his delegation found it difficult to identify legitimate leaders and those who could influence militant groups, prompting accusations that the government was intentionally undermining the peace process to justify military action.

The NDA eventually agreed to a ceasefire with the government on 20 August, suspending its attacks for the first time since January. However, another new militant organisation, the Niger Delta Justice Defence Group, began a pipeline-bombing campaign just as the NDA agreed to participate in a peace process. And, on 24 September, the NDA declared that it had attacked a major pipeline linked to Shell's Bonny terminal. The group blamed the breakdown of the ceasefire on government forces, which it accused of intimidating civilians in the region after the ceasefire began. By October, most militant groups in the Niger Delta had agreed to join new umbrella organisation the Pan Niger Delta Forum (PANDEF), under the leadership of Chief Edwin Clark, a former federal information commissioner and head of the Ijaw National Congress. As a consequence, the MEND negotiation team stepped aside to allow PANDEF to lead the dialogue with the federal government.

At a meeting in Abuja on 1 November, PANDEF presented Buhari with a list of 16 demands. The group called for the withdrawal of troops from the region, as well as the relocation of international oil companies' headquarters to the Niger Delta to provide jobs for young people and increased development funding.

Yet despite the ongoing dialogue, militant groups continued to attack pipelines and threaten further action in the region. By the end of the year, the president had dismissed the leadership of PANDEF, stating that the government would continue to search for credible leaders who could control the militants during a peace dialogue.

Growing military challenges

The Nigerian military struggled to maintain a balance between civil liberties and security in the Niger Delta. Although it made significant progress in arresting militants responsible for attacks on oil infrastructure, civilians often suffered human-rights abuses in security operations.

Launched in January, *Operation Awatse* led to the deployment of additional soldiers to protect waterways and oil installations. However, the new security measures failed to prevent the bombing on 28 January of Agip's pipelines in Brass local-government area, in Bayelsa State – an attack that led to a large-scale oil spill. The military subsequently detained the perpetrators of the bombing – who had also conducted several other attacks in the Niger Delta – but its increased presence exacerbated tension between civilians and the government. In April, Buhari issued another order to step up military deployments in the region, despite national and international criticism of the security forces' alleged human-rights abuses, particularly their use of intimidation tactics and excessive force. The government changed its approach in June, stating that the military presence would be reduced in favour of engaging with militant leaders to address their underlying concerns.

The previous month, the Nigerian Army arrested members of the NDA in connection with an attack on Chevron's oil pipelines. On 7 June, a spokesperson for the Nigerian Navy told journalists that the force had collaborated with other security agencies to arrest the coordinator of an attack on the Nigerian National Petroleum Corporation (NNPC) and Chevron oil facilities, who was suspected of illegally operating at least 35 abandoned oil-well heads and crude-oil pipelines in Warri South West local-government area. The spokesperson also announced that the militant leader responsible for the recent attack on Shell's Forcados pipeline had been arrested. The military declared that it had detained several high-profile members of the NDA, including a major logistics supplier and militants planning to assassinate a serving military officer in Kaduna State (the group denied that any of its members had been detained).

Another challenge for the military was the frequent emergence of new militant groups in the Niger Delta, which often set back its campaign due to a lack of intelligence on their capabilities and agendas. On 9 August, new militant group the Niger Delta Greenland Justice Mandate urged oil workers to evacuate their buildings within 48 hours, shortly (and less than 48 hours) before bombing a major pipeline belonging to the Nigerian Petroleum Development Company and Shoreline Natural Resources in Delta State. Ten days later, the group bombed another pipeline belonging to the former firm, in Delta State's Udu local-government area. The NDA carried out its first known attack on an energy installation in the southwest in July, bombing the NNPC's gas pipeline in Ogijo, in Ogun State. The militants had obtained access to the site by pretending to be NNPC officials on a maintenance patrol. The attack was one of at least 13 reported in the month.

The Nigerian Navy uncovered more than 100 illegal oil-bunkering sites in 2016. One of these was a major facility containing 500 drums of illegally refined petroleum, discovered in April at an abandoned warehouse in Ogbogoro area, in Rivers State. Additionally, the army destroyed 74 illegal refineries in September as part of *Operation Crocodile Smile*.

Environmental threats to local communities
International oil companies continued to face widespread criticism for the environmental effects of their operations in the Niger Delta. In January, six months after a spill at the Adibawa oilfields, Shell was accused of failing to deal with the accident properly and attempting to manipulate evidence during the official investigation that followed. The company consistently denied the allegations, insisting that members of the Edagberi community had thwarted its recovery

and remediation efforts by preventing representatives of industry regulatory agencies, the Rivers State Ministry of Environment and the company from accessing the spill site. The firm also said that it had immediately attempted to contain the spill – which followed the attempted theft of a well head – by constructing dykes, digging pits and deploying booms. However, further investigations to determine the cause and extent of the leak were prevented by locals, who reportedly demanded payments and employment in return for access. Shell also faced claims that it had failed to pay compensation to the victims of past accidents. In early March, the Artisan Fishermen Association of Nigeria requested a presidential investigation into an unpaid fine of US$3.6 billion for the Bonga oil spill of 2011, which had severely affected fishing communities in Delta, Bayelsa, Rivers, Akwa Ibom and Ondo states.

In August, there was a large oil spill at a Warri South West facility run by the NNPC. The following month, the leaders of ten local communities affected by the accident alleged that the NNPC had failed to adequately respond, signing a petition to the government that accused the company of negligence.

In February, the National Human Rights Commission had attempted to address the environmental threats to local communities by setting up a special panel to investigate oil spills and other pollution. According to the executive secretary of the commission, the panel had been established following an assessment of locals' petitions and complaints that a long-term solution to these environmental problems was necessary, as many oil spills had been either improperly cleaned up or ignored entirely, damaging aquatic life and farmland.

The long-awaited Ogoni clean-up and restoration programme was launched on 2 June, following recommendations from the UN Environmental Programme issued five years earlier. Although Buhari was due to inaugurate the effort, threats by militants in the Niger Delta prompted Vice-President Yemi Osinbajo to take his place. However, the programme encountered delays immediately after the inauguration, and did not begin in 2016. The government stated in November that it was in the process of ensuring that the project had a strong foundation and a clear organisational structure, so that there would be no problems with its budget, recruitment efforts or implementation.

The Ministerial Technical Audit Committee released in August a report that revealed substantial corruption in the Ministry of Niger Delta Affairs. Although the ministry had reportedly spent NGN700bn (US$2.2bn) on development contracts between 2009 and 2015, only 12% of the projects had been completed, with 18% stalled at the contract stage and 70% still under construction. The report also found that the process of awarding contracts violated the Public Procurement Act and often took place without the guidance of the procurement department.

The revival of militancy in the Niger Delta compelled the government to reassess its long- and short-term development goals there, and to explore ways for the military to establish a relationship with local communities that was not based on violence and fear. In launching 'Operation Red Economy', the NDA and other militant groups in the region demonstrated that they had the capacity to seriously harm all of Nigeria. Their activities caused disruption beyond the Niger Delta in an already fragile moment for the Nigerian economy. The challenge for federal and local governments was to draft a plan that not only addressed civilians' immediate concerns but also tackled the issues that had driven the conflict for so long. The PAP exit strategy was a crucial part of this plan, as it would affect 30,000 former militants who were on the government's side but might return to violence if they were pushed to the fringes of society again. The government also needed to fulfil Buhari's election promise to tackle corruption. The authorities had revealed little about how the organisational structure and practices of agencies such as the Niger Delta Development Commission, the Presidential Committee on the North East Initiative or the Ministry of Niger Delta Affairs would be reformed to ensure greater accountability and transparency in disbursement and procurement. Finally, the government needed to build greater cohesion and a sense of unity in the Niger Delta or else risk conflict between proliferating militant groups as they vied for influence.

Somalia

National elections dominated events in Somalia in 2016, with the country taking a step towards universal suffrage as its citizens chose a government for the first time since 1969. Yet the electoral process prompted al-Shabaab to focus its attacks on

Key statistics	2015	2016
Conflict intensity:	High	High
Fatalities:	4,000	3,500
New IDPs:		
New refugees:	31,000	

Mogadishu, in an attempt to demonstrate that Somalia was too insecure to conduct elections and to shake the international community's confidence in the undertaking. The vote also provided the group with new targets: election officials, delegates and candidates. In southern Somalia, the Somali National Army and the African Union Mission in Somalia (AMISOM) continued to push al-Shabaab back, while the Puntland authorities defeated a nascent affiliate of the Islamic State, also known as ISIS or ISIL, that had briefly captured the town of Qandala. Although the elections were not completed by the end 2016 as hoped, the process made significant progress in the last two months of the year and was poised to conclude in early 2017.

Military gains against al-Shabaab and ISIS

The main components of al-Shabaab's campaign were its battle for territory, particularly in southern states, and its disruption of the elections through violence in Mogadishu and the targeted killing of individuals such as delegates. The government maintained the pattern it established in 2015, driving al-Shabaab out of towns and cities while holding on to more territory than it had been able to in the past. The United States targeted al-Shabaab in strikes using unmanned aerial vehicles (UAVs) in the Shabelle and Juba regions, helping the army and AMISOM to win a series of victories against the group. However, the advance was slow and, on many occasions, the Somali government recaptured territory only temporarily. Despite the government's announcement in September that al-Shabaab had been militarily defeated, the group remained capable of conducting effective attacks and did not capitulate. Indeed, several times in the year, reports of al-Shabaab's imminent defeat circulated only for the government or AMISOM to request thousands of additional troops to help in the fight shortly thereafter. The military successes of pro-government forces increasingly created divisions between al-Shabaab's members – including its leaders – that further weakened the group.

However, any gains against al-Shabaab looked set to be short-lived if the state could not establish itself as the key provider of security. To do so, Somalia needed the help of the international community in building a government that was capable of fulfilling basic functions and seen as legitimate by the population. In December, AMISOM began *Operation Antelope* to rebuild major roads in Hir-Shabelle State, allowing security personnel greater access to the region and facilitating the transportation of consumer and agricultural goods there. Developmental efforts such as this also demonstrated progress to the Somali population and helped build confidence that the government was moving in the right direction.

Al-Shabaab's campaign to disrupt the elections through sustained attacks in Mogadishu was particularly striking as the group had only occasionally conducted operations in the city since withdrawing from it in 2011. Militants were able to carry out assaults in Mogadishu more easily in 2016 than 2015, partly because the Somali security forces were increasingly stretched by the additional security duties given to them during the elections. Moreover, a growing number of al-Shabaab fighters entered Mogadishu to escape American UAV strikes in the Shabelle and Juba regions, in the knowledge that the US was unlikely to target them in this way in an urban area. With many aspects of the elections either taking place within or being organised from the capital, attacks in the city sent a clear signal to both the Somali government and the international community that al-Shabaab had not been defeated and would threaten any new government.

Once the electoral process began, al-Shabaab implemented a strategy of assassinating individuals who had participated in the process. In September, the group threatened to attack polling stations across the country, stating that the elections was serving the interests of foreigners rather than the Somali people. A tribal elder who had been expected to act as a delegate in the elections was assassinated in Mogadishu in October, likely contributing to delegates' requests the following month that the government provide them with extra security. By the end of December, at least two more delegates had been killed. Groups other than al-Shabaab also participated in the violence – as seen in the brief battle between the Puntland authorities and self-proclaimed ISIS affiliates, as well as ongoing clashes between Puntland and Galmudug forces.

Despite having pledged allegiance to al-Qaeda in 2012, al-Shabaab continued to primarily focus on local issues rather than a worldwide jihad. Nonetheless, an al-Shabaab splinter group of around 20 militants who had pledged allegiance to ISIS claimed to have carried out its first attack in Somalia, detonating an improvised explosive device aboard a vehicle on the outskirts of Mogadishu in April 2016. The vehicle was destroyed, but it remained unclear whether anyone was killed in the incident – and AMISOM denied that those responsible were part of ISIS. A small group of ISIS militants took over Qandala in October, holding the town until it was retaken by the government in December. It was difficult to ascertain how many Somali militants had pledged allegiance to ISIS, but the killing of 30 of the fighters involved in the takeover of Qandala likely had a significant impact on the group's capabilities in Somalia.

Meanwhile, the dispute between the governments of Puntland and Galmudug led to violent incidents throughout the year. The fighting between the sides centred on the city of Galkayo, which straddles the border between the states. Following a series of attempts to reach a lasting peace agreement, tension between Puntland and Galmudug rose in October and, after a brief ceasefire, the clashes resumed in November. The United Nations reported that the violence caused the displacement of 75,000 people – some of them from camps for people who had been displaced by the wider conflict in Somalia. By the end of the year, the international community was encouraging both sides to find a peaceful solution to the dispute, but they had not signed an agreement.

Political progress

Somalia's elections adhered to a '4.5 power-sharing formula' as an interim step between appointed government and elections based on universal suffrage in 2020. Under this system, each clan elected or appointed delegates who then voted for members of parliament. Having been finalised in early 2016, the electoral process quickly encountered obstacles as Puntland challenged the 4.5 formula, arguing that districts should pick their own representatives for the lower house of parliament. The resulting delays continued into the final rounds of voting in December, when Puntland put forward its claim to the seats for the Sool and Sanaag regions allocated to Somaliland (the states had disputed ownership of the regions since Puntland's establishment in 1998). Ultimately, Puntland lost both of these challenges to the structure of the elections.

The international community continued to support the Somali authorities after they missed the initial August deadline for beginning the elections, despite having made clear that it would not extend the mandate of the president or parliament beyond August and September respectively. In October, Somalia's states began to put forward lists of candidates to be approved by the federal government. Candidates began to be elected in November, and most new members of the upper and lower houses of parliament were sworn in on 27 December. The presidential elections were postponed until the end of January 2017.

The elections proceeded more slowly than anticipated because of allegations of corruption, as well as a requirement that women make up a minimum of 30% of candidates and representatives in each state. The minimum threshold was controversial in many parts of Somalia; although the National Leadership Forum agreed to the measure, there was no law in place to enforce it. Some initial candidate lists failed to include enough female candidates, so had to be returned to the states concerned for revision before being approved. After the delegates voted, some states fell short of the threshold and were ordered to re-run their elections. By the end of the year, women made up 24% of new members of parliament – compared to 14% in the previous parliament.

The Somali auditor-general stated that the parliamentary elections had no credibility due to widespread vote-buying, fraud and intimidation. The Independent Electoral Disputes Resolution Mechanism (IEDRM) nullified the results of elections for 11 seats due to such irregularities in December. The move prompted the international community to express concern about how the organisation had made the decision after investigating 24 seats in all, and left the National Leadership Forum to decide how to deal with the nullified results. Lacking the legal authority to nullify the results, the IEDRM relied on the government to implement its recommendations. By the end of the year, the issue had not been settled: the National Leadership Forum announced that five of the 11 elections would be re-run, but the new parliament filed on 31 December a motion to prevent the decision from being implemented.

While the electoral process wore on, the Somali government finalised the creation of a federal system, which incorporated pre-existing governance structures such as those in Puntland. The final federal state to be created was Hir-Shabelle, in a merger of the Hiran and Shabelle regions shortly before the beginning of the elections. As a consequence, Somalia was divided into six federal states: Somaliland, Puntland, Galmudug, South West, Hir-Shabelle and Jubaland. However, Somaliland's government continued its efforts to secure independence for the state and, as such, refused to recognise the elections. Because of this, the vote for Somaliland's parliamentary representatives took place in Mogadishu.

Displacement and looming drought

Somali civilians faced significant threats from a continuing drought, which was particularly severe in parts of the north, and the Kenyan government's plan – declared in May – to close Dadaab refugee camp by November 2016. Announcing the closure, Kenyan Minister of the Interior Joseph Nkaissery said that the international community had acknowledged that it was time to shut down the camp, and that UN Secretary General Ban Ki-moon had acquiesced to the move in a private meeting with Kenyan officials. Nkaissery also claimed that Dadaab harboured terrorists. The declaration appeared to come as a surprise to both Somalia and the international community, especially as it ran counter to Kenya's international commitments to the UN and the US that all repatriations from the camp would be voluntary. Somalis comprised 95% of the estimated 300,000 refugees at the camp, the largest facility of its kind in the world.

Kenya's government contended that the camp posed a security threat to the country's citizens, citing high-casualty attacks by al-Shabaab on Kenyan soil and concerns that refugees in the facility had not been properly screened to prevent the entry of militants. The Kenyan authorities eventually realised that they would be unable to close the camp by the initial deadline without forcibly evicting the refugees. Meanwhile, the UN continued its voluntary repatriation programme, appealing in July for US$115.4 million to assist with the effort. However, there were reports of Somali refugees being turned away at the border by Somalia, which claimed that their security could not be guaranteed and that they would no longer receive humanitarian aid if they returned. The UN appointed in October a special envoy for Somali refugees, who was tasked with coordinating work to support them by the Somali government, host countries (Djibouti, Yemen, Uganda and Ethiopia, as well as Kenya) and the international community. In November, Kenya reaffirmed its commitment to closing the camp but extended the deadline to May 2017.

The drought persisted as an El Niño effect continued to hamper rainfall in Somalia and affected 1.4m people in Puntland and Somaliland, states experiencing their fourth consecutive rainy season that was relatively dry. However, the impact of the drought also began to be felt throughout the rest of the country. The UN estimated that due to the drought and the ongoing conflict, 5m people in Somalia were 'food insecure'. With substandard rainfall also predicted for 2017, it appeared that the drought would endanger civilians for some time to come.

South Sudan

Key statistics	2015	2016
Conflict intensity:	High	High
Fatalities:	3,500	3,000
New IDPs:	200,000	100,000
New refugees:	165,000	

Under concerted pressure from regional and international powers, the government of South Sudan and the Sudan People's Liberation Movement/Army–In Opposition (SPLM/A–IO) slowly implemented the peace deal between them in the first few months of 2016. Led by President Salva Kiir and Riek Machar respectively, the sides had signed the agreement in August 2015, but it seemed unpromising from the start. The deal maintained the political rivalry between Kiir and Machar, the two main figures in a civil war that began in December 2013, by returning to the pre-conflict political status quo and neglecting calls for accountability and justice. Although neither party genuinely committed to the ceasefire and the root causes of the war remained unaddressed, the agreement allowed for a decrease in violence.

However, by July 2016, observers had lost their cautious optimism about restoring peace and stability under the terms of the peace deal. The second half of the year saw further political and military fragmentation amid a worsening economic crisis, a shift towards greater ethnicisation of the conflict, an escalation in heavy fighting and a rise in insecurity across the country. Near the end of the year, Adama Dieng, the United Nations' special adviser on the prevention of genocide, warned that 'the signs are all there for the spread of this ethnic hatred and targeting of civilians that could evolve into genocide, if something is not done now to stop it.' The UN World Food Programme warned in December 2016 of an impending famine in South Sudan, due to 'unprecedented' levels of malnutrition and the fact that nearly four million people (one-third of the country's population) were 'severely food insecure'. Since the start of the conflict, around 3m people had been forced to flee their homes. Approximately 1.15m of them had fled to neighbouring countries, while 1.87m were internally displaced

– among whom more than 212,000 sought safety in UN Mission in South Sudan (UNMISS) Protection of Civilian (PoC) sites.

Collapse of a fragile peace process

In early 2016, South Sudan's politics was dominated by the stumbling attempts of the government and the armed opposition to form a Transitional Government of National Unity (TGoNU), as stipulated in the peace deal. Despite reaching a deadlock in the negotiations, Kiir issued in February a decree reappointing Machar as first vice-president (a position he had held until 2013). Machar said he would return to Juba and take up his post only after the capital had been demilitarised and government forces withdrawn, in line with the agreement. In April, following months of stalled negotiations on the formation of a TGoNU, around 1,300 SPLM/A–IO fighters travelled to Juba, along with some of the movements' senior leadership – including Deputy Chairman Alfred Lado Gore and Chief of General Staff Simon Gatwech Dual – in preparation for Machar's arrival. After several delays, Machar finally arrived in Juba on 26 April, and was sworn in as first vice-president shortly afterwards. South Sudan's newly appointed ministers joined the TGoNU on 29 April at a ceremony in the capital. But the fragile political and military arrangement did not last.

Some of the political obstacles to the implementation of the peace deal included the government's controversial decision in late 2015 to change the number of states in the country from ten to 28. Another was its continuous obstruction of the Joint Monitoring and Evaluation Commission (JMEC), responsible for overseeing the implementation of the peace deal, as well as the mandates and tasks of the TGoNU. Yet perhaps the most significant impediment to the agreement was the delay in demilitarising Juba. The presence of two parallel armies (the government's SPLA and the SPLA–IO) in the capital, outside their agreed cantonments, had a dramatic effect in such a socially, politically and economically volatile environment.

Growing tension between the forces sharply escalated on 2 July, when gunmen believed to be from South Sudan's military-intelligence agency shot and killed a senior SPLA–IO commander and his bodyguard in Juba. On 8 July, while Kiir and Machar were meeting at the presidential palace, government and SPLA–IO forces engaged in deadly clashes in the capital, beginning a wave of violence there that lasted until 11 July and prompted battles in other parts of the country. It was reported that approximately 300 people, including dozens of civilians, were killed on 8 July alone. The following day, South Sudan marked five years of independence with most of Juba under lockdown, as government and SPLA–IO forces fought across the city. Two Chinese UNMISS peacekeepers were killed and a further eight injured on 10 July, when their vehicle was hit by a mortar shell. Following another day of heavy clashes, on the evening of 11 July, Kiir and Machar both announced an immediate ceasefire.

The outbreak of violence had severe political, military and humanitarian repercussions. On 26 July, after Machar and his remaining SPLA–IO forces withdrew from the capital, Kiir replaced him as first vice-president with Taban Deng Gai, who was sworn in the same day. In response, a spokesman from Machar's SPLM/A–IO faction said the appointment was illegal and that the peace agreement had collapsed. Although the international community and the JMEC initially regarded the replacement of Machar as 'illegitimate', Deng secured international and regional support in the following months. But the new first vice-president failed to gain the backing of most of the SPLM/A–IO's various military groupings or of a social and political constituency across the country. Despite President Kiir's claim that the peace deal would still be implemented, the TGoNU had de facto collapsed due to the withdrawal of Machar, along with several of his senior supporters and members of the transitional government.

New geography of the war

Until July 2016, the conflict between the government and the SPLA–IO had been fought primarily in the Greater Upper Nile region, and with less intensity across the Greater Equatoria region. In the early months of 2016, there were sporadic armed clashes between government and opposition forces in various parts of Unity and Upper Nile states, including Koch, Rubkona, Mayom, Ulang and Nasir. Accordingly, the Ceasefire and Transitional Security Arrangements Monitoring Mechanism reported several violations of the ceasefire by government and opposition fighters. But the violence in Juba in July changed this dynamic by bringing the war to Greater Equatoria.

The conflict grew more intense from July onwards as its epicentre shifted, affecting urban and rural areas. Machar's predominantly ethnic Nuer SPLA–IO fighters joined forces with loosely aligned Equatorian SPLA–IO members in several parts of Greater Equatoria. Theretofore relatively free of violence, government-controlled Yei, in former Central Equatoria State, and the surrounding area became the scene of near-daily reports of criminality, extrajudicial killings and arbitrary arrests. Rural areas around Yei – but also Lainya, Morobo, Mundri, Maridi and Yambio in the Greater Equatoria region – were hit hard, with government forces allegedly conducting punitive attacks on civilians. Human Rights Watch reported on 23 November that civilians in and around Yei had been subjected to serious human-rights abuses, accusing the SPLA of conducting killings, rapes and arbitrary arrests, and the SPLA–IO of carrying out attacks and abductions. From August onwards, there was an escalation of violent incidents in the Greater Upper Nile region, reflected in steady reports of clashes between the government and the opposition. This confrontation took a heavy toll on the local population: there were many reports of the sides forcibly recruiting civilians, some of them children.

Throughout October, there was increasing polarisation across South Sudan – largely incited by political leaders. This exacerbated insecurity and strengthened the pre-existing, if less visible, ethnic dimension of the conflict. Government forces were widely perceived as being responsible for the violence across the Greater Equatoria region, which culminated on 31 October in the emergence of a new non-state armed group. Named the South Sudan Democratic Front and allegedly led by Lako Jada Kwajok, the new organisation called for the overthrow of the 'failed and illegitimate regime of the SPLM/SPLA Party through peaceful means, armed struggle or both'. The group was seemingly composed of fighters from the Greater Equatoria region; it called for an alliance with all other armed opposition groups, such as the SPLM/A–IO, the Agwelek forces of Johnson Olony, militias in Bahr el-Ghazal and the South Sudan Democratic Movement/Army–Cobra Faction (SSDM/A–CF). In September, senior opposition figure Lam Akol, a former minister in the TGoNU, declared the establishment of a new anti-government armed group, the National Democratic Movement. A few days later, Lieutenant-General Khalid Botrous, former deputy head of the SSDM/A–CF in Boma State, announced his defection from the government and stated that he would oppose President Kiir, claiming that the government had failed to honour the peace agreement signed with the SSDM/A–CF in 2014.

Threat of ethnic cleansing and genocide

The rising violence in South Sudan had enormous political, economic and humanitarian repercussions. The extended economic crisis in the country caused inflation of 835.7% in 2016. Many observers expected that the persistent political and economic emergency, as well as high levels of insecurity, would continue to force South Sudanese civilians to flee to neighbouring countries in 2017.

In the last few months of 2016, there was a proliferation of ethnic hate speech and incitement to violence, with growing criticism of President Kiir's government, seen as predominantly Dinka. Thus, anti-government sentiment spurred attacks on Dinka communities, which in turn fed a Dinka narrative of victimisation that justified acts of ethnic violence in self-defence. State and non-state actors increasingly targeted civilians for their ethnicity. On 13 October, Information Minister Michael Makuei Lueth cautioned communities in the Greater Equatoria region not to support 'terrorists', or rebel groups, threatening them with repercussions if they did so. His remarks reflected a further escalation of ethnic violence and targeting, implicitly acknowledging the SPLA's strategy of collectively punishing civilians who were seen to be taking a side.

By October, the international community and domestic political and civil-society actors were warning against the growing use of ethnic language and violence. On 12 October, senior members of the SPLM–Former Detainees stated that South Sudan would face the possibility of genocide if the security situation failed to improve, referring to 'the ills of ethnic cleansing and fragmentation'. Two weeks later, Zeid Ra'ad al-Hussein, the UN's human-rights chief, also denounced the political exploitation of ethnic language. The UN Panel of Experts on South Sudan echoed these warnings in November. Having concluded a ten-day visit to the country on 1 December, the UN Commission on Human Rights announced that 'many of the warning signals of impending genocide' were evident in South Sudan. Yasmin Sooka, the commission's chairwoman, declared that violence and ethnic tension there had reached unprecedented levels, warning 'there is already a steady process of ethnic cleansing underway in several areas of South Sudan using starvation, gang rape and the burning of villages.' She also referred to the forced recruitment of children and other civilians, sexual and gender-based violence, and the 'total breakdown of law and order'. Dieng spoke in November of the 'extreme polarization of some ethnic groups', adding that 'genocide is a process. It does not happen overnight. And because it is a process and one that takes time to prepare, it can be prevented.' He called on the international community to address the short- and long-term causes of the violence.

Ineffective international response

Although it hosted more than 212,000 people in its PoC sites across the country, UNMISS often failed to protect civilians in 2016 – most notably on 17–18 February, when unidentified gunmen opened fire on people sheltering in its site in Malakal, in former Upper Nile State. The fighting intensified after government troops entered the camp, sparking a conflict between militias from the three main ethnic groups there. Médecins Sans Frontières confirmed that the combatants killed at least 18 people – including two of its staff – and injured approximately 100 others, while burning down much of the camp, which had housed 47,000 people. UNMISS was also accused of failing to protect civilians, including humanitarian workers, during the July clashes in Juba.

Despite the warnings and the seriousness of the humanitarian crisis in South Sudan, the regional and international response was painstakingly slow in 2016, especially after July. Throughout the year, the international community focused its efforts on establishing the TGoNU and implementing other parts of the August 2015 peace agreement, even after it became clear that the deal had collapsed with Machar's withdrawal from Juba. Neighbouring countries went along with the government's narrative that the TGoNU was functioning as expected and the peace deal being implemented as planned. As the year wore on, President Kiir convinced Sudan, Ethiopia and Kenya to further politically isolate Machar and the SPLM/A–IO. By doing so, these regional powers overlooked the fact that the faction had significant popular support in South Sudan and effectively reduced the likelihood of a political solution to the crisis.

The international community responded to the July crisis in Juba by focusing on a new, robust peacekeeping mandate, as well as the need for an arms embargo and additional targeted sanctions, suggesting that it lacked a political strategy for addressing the situation. On 12 August, the UN Security Council agreed to send 4,000 regional peacekeepers to South Sudan. This Regional Protection Force would be authorised to 'promptly and effectively engage any actor that is credibly found to be preparing attacks, or engages in attacks, against United Nations protection of civilians sites, other United Nations premises, United Nations personnel, international and national humanitarian actors, or civilians'. A UN Security Council delegation led by US Ambassador Samantha Power visited South Sudan in early September, but it failed to achieve concrete results and, by November, the government was still blocking the deployment of the Regional Protection Force. On 18 November, the United States circulated a draft resolution at the UN Security Council calling for an arms embargo on South Sudan, as well as sanctions targeting several individuals, including Machar, Makuei and SPLA Chief of General Staff Paul Malong Awan. However, Russia and China rejected the resolution, a move welcomed by the government of South Sudan.

As indicated by the warnings of multiple actors and UN bodies, there was little prospect of peace in 2017. Indeed, it was highly likely that the onset of the dry season – when it becomes easier to move around South Sudan – would see a further deterioration of the political, security and humanitarian situation in the country.

Sudan (Blue Nile, Darfur and South Kordofan)

Despite completing its National Dialogue initiative in October 2016, Sudan was troubled by conflict in Darfur, South Kordofan and Blue Nile states throughout the year, leading to further decline in the country's humanitarian situation. Although the

Key statistics	2015	2016
Conflict intensity:	Medium	Medium
Fatalities:	3,000	3,500
New IDPs:	400,000	200,000
New refugees:	21,000	

warring sides met for peace talks in Addis Ababa in August, the process rapidly collapsed. In South Kordofan and Blue Nile states, ongoing battles between government forces and the Sudan People's Liberation Movement–North (SPLM–N) resulted in the displacement of civilians in government-controlled areas, and continued to prevent international aid organisations from reaching territory held by the rebels.

Another failed peace process

After the sides failed to reach an agreement on a ceasefire or humanitarian access to rebel-controlled areas in Blue Nile and South Kordofan in November 2015, the new year began with tentative discussions on a road map for peace negotiations between the government and the armed opposition. The African Union High-Level Implementation Panel (AUHIP), led by former South African president Thabo Mbeki, mediated the talks. In January, AUHIP discussed the road map separately with the SPLM–N, in relation to Blue Nile and South Kordofan, and the two main Darfuri armed groups, the Justice and Equality Movement (JEM) and the Sudan Liberation Movement–Minni Minnawi (SLM–MM), in relation to Darfur. But these negotiations took place against a backdrop of clashes in South Kordofan and Blue Nile states, as well as a sharp escalation in fighting in the Jebel Marra region of Darfur.

On 18 March, government representatives and armed opposition groups from South Kordofan, Blue Nile and Darfur began a strategic consultation meeting in

Addis Ababa. The government signed the AU-sponsored Roadmap Agreement on 21 March, but several opposition groups – the National Umma Party (NUP), the SPLM–N, the JEM and the SLM–MM – refused to follow suit, despite international pressure to do so. In the first half of the year, there were three distinct dimensions to the seemingly endless discussions on potential peace processes. The first of these was whether the coalition of armed and non-armed opposition groups known as 'Sudan Call' would sign the Roadmap Agreement. The second was whether the 2011 Doha Document for Peace in Darfur could be reopened for negotiation to groups that were not among its initial signatories. The third dimension was the potential amalgamation of the Qatari and AU peace initiatives. The international community – including the AU, the United Nations, the European Union and the 'troika' of Norway, the United Kingdom and the United States – continued to pressure Sudanese opposition groups, to no avail. The Sudan Call alliance maintained its rejection of the Roadmap Agreement until 6 July, when the leader of the NUP, Sadiq al-Mahdi, announced that the alliance was likely to sign the Roadmap Agreement. As a result, Sudan Call entered into discussions on the Roadmap Agreement in Paris on 19 July.

August saw the initiation and swift failure of another round of peace negotiations in Addis Ababa between the government of Sudan and armed opposition groups from Darfur, Blue Nile and South Kordofan. On 7 August, members of Sudan Call travelled to Addis Ababa to engage in the AU-sponsored peace talks with the government. Mbeki also met with four members of Sudan Call: the SPLM–N, from South Kordofan and Blue Nile states; the SLM–MM from Darfur; the JEM, also from Darfur; and the NUP, the most influential political opposition party in Sudan. On 8 August, Sudan Call signed the Roadmap Agreement. The deal included arrangements for a ceasefire in South Kordofan, Blue Nile and Darfur, and for a peace process involving the armed opposition in the National Dialogue convening in Khartoum. However, two days later, the SPLM–N delegation accused the government of defaulting on the framework agreement. By 14 August, peace talks to secure a long-term ceasefire in the three warring regions had collapsed.

The government initially postponed the peace process indefinitely, before announcing on 20 August that it would resume talks with the armed opposition groups the following month. On 15 September, members of Sudan Call were preparing to meet in Addis Ababa to coordinate their positions, ahead of discussions with the government over a cessation of hostilities and humanitarian access, as well as a subsequent national constitutional dialogue conference. But Khartoum continued to reject the SPLM–N's demand for delivery of humanitarian aid to Blue Nile and South Kordofan via the Ethiopian town of Asosa. Sudan Call announced on 30 September that, unless the government declared a ceasefire that allowed for the delivery of humanitarian aid to Darfur, Blue Nile and South Kordofan, it would boycott the rescheduled preparatory meetings in Addis Ababa. Despite the deadlock on the AU-brokered peace talks and ongoing clashes that continued to force people to flee their homes, Sudanese President Omar al-Bashir declared on 7 September that peace was returning to Darfur. By the end of the year, Ugandan President Yoweri Museveni had joined the effort to reinvigorate the stalled peace talks, meeting the leaders of the SLM–MM and the JEM, Minni Minnawi and Gibril Ibrahim respectively, in Entebbe.

While 2016 saw little progress in achieving a comprehensive political solution for South Kordofan, Blue Nile and Darfur, Central Darfur Governor Jaafar Abdel Hakam and a breakaway faction of the Sudan Liberation Movement led by Abdel-Wahid al-Nur (SLM–AW) signed the Korona Peace Agreement for western Jebel Marra on 8 November. The main SLM–AW faction rejected the agreement. By 13 November, President Bashir had ordered the implementation of the Korona Peace Agreement with breakaway faction the Sudan Liberation Movement–General Leadership, comprising several commanders and around 2,000 troops.

National Dialogue: maintaining the political status quo

Having concluded the National Dialogue initiative, the government signed the National Document on 10 October. The document was designed to resolve Sudan's social, economic and political problems, and to serve as a springboard for drafting a permanent constitution. It recommended a series of democratic reforms, including the appointment of a prime minister and direct elections for the president, members of the parliament, the Council of States and local commissioners.

But the NUP, the JEM, the SPLM–N and other factions refused to sign the National Document, saying that it was part of a deceptive attempt to maintain the political status quo. The NUP called the forum 'mere play' and argued that any dialogue should start by stopping the war, allowing the delivery of humanitarian aid and organising discussions in a neutral forum. Sudan Call and Future Forces of Change, another alliance of opposition groups that had refused to participate in the National Dialogue, agreed on 7 October to work together to establish a unified national political project. Other opposition groups, such as the Popular Congress Party, were also sceptical of President Bashir's assurances that his government would implement the recommendations of the National Dialogue committees. Although President Bashir continued to encourage opposition groups to sign the National Document, Sudan Call warned on 23 October that, unless the ruling National Congress Party government launched an internal dialogue and a more inclusive political process, its members would continue their efforts to overthrow the regime.

The US maintained its sanctions on Sudan in 2016 due to the government's lack of commitment to human rights and freedom of speech. In November, as protests against rising fuel and electricity prices broke out across the country, the regime responded by arresting several opposition leaders and closing media outlets. The National Intelligence and Security Services continued to regularly confiscate copies of newspapers that reported on anti-government demonstrations.

Air raids and armed clashes

Reports of government aerial and ground offensives, human-rights abuses and violations of international law in Darfur, Blue Nile and South Kordofan persisted throughout the year. The violence followed an established pattern:

fighting intensified in the dry season and lapsed in the rainy season, as it became difficult to travel on roads and in rural areas.

On 29 April, the Sudanese Revolutionary Front, an umbrella group of opposition militias, declared a unilateral six-month cessation of hostilities. This came five days after the SLM–MM, the JEM and the SPLM–N announced a joint cessation of hostilities in Darfur, Blue Nile and the Nuba Mountains, in South Kordofan. On 17 June, President Bashir also declared a unilateral six-month ceasefire, which initially covered Blue Nile and South Kordofan, and was later extended to Darfur. Although the security situation showed some improvement in July, the Sudanese Air Force intensified its air raids the following month, using barrel bombs and other munitions across Jebel Marra, Blue Nile and South Kordofan.

Reflecting the deep distrust between the parties to the conflict, the SPLM–N's chief of general staff said on 19 July that President Bashir's ceasefire announcement was 'a ploy' to allow for the reorganisation of government troops in South Kordofan and Blue Nile. The government purportedly extended its ceasefire for another two months on 10 October and, a few weeks later, the SPLM–N, the SLM–MM and the JEM extended their cessation of hostilities for six months, citing humanitarian concerns. However, as in previous years, the ceasefires and cessations of hostilities were not fully implemented. By 16 November, government forces and SPLA–N fighters were battling each other in Al-Azarak, one of the most productive agricultural areas in the Nuba Mountains.

As the under-reported war in South Kordofan and Blue Nile entered its fifth year, continued aerial attacks on civilians and severe fighting between the Sudan Armed Forces (SAF) and the SPLA–N caused hundreds of fatalities. Entering its 13th year, the conflict in Darfur generated persistent reports of killings, rapes and aerial attacks, especially in the Jebel Marra region. The SAF and its paramilitary Rapid Support Forces launched an intense campaign against armed opposition groups – especially the SLM–AW – in the region. After weeks of clashes, the Sudanese government declared on 28 January that its forces had defeated the SLM–AW in the region, and announced on 1 March that they controlled all areas north of Jebel Marra in Central Darfur. Yet the SLM–AW denied these claims, insisting that its forces had killed dozens of government soldiers and militia fighters, and that it had seized the Katrom area of Jebel Marra. Fighting between the SLM–AW and the SAF affected 159,000–194,000 people between January and early September.

In a report published on 29 September, Amnesty International accused the government of Sudan of using chemical weapons in Jebel Marra. The organisation estimated that 'between 200 and 250 people may have died as a result of exposure to chemical-weapons agents, with many – or most – being children'. It also reported that hundreds of other people had been injured in the attacks. The government denied the allegations, but the EU, the UK and France called for an international investigation into the allegations. Amnesty International concluded that Darfur has 'been stuck in a catastrophic cycle of violence for more than 13 years. Nothing has changed except that the world has stopped watching.'

In late 2016, there was also a spate of violent clashes between farmers and herders across Darfur. The UN reported on 10 October that tribal tensions between these groups had led to the displacement of 1,600 people in Tawila, in eastern Jebel Marra.

Critical humanitarian situation

In the context of unsuccessful peace talks, a political process that failed to engage with key opposition groups and continuous fighting across Darfur, Blue Nile and South Kordofan, the humanitarian situation deteriorated throughout 2016. International aid organisations continued to face bureaucratic impediments in accessing the communities most affected by the violence, especially those in Blue Nile and South Kordofan.

In May, the government, the AU and the UN began discussing the development of an exit strategy for the AU–UN Mission in Darfur (UNAMID). However, in a June briefing on Sudan to the UN Security Council, UN peacekeeping head Hervé Ladsous emphasised that the nature of the conflict in Darfur was unchanged and that 2.6 million people in the region remained displaced. Consequently, the Security Council extended UNAMID's mandate until 30 June 2017, noting that the situation in western Sudan continued to constitute a threat to international peace and security.

Indeed, the UN Office for the Coordination of Humanitarian Affairs reported in March that 129,000 people had been displaced by the fighting in Jebel Marra since the start of the year. Although the Sudanese government denied the existence of rebel groups in Jebel Marra, the UN disputed this claim and the Sudanese Air Force continued to carry out airstrikes in the region. Many displaced people rejected the government's calls for them to return to their homes in North Darfur, saying they would only do so after the security situation improved and they could gain access to services there.

The conflict in South Sudan also had an impact on human security in Sudan, as people crossed the border seeking safety. By mid-November, there were more than 263,000 South Sudanese refugees in Sudan, the majority of them in Darfur, South Kordofan and near the border with Renk, in former Upper Nile state. Despite the severe humanitarian crisis in Sudan, relations between international aid organisations and the government were strained. On 21 September, the Sudanese authorities forbade foreign organisations from working in refugee camps in East Darfur, Blue Nile and West Kordofan, with interior minister Babiker Digna stating that all assistance would be distributed by the UN High Commissioner for Refugees and local humanitarian organisations.

Given the failure of the August peace talks and the lack of genuine commitment to political reforms, it appeared that civilians across Darfur and the Nuba Mountains might face another destructive fighting season, including aerial raids, in early 2017. Although there were likely to be temporary breaks in the violence, the conflict looked set to persist until its root causes were addressed and significant political change occurred in Khartoum.

Chapter Five

South Asia

Afghanistan

Key statistics	2015	2016
Conflict intensity:	High	High
Fatalities:	15,000	16,000
New IDPs:	300,000	600,000
New refugees:	23,000	

The international community had little confidence in Afghanistan's prospects for 2016. In his final report to the UN Security Council on 15 March, Nicholas Haysom, head of the UN Assistance Mission in Afghanistan, said that 'survival will be an achievement' for the Afghan government in 2016, given the country's strengthened insurgency, troubled peace process, low economic growth, political divisions and reliance on financial donors. Nonetheless, the government did survive – albeit amid rising fatalities, injuries and displacement (including a record number of civilian casualties) and while losing control of parts of the country.

The US special inspector general for Afghanistan reconstruction (SIGAR) estimated that 6,785 members of the Afghan National Security Forces (ANSF) were killed and 11,777 others injured between 1 January and 12 November 2016. By the end of the year, the government controlled or influenced only 57% of the country's 407 districts – 15% less than in 2015 – while insurgents controlled or influenced 10%, and the sides contested the remaining 33%. There were an

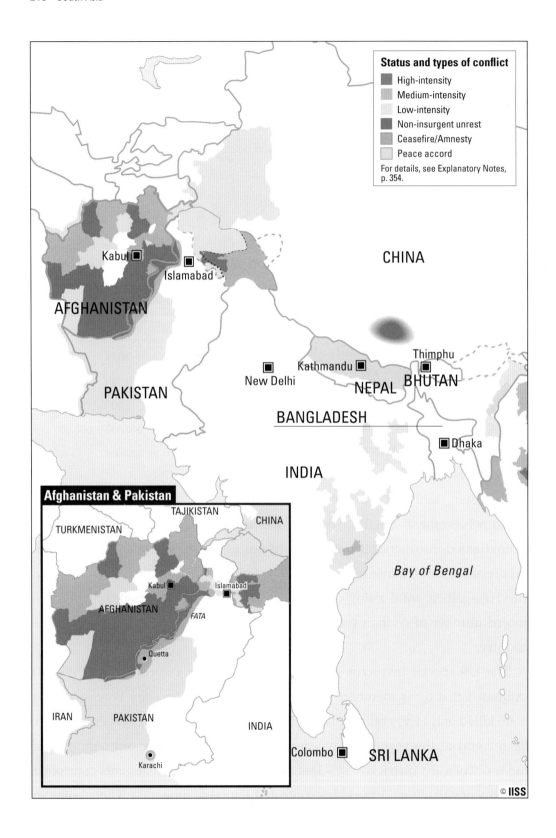

estimated 45,000 anti-government fighters in Afghanistan, 20–25% of them foreigners. The Afghan Taliban tried to capture provincial capitals at least eight times in the year, although pro-government forces eventually repelled each assault. The group funded its efforts partly through an increase in opium-poppy cultivation: according to the UN Office on Drugs and Crime, in 2016 the amount of Afghan land given over to such activity increased by 10%, and opium production rose by 43%, compared to the previous year.

In the run-up to the 8–9 July NATO summit in Warsaw, then US president Barack Obama announced that 8,400 American troops would remain in Afghanistan in 2017 – 2,900 more than previously announced. At the summit, members of the Alliance agreed to continue funding the ANSF until 2020, at an annual cost of around US$4.5 billion.

Stuttering peace process and a new Taliban leader

The peace process between Kabul and the Taliban appeared to regain some momentum on 11 January, when Afghanistan, Pakistan, China and the United States met for the first time as the Quadrilateral Coordination Group, in Islamabad. By the end of February, the parties had met three more times to discuss the issue. The group invited 'all Taliban and other groups' to participate in direct peace talks with the Afghan government, slated to begin in the first week of March. However, the Taliban's Qatar-based political office said that it was 'unaware of plans for talks' and that its preconditions for participation had not changed. The Taliban rejected talks with Kabul as 'futile' and 'misleading' on 5 March.

Sartaj Aziz, foreign-affairs adviser to the Pakistani prime minister, publicly acknowledged in early March that Pakistan had 'some' influence over the Afghan Taliban because the group's leadership was based in Pakistan. Islamabad had previously denied such claims. On 8 March, the Pakistani media reported that Taliban leader Mullah Mohammad Akhtar Mansour had urged his supporters in Pakistan to move to Taliban-controlled areas of Afghanistan to evade pressure from the Pakistani government. Yet in an address to a joint session of parliament on 25 April, Afghan President Ashraf Ghani said that he no longer expected Islamabad to bring the Taliban to the negotiating table. Shortly afterwards, Richard Olson, US special representative for Afghanistan and Pakistan, stated: 'we believe Pakistan has not taken action against terrorist groups that threaten its neighbours'.

On 21 May, the US killed Mullah Mansour using an unmanned aerial vehicle (UAV) in the Ahmad Wal area, part of the Pakistani province of Balochistan. A Pentagon spokesman said two days later that because Mansour had been plotting attacks on US and coalition personnel, the operation qualified as a 'defensive strike' under the rules of engagement for US forces in Afghanistan. Although Afghan Chief Executive Officer Abdullah Abdullah described Mansour as 'the main figure preventing the Taliban from joining the peace process', the undertaking made no progress until September.

The group announced on 25 May that it had appointed Haibatullah Akhundzada as Mansour's replacement. According to Taliban sources, Akhundzada had been chosen over the increasingly popular Sirajuddin Haqqani, leader of the Haqqani network and deputy head of the Taliban, due to concern that the appointment of the latter would precipitate an increase in pressure from the US military.

Akhundzada established his control over the group by co-opting rivals – and did so more quickly than his predecessor had been able to. The Taliban announced on 14 August that the deputy commander of the Mohammad Rasool splinter faction had rejoined the main group, as had a commander in Uruzgan province. Mullah Nanai, a former supporter of Mansour, was appointed chief justice in August.

Akhundzada seemed to be open to peace talks. He said on 11 September that the group continued to make diplomatic efforts, adding: 'we want to have relations with the world and answer their questions and mitigate their concern, so that we will save our country from others' harms in future and that others will also not be harmed from our country'. Akhundzada appeared to be referencing the initial US objective in Afghanistan in the wake of 9/11: preventing terrorists from using the country as a base from which to launch attacks on the West.

Tayyab Agha, former head of the Taliban's political office, said in October that the group did not have strict preconditions for peace talks, and that progress was possible. Although it was unclear whether he retained influence on the insurgency, he may have been urging the Taliban to soften its stance while positioning himself as the acceptable face of the group in the West.

In September and October, the Afghan government and the Taliban held secret talks in Qatar. The discussions reportedly involved a US diplomat but no Pakistani representatives, a fact that seemed to unnerve Islamabad. On 12 October, six days before the talks became public knowledge, the Pakistani authorities arrested three senior members of the Taliban in what one Western diplomat described as an attempt to 're-establish control' over the process and prevent the group from acting independently.

Meanwhile, Afghanistan–India relations continued to improve, despite militant attacks on the Indian consulates in Mazar-e-Sharif and Jalalabad. New Delhi continued to back various infrastructure projects in Afghanistan, pledging US$1bn in economic assistance to the country. General John Nicholson, commander of the US-led *Operation Resolute Support* in Afghanistan, visited New Delhi on 10 August to encourage Indian leaders to provide Kabul with further military aid. On 26 November, India delivered the last of four Russian-made Mi-24 helicopters to the Afghan Air Force.

Russia also became more active in Afghanistan, delivering its first major military-aid package in 15 years – comprising 10,000 AK-47 assault rifles – to Afghan forces in February. Russian officials publicly acknowledged that 'the Taliban is a potent political armed force', and confirmed that the Kremlin had maintained regular contact with the group since 2009 to protect Russians in Afghanistan. The Iranian ambassador in Kabul also acknowledged his country's contacts with the Taliban. Russia then hosted China and Pakistan for a trilateral meeting concerning the security situation in Afghanistan, which concluded on 27 December. This was the third such meeting, but the first to produce a public statement. The event reaffirmed Russia's emphasis on combating Khorasan Province (ISIS–KP) – the local affiliate of the Islamic State, also known as ISIS or ISIL – while taking a pragmatic stance on the Taliban. After Kabul accused Russia and Iran of cooperating with the Taliban against ISIS–KP, the parties to the discussions invited Afghan representatives to attend any future meetings.

Senior Afghan officials made several visits to China, reflecting the country's growing importance to the Afghan peace process. Afghan Minister of Foreign Affairs Salahuddin Rabbani stated in January that Afghanistan was willing to 'actively participate' in the Belt and Road Initiative, Beijing's massive regional economic project. China pledged more than US$70 million in unspecified military assistance to Afghanistan in February. A Taliban delegation visited China on 18–22 July, reportedly to discuss the peace talks. In November, Beijing denied reports in the Indian media that Chinese troops had conducted patrols in Badakhshan province.

Kabul signed a peace agreement with rebel group Hizb-e-Islami on 29 September, following around six months of negotiations. Having apparently been militarily inactive for some time, the group committed to a permanent ceasefire and to ending its support for terrorist groups. In return, the government agreed to request the repeal of international sanctions on Hizb-e-Islami, and to offer immunity to the group's leaders for crimes committed during the Afghan civil war in the 1990s.

Despite this achievement, Afghanistan's national unity government remained anything but unified. There was escalating tension between Ghani, Abdullah and Vice-President Abdul Rashid Dostum. Dostum repeatedly complained that he had been 'sidelined', accusing Ghani and Abdullah of 'sharing power among themselves'. Both Dostum and acting Balkh governor Atta Mohammad Noor personally led anti-Taliban operations in March, seemingly to burnish their credentials as military commanders and providers of security. On 20 June, parliament approved Masoom Stanekzai and General Abdullah Khan Habibi as head of the National Directorate of Security (NDS) and defence minister respectively. Afghanistan had been without a defence minister since the inauguration of the unity government, with Ghani and Abdullah unable to agree on a candidate. On 11 August, Abdullah publicly described Ghani as 'not fit for the presidency'.

Around two months later, Dostum accused the president of favouring his fellow Pashtuns and called Abdullah 'incapable'.

The government also suffered damage to its legitimacy on 29 September, when it missed the two-year deadline for achieving several of its inaugural objectives. It had failed to hold elections, enact electoral reforms and establish a constitutional Loya Jirga (a gathering of Afghan elders and other leaders) to discuss replacing the post of chief executive with that of prime minister. Moreover, parliament voted on 12–15 November to dismiss seven cabinet members, including Rabbani, an ally of Abdullah. Ghani instructed the ministers to remain in their jobs while they awaited a Supreme Court ruling on the issue.

Fierce Taliban offensive

The Taliban continued its assaults on district and provincial centres, while attempting to control strategically important highways. Kunduz, Helmand, Farah, Faryab and Uruzgan were among the worst-hit provinces in 2016. The year began badly, with SIGAR reporting that the Afghan government lost control or influence over 19 districts between January and May. General Nicholson revealed that ANSF casualties in the first half of 2016 were around 20% higher than they had been in 2015. According to a confidential report by the Afghan interior ministry that was leaked in February, one-quarter of Afghanistan's 144,000 police officers left their jobs in 2015, many by deserting. The report also said around 200 police officers had been killed in January 2016 alone.

By the end of the year, the government controlled only two of Helmand's 14 districts. Six of the districts were contested, and the remainder largely controlled by the Taliban. The head of Helmand's provincial council said on 10 January that around 40% of registered security personnel existed only on paper, and that the group controlled approximately 65% of the province. The insurgents continued their attacks in Musa Qala, Marjah, Nahr-e-Saraj and Sangin districts as part of a plan to encircle the provincial capital, Lashkar Gah. A US military spokesman said on 25 January that the Afghan army in Helmand was being 'rebuilt' because of a combination of 'incompetence, corruption and ineffectiveness'. In mid-February, a joint US–Afghan operation in Sangin led to the death of one police officer and the arrest of 30 others. The authorities suspected that the detained officers had provided weapons and ammunition to the insurgents, and planned to surrender to them.

The US confirmed on 8 February that it would redeploy several hundred soldiers to Helmand from other parts of Afghanistan by the end of the month, in support of the 215th Corps of the Afghan National Army (ANA) and to protect American forces already stationed there.

The Taliban briefly captured Khan Neshin district in mid-March, before the security forces reported retaking it on 27 March. This came shortly after the ANSF abandoned its bases in Musa Qala and Now Zad on the recommendation of *Resolute Support* officials aiming to secure strategically important locations in Helmand such as Lashkar Gah, a national highway and the Kajaki Dam. The defence ministry formally announced the change of strategy on 26 March, stating that a reduction of security checkpoints in vulnerable areas would free up forces for offensive operations. US officials in Kandahar urged a similar course of action there, calling on the ANSF to go on the offensive. However, in May and June, the reduced security presence on highways led to a spike in Taliban attacks on civilian travellers, including abductions. The US sent Brigadier-General Andrew Rohling to command the train, advise and assist force in Helmand – a task normally given to a less senior officer.

The Taliban announced the launch of its spring offensive, named *Operation Omari* after its late leader, on 12 April. Four days later, the ANSF had begun a counter-offensive entitled *Operation Shafaq* and was fighting in 25 districts across 12 provinces.

The Taliban also continued to carry out complex suicide attacks in urban centres. On 19 April, the group killed at least 64 people and injured 347 others in Kabul by detonating a large truck bomb next to a building housing close-protection services for senior government officials. Following the explosion, three attackers wearing uniforms entered the compound, engaging in a gun battle before dying themselves. Most of the casualties were civilians. On 8 May, the government responded to the attack by executing six convicted terrorists at Pul-e-Charkhi prison in Kabul. The Taliban retaliated by conducting several assaults on the judiciary across Afghanistan.

On 15 April, the group attacked several districts in the surroundings of Kunduz city, reportedly taking control of several highway checkpoints. The government pushed back the assault with the help of US special forces. The Afghan Air Force, particularly its recently delivered *Super Tucano* light attack aircraft, played an important role in repelling Taliban assaults, while the police, the military and the intelligence services demonstrated that they had greatly improved their coordination since the fall of Kunduz city in 2015.

On 12 May, the ANSF announced that it would conduct *Operation Khanjar* to push the Taliban out of the Lashkar Gah area. Nine days later, the armed forces reportedly cleared Yakhchal, in Nahr-e-Saraj district, along with most of Marjah district. On 29–30 May, the Taliban killed at least 57 police officers in assaults on highway checkpoints in Nad-e-Ali and Nahr-e-Saraj. The group made a similar push to encircle Uruzgan's provincial capital, Tarin Kot.

By June, the Taliban controlled at least nine districts: Baghran, Dishu, Musa Qala and Now Zad, in Helmand province; Warduj and Yamgan, in Badakhshan; Kohistanat, in Sar-e-Pul province; Nawa, in Ghazni province; and Khakar, in Zabul province. With the group controlling or influencing more territory than at any time since 2001, the US introduced more lenient rules of engagement. These allowed American forces to target the Taliban with airstrikes, provide close air support to ground troops and operate with conventional ANA units (rather than just Afghan special forces) if 'their engagement can enable strategic effects on the battlefield'.

Militants carried out two major suicide bombings in Kabul in late June. The first of these killed 14 Nepalese guards and wounded eight other people, some of them civilians, at the Canadian Embassy. Both the Taliban and ISIS–KP claimed responsibility for the assault. In the second attack, two Taliban fighters targeted a convoy of buses carrying police cadets, killing 33 people, most of them Afghan security personnel, and injuring 40 others.

The same month, the group temporarily gained control of three districts in Kunduz province, outnumbering the ANSF in the province. In Faryab province, 70 ANSF personnel had to be rescued from Almar district after the Taliban besieged the area. The group seized yet another district in Helmand – Khan Neshin – on 30 July, killing at least 24 police officers in the process. Yet on 4 July it reportedly lost around 67 fighters during a large-scale attack on Musa Khel district, in eastern Khost province, after being repelled by the ANSF.

The head of Helmand's provincial council claimed on 6 August that at least 586 ANSF personnel and 250 civilians had been killed or wounded in the preceding weeks. Around 30,000 villagers in the province had fled their homes and were unable to return because of the threat posed by the Taliban's roadside bombs. Displaced people continued to stream into Lashkar Gah from the countryside. Nawah-e-Barakzai district, normally one of Helmand's safest areas, fell briefly to the Taliban before being retaken on 11 August. A similar course of events took place in Garmser district two days later. The insurgents' failure to overrun Lashkar Gah largely resulted from a significant increase in airstrikes conducted by the US, which deployed B-52H *Stratofortress* strategic bombers to the Afghan theatre for the first time in ten years. The Pentagon confirmed that by 22 August, up to 130 US troops had arrived in Lashkar Gah to help the ANSF defend the city. These troops were to be stationed there for an unspecified length of time in an 'advisory capacity'. While besieging Lashkar Gah, the Taliban made another attempt to capture Kunduz city, employing its established strategy of encirclement.

The group killed 30 ANSF personnel in a successful two-week offensive to capture Jani Khel district, in Paktia province. The insurgents held the district for around a week and a half, before the government recaptured it on 5 September. The same day, the Taliban began its operation to seize Tarin Kot, while conducting a suicide bombing at the Ministry of Defense in Kabul that killed 41 people, including an ANA general and senior police officers, and wounded around 110 others, many of them civilians. Although the group came close to seizing Tarin Kot on 8 September, it was pushed back the same day. However, some reports noted that the Taliban had overrun or destroyed all checkpoints around Tarin Kot, and that fighting near the city continued. As the group resumed its assault on Tarin Kot on 22 September, an ANA commander accused the police of handing over 89 checkpoints to the militants. The interior ministry denied the claim, arguing that 'tribal issues' had complicated the authorities' response.

On 2–3 October, the insurgents launched simultaneous attacks on the capitals of Farah, Kunduz, Baghlan and Uruzgan. The Taliban once again entered Kunduz city, targeting symbolically important administrative buildings there. As in 2015, there were reports that the police had abandoned their posts when faced with the Taliban advance. The ANSF regained control of most of the city the following day, partly due to at least three US airstrikes. With the insurgents reportedly forcing people from their homes in an attempt to create a large number of internally displaced persons, the UN estimated on 14 October that 39,000 people had fled from Kunduz in the preceding two weeks.

The Taliban maintained its encirclement of Lashkar Gah. To the west of the city, the group captured Nawah-e-Barakzai district and killed its district police chief on 3 October. By 14 October, the Taliban had killed more than 200 ANSF personnel and 45 civilians in its latest offensive on the city (which it remained unable or unwilling to enter). Three days earlier, it had ambushed and killed 100 soldiers and police officers who were attempting to escape a siege at nearby Chah-e-Anjir. Up to 150 other troops surrendered, while the Taliban seized 22 Humvee armoured vehicles, dozens of trucks and hundreds of firearms. The government deployed around 300 special-forces personnel as regular infantry to help defend Lashkar Gah.

The Taliban's 11-day siege of Farah city also failed. But the governor of Farah province said that only US airstrikes, which killed at least 27 insurgents, prevented the group from taking the city. *Resolute Support* special forces had also deployed to Farah city. The Taliban briefly entered Tarin Kot on 3 October. By the end of the month, three ANSF bases in Uruzgan had fallen to the insurgents, suggesting an alarming lack of coordination, supplies and morale among security forces in the province.

General Nicholson said on 21 October that the Taliban had failed to achieve its military objectives. This was due not only to US airstrikes – around 700 of which took place in Afghanistan between 1 January and 26 October 2016, compared to 500 in all of 2015 – but also the resilience of the ANSF, which lost a large number of troops. Yet the group's capacity to carry out massive coordinated operations remained a major cause of concern.

Around the same time, there was a surge in attacks on Western targets, with three American soldiers dying in two separate bombings in early October. On 19 October, an Afghan soldier killed two US citizens in an apparent insider attack at Camp Morehead, near Kabul. On 12 November, a Taliban suicide bomber killed two US soldiers and two US contractors at Bagram air base, in Parwan province. Two days earlier, a Taliban fighter had detonated a truck at the German consulate in Mazar-e-Sharif, killing at least four Afghan civilians and wounding 129 others. German forces killed another two civilians following the explosion after they refused to stop at a checkpoint. Nonetheless, six days later Berlin committed to maintaining its 980-strong troop presence in Afghanistan.

There was a decline in clashes across Afghanistan following the Taliban's October offensive, suggesting that a stalemate was developing as winter began. On 21 November, the ANSF launched *Operation Shafaq* 2 in Nangarhar province with the aim of cutting insurgent supply routes from Pakistan during winter, later expanding the mission to Ghazni, Logar, Paktia and Paktika provinces.

Campaign against Khorasan Province and al-Qaeda

On 14 January, the US government formally designated ISIS–KP as a foreign terrorist organisation, making material support for the group illegal. Separately, the White House expanded the legal authority for the US Department of Defense to target the group. By 20 February, airstrikes and ANSF ground operations had dislodged ISIS–KP from several of its strongholds in Achin district, in Nangarhar. In the first three months of 2016, US forces in Afghanistan carried out around 100 airstrikes against ISIS–KP and al-Qaeda, approximately 70–80% of which targeted the former group in Nangarhar. Afghan officials said in March that these operations had forced many ISIS–KP militants to relocate to Kunar province.

General Nicholson announced on 12 July that US troops had shifted their operational focus to Nangarhar after being assigned new rules of engagement. This soon had an effect: on 26 July officials declared that Afghan forces, with US air support, had expelled ISIS–KP from Kot district. Reportedly, the operation killed 122 militants, including Saad al-Emarati, the group's deputy military commander. Five US special-forces personnel were also injured in the fighting.

On 23 July, an ISIS–KP suicide bombing killed at least 80 Shia ethnic Hazaras and wounded 231 others at a demonstration in Kabul, the group's first attack in the capital. As this was the deadliest single attack there since 2001, the Ghani administration received harsh criticism for its apparent failure to provide adequate security. An ISIS–KP spokesman said the group would continue to target Hazaras 'unless they stop going to Syria and stop being slaves of Iran' (Hazaras had fought on the side of the Syrian government in the country's civil war).

The same month, the US killed ISIS–KP leader Hafiz Saeed Khan in a UAV strike. Abdul Haseeb, an Afghan and a former member of the Taliban, replaced Khan. General Nicholson said on 10 August that around 25% of ISIS–KP's 1,200 fighters had been killed in recent operations (70% of these fighters were thought to be former members of Tehrik-e-Taliban Pakistan). Despite having agreed to a series of local ceasefires over the preceding two months while they fought back against counter-insurgency operations in eastern Afghanistan, the Afghan Taliban and ISIS–KP had by late August begun to battle each other once again.

The ANSF initially succeeded in its high-profile *Operation Qahar Selab* in early September, but neglected to hold areas that it had recaptured from ISIS–KP. As a consequence, militants returned to these areas just weeks later. Nevertheless, *Resolute Support* declared in October that the group had been reduced to about 1,000 fighters in three districts of Nangarhar province. On 30 November, the security forces reported that they had expelled ISIS–KP from Pachir Wa Agam, after it attempted to gain a foothold in the Tora Bora mountains.

The group claimed responsibility for a gun attack on Kabul's Sakhi Mosque on 11 October, during the feast of Ashura, in which at least 18 Shi'ites were killed and 62 others were wounded. The following day, an ISIS–KP suicide bombing at an Ashura gathering in Balkh district killed 15 people and injured 30 others. Six people died in another such bombing in Kabul on 16 November. Five days later, an ISIS–KP suicide bomber killed 32 people and injured around 50 others at a Shia mosque in Kabul.

General Nicholson said in April that the leadership struggle within the Taliban had increasingly led to 'overt cooperation' between the group and al-Qaeda, as the latter provided unspecified 'capabilities and skills' to the former. There were thought to be 200–500 members of al-Qaeda in Afghanistan. On 20 December, the US confirmed that three of the group's leaders – Farouq al-Qahtani, Bilal al-Utabi and Abd al-Wahid al-Junabi – had died in an airstrike in Kunar province on 23 October. Qahtani was al-Qaeda's 'emir' in Afghanistan, while Utabi was one of his lieutenants. Another six members of the group died in UAV strikes or ground operations in 2016, while Afghan intelligence operatives dismantled a camp belonging to it in Mizan district, in Zabul province, on 19 September.

Proliferating threats to civilians

The United Nations announced that the conflict killed 3,498 civilians and injured 7,920 others in 2016, the highest annual number of civilian casualties ever recorded by the organisation in Afghanistan. Those who had been killed or injured included 3,512 children – another record. Helmand province alone saw 891 civilian deaths. Anti-government insurgents were deemed responsible for 61% of civilian casualties, and pro-government forces for 24% (with 10% attributed to both sides, and 5% to unidentified ordnance). Although several oft-used insurgent tactics – such as suicide attacks and the deployment of improvised explosive devices – caused many of the casualties, most resulted from conventional ground engagements.

Human Rights Watch reported in February that the Taliban had increased its recruitment of child soldiers since mid-2015. Indeed, intelligence operatives arrested a 14-year-old would-be suicide bomber from the group in Kabul on 28 February. But pro-government forces also employed minors. In February, the Taliban killed a ten-year-old boy who had fought with local police in Tarin Kot, in what appeared to be a targeted assassination.

Afghanistan remained one of the most dangerous places in the world to be a journalist. The Taliban killed seven employees of Moby Group, a media company, in a suicide attack in Kabul on 20 January. At least 25 others were injured in the assault. The Taliban had previously threatened the firm's subsidiaries due to their perceived bias in reporting on the insurgents' capture of Kunduz in September 2015.

The Taliban's attempted control of highways around the country also endangered civilians. On 31 May, the insurgents abducted 185 people from buses at a roadblock in Aliabad district, in Kunduz province. At least 12 people died in the incident; although the group freed 160 of the victims shortly thereafter, some remained in captivity. The Taliban also appeared to be behind a series of abductions of, and other attacks on, around 120 travellers in June, which led to the deaths of approximately 12 people. Its objectives appeared to be to find and kill government employees or military personnel, and to demonstrate Kabul's inability to provide security. On 25 October, militants captured dozens of civilian hostages during an attack on Firozkoh, the capital of Ghor province. After hearing that one of their commanders had been killed in a clash with security forces, they executed at least 26 – but possibly as many as 42 – of the hostages.

The acting police chief of Faryab province said that local militias participated in 50% of operations launched by the security forces there. Such groups often proved to have shifting allegiances, little training and a propensity to abuse civilians.

Airstrikes also caused civilian casualties. On 6 April, a government official claimed that two American air operations in Paktika province had killed 21 civilians (the US said the strikes were aimed at insurgents, but provided no further details). According to the UN Assistance Mission in Afghanistan, another US airstrike killed at least 15 civilians in Achin district on 28 September. Afghan and US officials insisted that ISIS–KP militants had been targeted in the operation, and that it might also have killed up to six militants. One month later, a strike by American

UAVs reportedly injured at least 12 civilians in Sherzad, in Nangarhar. On 3 November, a failed US and Afghan special-forces operation in Kunduz city led to the deaths of 32 Afghan civilians, four Afghan security personnel and two US soldiers. Another four US soldiers and 19 civilians were wounded in the operation.

Due to these threats, Afghans constituted one of the largest groups among the refugees and other migrants travelling to Europe. The UN Office for the Coordination of Humanitarian Affairs reported on 12 December that around 600,000 people had been internally displaced in Afghanistan in 2016. As a consequence of its deteriorating relationship with Kabul, Islamabad repatriated at least 600,000 Afghan refugees in Pakistan, many of whom had been there for decades and as such lacked homes in Afghanistan. President Ghani told Afghanistan's parliament on 27 September that he expected more than 1m refugees to return to the country in 2017.

Thus, Afghanistan itself took in more Afghan refugees than any other country, particularly as some also returned from Iran – and from Europe, under a deal signed in the run-up to the Brussels Conference on Afghanistan in October. The agreement allowed the EU to deport an unrestricted number of Afghan asylum seekers whose applications had been rejected, and obliged Kabul to accept them. This measure was likely a prerequisite for the US$15.2bn in non-military assistance that donor countries pledged to Afghanistan at the conference.

India (Assam)

The marked improvements in Assam's security that began in 2015 continued the following year, with positive developments in the peace process between the federal government and the pro-talks faction of the

Key statistics	2015	2016
Conflict intensity:	Low	Low
Fatalities:	60	100
New IDPs:		
New refugees:		

United Liberation Front of Asom (ULFA–PTF). Meanwhile, the security forces continued to achieve substantive results in operations against hardline organisations such as the National Democratic Front of Bodoland–Songbijit (NDFB–S) and the Karbi People's Liberation Tigers (KPLT). Although the annual number of conflict-related fatalities rose from around 60 in 2015 to approximately 100 in 2016 as a result of both militant attacks and security operations, it remained significantly lower than the 305 recorded in 2014. However, resurgent militancy by the United Liberation Front of Asom–Independent (ULFA–I), the formation of a new Karbi insurgent group and the persistence of identity politics in Assam suggested that the state still faced significant challenges.

Change of state government

The most significant political event in Assam in 2016 was the April elections for the state assembly, which pitted Congress-party stalwart and long-serving chief minister Tarun Gogoi against Bharatiya Janata Party (BJP) candidate Sarbananda Sonowal. Congress had been criticised in 2015 for failing to deal with persistent illegal migration into the state, an issue that played a crucial role in civil unrest and the armed conflict in Assam. Sonowal capitalised on the resulting discontent in a campaign driven by anti-migration sentiment, branding illegal migrants 'enemies of the state' in one particularly fiery speech on 11 February. The approach worked: the BJP secured a clear majority of 86 seats in the 126-seat assembly, bringing Gogoi's 15-year tenure to an end and the state assembly into political alignment with the central government. The new administration quickly announced that it would step up the construction of fences along the border with Bangladesh, which the federal Ministry of Home Affairs ordered to be completed by the end of 2017. The urgency surrounding the fences became intricately linked with national security in July, when Jamaat-ul-Mujahadeen Bangladesh attacked a café in Dhaka. Given the group's links with international jihadist movements, the incident sparked fears that terrorist organisations would exploit Assam's porous borders to conduct operations in India.

Nonetheless, the political alignment of the federal and state governments did not always ensure the smooth implementation of policy. The BJP's Assam wing came under considerable pressure from its counterparts in New Delhi on 19 July, when they tabled a bill proposing that Hindu, Buddhist, Jain, Parsi and Sikh refugees from Pakistan, Afghanistan and Bangladesh be offered citizenship. Running against the anti-migration provisions of the Assam Accord, the move prompted rallies in opposition by political parties such as Congress and Asom Gana Parishad, as well as militant organisations such as the ULFA–I.

Indeed, these alterations to citizenship laws slowed talks between the central government and the ULFA–PTF. In early 2016, the ULFA–PTF announced that influential leader Anup Chetia, released from prison on bail in December 2015, would join and head its peace talks with the government. Chetia's ULFA–PTF immediately met with Gogoi and signalled its commitment to the peace negotiations by condemning violence, reaching out to other former militant groups, such as Dima Halam Daogah, and advocating better living standards for former ULFA militants who had surrendered. On 22 June, the ULFA–PTF also encouraged civil-society organisations to become involved in the peace process. However, Chetia announced on 4 October that peace talks with the government had been placed on hold due to the uncertainty surrounding the citizenship bill.

Meanwhile, the government was making little progress in its peace process with the National Democratic Front of Bodoland–Progressive (NDFB–P), with a spokesperson for the group revealing in September that there had been no talks between the parties since June 2015. On 26 October, following a meeting between representatives of the Assam government, New Delhi and the NDFB–P, the sides announced that the talks would resume in late November. Yet by the end of the year, it remained unclear whether the discussions had taken place.

Sustained attrition

The federal government continued to pursue a relentless counter-insurgency campaign in 2016, putting groups that had conducted major attacks in recent years on the defensive. For example, the NDFB–S indicated its willingness to engage in a ceasefire at the beginning and end of the year, only for the government to publicly reject these overtures on 21 January and 16 December respectively. Most of the 22 killings of NDFB–S fighters during the year resulted from the government's persistent security operations.

The decline in the group's capabilities became particularly apparent in the third quarter of 2016, when 12 of these fatalities occurred. Following a series of army and police operations in July, NDFB–S militants launched on 5 August one of the deadliest attacks in Assam since the ethnic violence of December 2014, shooting to death 14 civilians in a marketplace in Balajan Tiniali. There, in contrast to its previous attacks, the NDFB–S chose not to target a specific ethnic group and instead fired indiscriminately into the crowd, resulting in the deaths of several Bodos. Although the militants who conducted the assault were members of the NDFB–S, the group itself denied involvement, possibly in a bid to deflect further pressure from counter-insurgency operations and, potentially, to avoid a backlash from the Bodo community. Two days after the incident, the state government declared a 'zero-tolerance' policy towards the group, bringing in three more companies of federal paramilitary personnel to bolster counter-insurgency operations and prevent further attacks. By the end of the year, with only around 150 fighters and no prospect of gaining a ceasefire deal, the group appeared to be on the verge of collapse. The KPLT also faced intense pressure from the security forces throughout the year, forcing it to declare a unilateral six-month ceasefire on 24 October (the group had long 'prayed' for a ceasefire agreement with the government, according to state police).

Despite this progress, the insurgency in Assam still presented significant challenges. The apparently imminent collapse of the NDFB–S and the KPLT in the state prompted I.K. Songbijit – former leader of the NDFB–S – to create on 27 October the People's Democratic Council of Karbilongri, an organisation that sought to fill the gap left by the KPLT's withdrawal from Karbi Anglong district. Although the new group did not engage in any significant activities in 2016, the fact that a former NDFB–S leader had established it reflected the increasing cooperation between militants in the northeast. Indeed, statements of approval from groups such as the ULFA–I and the National Socialist Council of Nagalim–Khaplang suggested that it was likely to become part of umbrella organisation the United Liberation Front of Western South East Asia (UNLFWSEA), which had given them valuable support in recent years.

With an estimated strength of 120 fighters, the ULFA–I was likely weaker in late 2016 than it had been at any point since its foundation, in 1979. Nonetheless, the group had maintained its presence in Assam through a series of attacks. One of these came on 4 April, shortly after the conclusion of the state elections, when it killed three civilians by detonating a bomb in Dudhnoi, in Goalpara district. Militants from the group kidnapped the son of a Tinsukia district BJP politician from the Assam–Arunachal Pradesh border area on 1 August, releasing him on 10 September after declaring that he was not a government spy. The group also carried out a flurry of assaults around the time of India's Independence Day. On 11 August, it detonated a bomb near a tea estate in Tinsukia district, although no one was injured in the blast. The group also killed two civilians in a gun attack in Bamungaon, in Tinsukia district, on 13 August, and murdered five other people across Upper Assam on 15 August. After New Delhi announced on 8 November that it would demonetise all 500-rupee and 1,000-rupee notes in a bid to fight corruption, the ULFA–I became increasingly engaged in robbery and extortion, firing on a commercial vehicle and killing one civilian in Tinsukia district a week later. Although the government's demonetisation policy appeared to have a significant effect on the NDFB–S, the KPLT and the ULFA–I, there were reports on 15 December that the ULFA–I had converted approximately 50% of its cash into legal currency. This increased the perceived threat posed by the ULFA–I, which, with the assistance of its allies in the UNLFWSEA, had killed three soldiers after ambushing their convoy in Tinsukia district on 19 November.

Desperate to remain relevant while under severe pressure from the security forces, militant groups in Assam conducted a series of recruitment efforts, extortion drives and attacks against civilian targets. This growing activity was particularly evident among NDFB–S and ULFA–I fighters. Extortion appeared to have a marked effect on the economy: truck drivers operating on Highway 37 revealed on 25 September that the price of goods had increased due to a renewal of the practice along the route. Meanwhile, counter-insurgency operations also affected civilian life. The army responded to the ULFA–I's 19 November attack by conducting a large-scale search operation in Pengaree, in Tinsukia district. The undertaking hampered agricultural activities in the area, reflecting the long-standing predicament of civilians caught between the militants and the security forces.

India (CPI–Maoist)

Key statistics	2015	2016
Conflict intensity:	Low	Low
Fatalities:	300	400
New IDPs:		
New refugees:		

The Naxalite conflict intensified in 2016, albeit while remaining significantly less violent than at its peak in 2011. The balance of power continued to shift in favour of the government, at the expense of the Maoist insurgency. Federal and state security forces consistently defeated the Communist Party of India–Maoist (CPI–Maoist) and its splinter groups on the battlefield, with the insurgents incurring more than three times as many fatalities as their opponents. Clashes between state security forces and Maoist militants remained the main cause of fatalities throughout 2016. As in 2015, the authorities carried out security operations in tandem with economic-development projects and other forms of community outreach. There were signs that this strategy had reduced support for the insurgency among civilians, a trend compounded by an apparent crisis of leadership and serious morale problems within the CPI–Maoist. Yet despite their battlefield losses, the Maoists often killed civilians suspected of cooperating with the government. The security forces were also implicated in several severe human-rights violations in the year, prompting the judiciary to open a series of investigations into the incidents.

Coordinated development and security efforts

The Indian government continued to implement a dual policy of security measures and economic development in its counter-insurgency campaign against the CPI–Maoist. This entailed using the security forces to protect, and sometimes build, infrastructure and business projects in rural areas previously cut off from the government and its services. The top commanders of several states' anti-Naxalite security wings expressed confidence in this approach, attributing recent successes against the insurgency to its implementation. Indeed, the government supported a diverse array of development projects in the Red Corridor, the region of India affected by Naxalite operations, in 2016. One of the largest of these projects took place in Odisha, where the state government partnered with the Asian Development Bank to finance a major road-construction project in Naxalite-affected districts. In another, the Bihar security forces began helping villages in Jamui and Munger districts build poultry farms. The district governments of Dantewada and Gadchiroli, in Chhattisgarh and Maharashtra respectively, began administering community-based agriculture and eco-tourism projects to boost employment and economic output. Anti-Naxalite forces also initiated community-outreach programmes, such as those in which they shared well water at their outposts with nearby villages.

It remained unclear whether development projects – as opposed to government successes on the battlefield, the availability of amnesty packages or other factors – were turning civilians away from the insurgency. However, it was increasingly apparent that many local communities were turning against the CPI–Maoist. More than 2,000 people in excess of a dozen 'liberated' villages, or communities that once recognised the legitimacy of the CPI–Maoist and submitted to its authority, cut their ties to the group in 2016. The changes were especially clear in Malkangiri district, in Odisha. These villages formally accepted the authority of the state in public ceremonies, during which all residents, including armed former fighters, appeared before police to surrender and accept rehabilitation packages. It was possible that the villagers participated in the ceremonies for the sole purpose of receiving the cash disbursements included in the rehabilitation packages.

However, the Maoists' targeted killing of civilians suspected of collaborating with the state in the Red Corridor continued throughout the year, likely undermining popular support for the insurgency. Thus, because publicly surrendering to the security forces carried with it the risk of retaliation, many of the villagers who surrendered probably did so sincerely.

Several villages took even more drastic steps than mass surrender. In one instance, the residents of Nama village, in Chhattisgarh's Sukma district, armed themselves with traditional weapons and carried out night-time patrols to deter Maoists from entering their territory. Several neighbouring villages joined the effort, prompting the Maoists to kill one of their leaders in November. In an even more drastic act of resistance, villagers in Malkangiri district gave poisoned food to a passing squad of Maoists, leading to the death of one of the militants. And, in Gadchiroli district, a local non-governmental organisation (NGO) used donated materials and volunteer labour to build a bridge over the Jui River, connecting more than a dozen largely inaccessible villages to the relatively developed Bhamragarh area. The organisation proceeded with the project despite threats from local Maoists, who worried that the bridge would make it easier for the security forces to reach the villages. The successful completion of the project in December suggested that there was strong local demand for development, at least in some areas. This lent credence to Indian commanders' assessment that the development approach successfully undermined civilian support for the CPI–Maoist, thereby weakening the insurgency.

Insurgent losses on the battlefield

Federal and state security forces clashed with CPI–Maoist militants in more than 100 separate engagements, primarily in the states of Chhattisgarh, Odisha, Andhra Pradesh and Jharkhand. These skirmishes, which occurred almost exclusively in small villages and remote forested areas, increased the annual death toll among government soldiers (by 24%) and militants (by 91%). By the end of 2016, the security forces had killed at least 193 militants while sustaining 57 fatalities. This reflected a significant increase in the battlefield effectiveness of government forces, as militants had died at approximately twice the rate of security personnel in such clashes in 2015 (the first year in which government forces inflicted substantially more losses on their opponents than they incurred since 2006, when the CPI–Maoist formed). The trend marked a striking reversal of fortune: the fatality ratio was around 2:1 in the CPI–Maoist's favour as recently as 2012.

The most important battle between the government and the CPI–Maoist occurred in late October, when security forces based in Andhra Pradesh raided a Maoist meeting near Balimela reservoir, in Malkangiri district. Most high-ranking leaders of the Andhra–Odisha Border Special Zonal Committee (AOBSZC), one of the CPI–Maoist's most powerful organisations, attended the meeting – and at least 13 of them died in the operation. Bakuri Venkata Ramana, the last high-ranking tribal member of the CPI–Maoist, was among those killed. His death seemed likely to weaken relations between the CPI–Maoist and tribal communities of Andhra Pradesh and Odisha, making it more difficult for Maoists to recruit and conduct operations in these areas. The security forces also believed that Ramakrishna – who led the AOBSZC until at least August 2016 – attended the meeting, and may have been wounded when he fled. The death or capture of so many insurgent leaders almost certainly impeded the CPI–Maoist's activities in Andhra Pradesh and Odisha in autumn 2016, and appeared likely to disrupt its command and control structures in the long term.

The fallout from the raid contributed to an apparent crisis of leadership in the CPI–Maoist. Kudumula Venkata Rao, the highest-ranking leader of the People's Liberation Guerrilla Army, the CPI–Maoist's armed wing, died of severe jaundice in April. In August, the insurgent group reportedly replaced Ramakrishna with Gajarla Ravi as head of the AOBSZC. Although Ramakrishna led a Maoist rally in Malkangiri district in early October (which several thousand villagers were forced to attend), it was unclear whether he was acting on his own or under Ravi's authority. If the former, Ramakrishna may have created a major split in the CPI–Maoist. Moreover, there were also reports that the group experienced leadership problems outside Odisha and Andhra Pradesh. One former Maoist leader claimed in late October that some local commanders in Jharkhand had rejected the authority of Sudhakaran, a high-level commander sent to the state from Andhra Pradesh by the Maoist central committee. Sudhakaran had been redeployed to Jharkhand to reorganise the leadership structure there.

Alongside the tension within the leadership, the CPI–Maoist was also contending with a sharp decline in morale. At least 2,281 Maoist supporters or militia fighters surrendered to the authorities, often in large groups, in 2016, accelerating a trend that had begun the previous year. Although most of the surrenders occurred in Chhattisgarh and Odisha, militants also gave up their arms in Jharkhand, Telangana, Andhra Pradesh and Maharashtra. In addition to civilian sympathisers and low-ranking fighters, more than a dozen commanders, several of whom were key leaders in their respective areas of operation, also surrendered. Dinesh, a high-ranking commander in the CPI–Maoist's Bihar–Jharkhand–North Chhattisgarh Special Area Committee, surrendered to police in Jharkhand's Latehar district in mid-April. In June, one of the CPI–Maoist's top leaders in the Andhra–Odisha border region surrendered in Visakhapatnam district, while the senior leader of the group's Rayalaseema committee surrendered in Kurnool district. These two commanders cited rampant sickness, especially among female insurgents such as themselves, as a major factor in their decision to quit. Coupled with thousands of desertions, these developments almost certainly eroded the Maoists' ability to control territory or challenge the security forces in battle.

Nonetheless, the CPI–Maoist continued to pose a threat to civilians, killing at least 83 suspected police collaborators, or others who voiced opposition to the Naxalite movement, in 2016. Although this was less than half the number in 2012 – the year in which security forces were least effective at combating Maoists on the battlefield – it was nearly unchanged from 2014 and 2015, when the CPI–Maoist killed at least 84 and 82 civilians respectively. It appeared that the government's increasing effectiveness on the battlefield had not yielded proportional improvements in civilian safety. This was particularly apparent in December, when the Border Security Force failed to prevent a Maoist raid on Kudmulgumma village, which had renounced the CPI–Maoist.

Alleged human-rights abuses by the security forces

As in previous years, federal and state security forces were implicated in a series of human-rights abuses in states affected by the conflict. Some of these incidents likely began as accidents – as seen in February, when a unit of Greyhounds, Andhra Pradesh's elite anti-Naxalite security forces, killed two young men by firing on a traditional hunting party near Vakalapudi village, in Visakhapatnam district. While the Greyhounds probably believed they were attacking Maoists, it subsequently emerged that the hunters had no affiliation with any Naxalite group and were only armed with bows, arrows and other traditional weapons. Despite overwhelming evidence to the contrary, Greyhounds officials continued to portray the incident as an exchange of fire between the security forces and armed Maoists, suggesting an attempt to cover up the killings to avoid legal repercussions.

The security forces were also accused of deliberate acts of violence against civilians throughout the year. Some of these abuses may have been politically motivated, especially in districts in which new mining licences had been granted despite fierce opposition from tribal residents. In March and April, representatives of the Dongria Kondh tribe claimed that the police used arbitrary detention, torture and intimidation to silence opponents of mining projects near the Andhra Pradesh–Odisha border.

In April, police officers illegally detained and beat five suspected Maoists in Gadchiroli district. Two months later, another villager in the district accused the police of illegally detaining and torturing him, before forcing him to don a Maoist uniform. There was also evidence that police officers raped and murdered Madkam Hidme – a young woman from the Gond tribe – before portraying her as a militant whom they killed in an exchange of fire in Sukma district.

Moreover, the security forces allegedly harassed activists and politicians investigating human-rights violations. For example, Bela Bhatia, a human-rights activist and professor of the Tata Institute of Social Sciences, in Mumbai, claimed that police had harassed her as she helped to document incidences of gang rape committed by security forces in Sukma and Bastar districts in 2015. Soni Sori, another activist and the leader of the Aam Aadmi Party, claimed that the police illegally detained and tortured members of her retinue in Sukma district in May, aiming to extract forced confessions that would incriminate her as a Maoist sympathiser. In December, the police arrested the members of a fact-finding team (comprising lawyers, students, journalists and university professors) attempting to investigate human-rights abuse in Chhattisgarh. Civilians in the Red Corridor often demanded that the security forces be subjected to greater scrutiny and accountability. Meanwhile, local NGOs were instrumental in pressuring the federal and state governments to launch inquiries into incidents such as that involving the Greyhounds in Andhra Pradesh and an attack on five members of the same family in Odisha. In October, two men appeared before the High Court of Chhattisgarh to submit a complaint that the police had executed their teenage sons near Garda village, in Dantewada district. The alleged rape and execution of Madkam Hidme in Chhattisgarh led to protests by activists outside the Red Corridor, who picketed Chhattisgarh Tourism Board offices in Kolkata, in West Bengal. Civilians also demanded that the government investigate historical human-rights violations. For instance, in November, one woman pressed the High Court of Kerala to open an investigation into the February 2015 death of her husband, a photographer, in Palakkad district, after submitting forensic evidence that the Thunderbolts, Kerala's anti-Naxalite police force, executed him. The judiciaries of both the federal and state governments had often rewarded such demands by opening official investigations. In October, the Central Bureau of Investigation indicted seven police officers from Chhattisgarh State for setting fire to 160 houses in several villages in Dantewada district in March 2011. Such cases marked an important departure from the status quo, and had the potential to affect the dynamics of the conflict in the long term. Prosecutions of security personnel had theretofore been extremely rare.

Such abuses threatened to undermine the authorities' counter-insurgency strategy by losing the battle for public support. Although there was little evidence that this was happening yet – for instance, tribal communities in Odisha actively defied Maoist orders while simultaneously protesting against alleged abuses by the security forces – it was possible that the authorities would eventually exhaust civilians' patience. Given its proven capacity to gain recruits and support by exploiting popular grievances against the state, the CPI–Maoist may use such human-rights violations to recapture the allegiance of erstwhile supporters.

India (Manipur)

Manipur experienced significant civil and political unrest throughout 2016, creating an environment favourable to the operations of non-state armed groups. There was an increasingly tense relationship between valley politics, centred on ethnic Meiteis, and hill politics, revolving around tribal communities such as the Kukis and the Nagas. This was partly due to the political manoeuvring of the Meitei-dominated state government, which came under pressure from its main constituencies and civil-society organisations in the hill districts. By the end of the year, the dynamic appeared to pose a significant threat to the peace process between the federal government and the National Socialist Council of Nagalim–Isak Muivah (NSCN–IM), generating friction between New Delhi and the Manipur authorities.

Key statistics	2015	2016
Conflict intensity:	Low	Low
Fatalities:	100	45
New IDPs:		
New refugees:		

Escalating inter-communal tensions

Protests in favour of implementing the Inner Line Permit system remained a prominent feature of Manipur's political landscape, as did changes to land-ownership laws. Activist organisations such as the Joint Committee on Inner Line Permit System (JCILPS) – which primarily draws its support from the Meitei community – advocated the adoption of the bills, which would restrict migration to the state by both Indian and foreign nationals. Members of the JCILPS continued to pressure the state government through vigilantism: the group and its affiliates apprehended 375 non-local labourers in July alone. Some of the demonstrations turned violent – as seen in September, when several police officers and 19 pro-JCILPS student protesters were injured in and around Imphal. Tribal communities in the hill districts remained wary of attempts to alter land-ownership legislation, seeing them as designed to reduce their rights in these areas. Consequently, the JCILPS and the Joint Action Committee against Anti-Tribal Bills (JACAATB) launched *bandhs* (strikes), economic blockades and counter-protests in the valley and hill districts. These efforts intensified on 10–20 June, as JCILPS imposed an economic blockade on the hill districts – leaving 400 vehicles stranded – and the JACAATB imposed a counter-blockade on the valleys. On 31 August, the United Naga Council (UNC) condemned the state government for its perceived apathy towards Naga and tribal populations in Manipur, largely maintaining the dynamics of the political unrest there in 2015, which revolved around the same issues.

Inter-communal tensions, long a feature of Manipur's fragile politics, grew worse in late 2016, in the run-up to state-assembly elections scheduled for spring the following year. Chief Minister Okram Ibobi Singh declared on 21 October that no part of Manipur could be integrated into 'Greater Nagalim' – an entity proposed by the NSCN–IM as part of the peace process, which would incorporate Naga-dominated territories across Assam, Nagaland and Arunachal Pradesh. The group reacted violently three days later, firing on Singh's helicopter at a helipad in Ukhrul district, one of the proposed areas of Greater Nagalim. The state government subsequently announced that it would create the new administrative districts of Sadar Hills and Jiribam, despite condemnation from Naga civil-society organisations and militant groups such as the NSCN–IM. Ultimately, this provoked the UNC to launch on 1 November a crippling, indefinite economic blockade on the valleys of Manipur. The move severely limited access to commodities and fuel in the state, provoking counter-blockades in the valleys that led to looting and violent attacks on vehicles travelling towards the hill districts. The state government responded by creating on 8 December seven new administrative districts, many of which bifurcated what the UNC and other Naga political organisations considered to be their ancestral lands. The move hardened the resolve of the groups maintaining the economic blockade.

Decline in fatalities

Despite the deteriorating political situation, there was a considerable decline in fatalities, with around 45 conflict-related deaths in 2016, compared to 100 in 2015 and 55 in 2014. Among the 27 fatalities that occurred in the second quarter of 2016, 14 occurred during and in response to a 22 May ambush on an Assam Rifles patrol in Hengshi, in Chandel district. There, militants operating under the banner of umbrella organisation the Coordination Committee killed six Assam Rifles personnel. This was the deadliest incident in northeast India since June the previous year, when the National Socialist Council of Nagalim–Khaplang (NSCN–K) launched an attack in the same district. Bordering Myanmar, Chandel district is, like many areas of Manipur, acutely vulnerable to cross-border militant operations. In line with its established approach, the Indian military reportedly responded to the 22 May attack by conducting a cross-border operation against the Coordination Committee, killing eight members of the group in Myanmar. Although Indian officials did not confirm these reports due to the sovereignty issues they might raise, the incident suggested that retaliatory cross-border strikes remained a key part of the security forces' approach to combating militancy in the northeast.

Manipur's authorities were also forced to contend with militant activity in Naga-dominated areas adjacent to Nagaland State. This was particularly problematic given that the NSCN–IM's ceasefire arrangement with the Nagaland government did not formally prohibit attacks in Manipur. The group reportedly killed a civilian in Imphal on 1 March, a militant from rival group the Zeliangrong United Front on 13 September and another civilian, accused of assisting the NSCN–K, in Tamenglong district on 29 November. The escalating tension between Manipur's valley and hill communities appeared to prompt the

NSCN–IM to conduct a series of limited, albeit politically significant, attacks in the state. The first of these came on 24 October, with the assault attack on Chief Minister Singh's helicopter. On 15 December, the group killed three police officers in two separate attacks, and carried out a raid on an India Reserve Battalion post. Staged in the newly created Tengnoupal and Noney districts respectively, the assaults demonstrated the NSCN–IM's willingness to go beyond clashes with rival armed groups by directly attacking the security forces.

Threat to the Naga peace process
The Manipur government attempted to placate the JCILPS by discussing the introduction of an Inner Line Permit system throughout the year, but it did so without taking substantive action on the issue. Combined with the creation of seven new administrative districts on 8 December, this compounded hostility towards the government among Nagas, including civilian political activists and members of non-state armed groups. Facing a significant economic crisis in Manipur, New Delhi deployed in November and December 17,500 federal paramilitary personnel to alleviate the blockade on major transport arteries. On 27 December, the federal government urged the Manipur authorities to utilise these forces properly, accusing the state's leaders of failing to uphold their constitutional responsibilities amid what it described as a 'humanitarian crisis'. The following day, the NSCN–IM condemned the deployment of paramilitary personnel, arguing that the move had allowed the state government to undermine trust in the peace process. Meanwhile, political organisations based in the valley accused the UNC of working to fulfil the interests of the NSCN–IM. The growing dispute indicated that the highly secretive peace process between the federal government and the NSCN–IM had failed to address the competing political interests of both Nagas and valley-based political groups in Manipur. The resulting hostility between the parties threatened to exacerbate tensions created by the 2017 elections, presenting challenges to the peace process that could have serious ramifications for not only Manipur, but also Nagaland, Assam and Arunachal Pradesh.

Despite the decline in conflict-related fatalities, civilians in Manipur remained vulnerable to chronic security threats, particularly violent extortion by militants. Labourers and other civilians were kidnapped and often killed: a truck driver abducted in Ukhrul district on 3 April was found dead two weeks later, while on 21 December unidentified assailants kidnapped a project manager in Noney district. Militants frequently issued threats against organisations such as Imphal's Regional Institute of Medical Sciences and Jawaharlal Nehru Institute for Medical Sciences – typically by placing grenades in or around their residences or facilities. Armed groups also attempted to intimidate civil-society organisations: on 26 December, members of the Kangleipak Communist Party placed a grenade outside a newspaper office in Wahengbam Leikai, in Imphal West district.

On 26 July, human-rights activist Irom Sharmila ended her 16-year hunger strike against the imposition of the Armed Forces (Special Powers) Act in Manipur. Although the federal government confirmed on 3 August that all districts in the state, barring the Imphal municipal area, would continue to be classified as 'disturbed', Sharmila vowed to maintain her opposition to the legislation by running in the state elections as part of newly formed political party the People's Resurgence and Justice Alliance. However, given the ongoing imposition of the act and the increased deployment of security forces in Manipur, human-rights movements such as Sharmila's faced substantial challenges in their attempts to influence mainstream politics.

India (Nagaland)

Throughout 2016, developments in various peace processes between the Indian government and Naga non-state armed groups, particularly the National Socialist Council of Nagalim–Isak Muivah (NSCN–IM), shaped the conflict in Nagaland. Although the organisation appeared to reach the final stages of a settlement with the government in August 2015, the sides did not conclude the process the following year. By the end of 2016, the prospective settlement appeared to be threatened by escalating political tension between Nagas residing in the hill districts of Nagaland and Meiteis living in the valleys of neighbouring Manipur.

Key statistics	2015	2016
Conflict intensity:	Low	Low
Fatalities:	100	20
New IDPs:		
New refugees:		

Due to the NSCN–IM's strong presence in the hill districts of Manipur, and its status as a non-signatory to the ceasefire in the state, relations between the group, the central government and the state government became increasingly complicated. While there were few major armed clashes between Naga militants and the security forces, and few direct conflict fatalities, in Nagaland, all Naga armed groups maintained complex extortion and recruitment networks that fuelled rivalries between them in the state and beyond – a dynamic that had a negative impact on human security.

Declining violence

Despite abrogating its 14-year ceasefire with the government in April 2015, the National Socialist Council of Nagalim–Khaplang (NSCN–K) was unable to cause a significant number of casualties in 2016. This was reflected in the drop in total fatalities related to the Naga conflict, from 100 in 2015 to 20 in 2016 (only five more than in 2014). Although the group frequently clashed with army and Assam Rifles personnel, it often did so when forced on to the defensive by counter-insurgency operations. The army killed three NSCN–K militants in Lohit district, in Arunachal Pradesh, on 16 February; two days later, the Assam Rifles killed another member of the group in Nagaland's Dimapur district. Six people were injured in August and September as NSCN–K fighters clashed with security-forces personnel in Mon and Zunheboto districts, in Nagaland, as well as Changlang district, in Arunachal Pradesh. The injury of four members of the Assam Rifles in the Zunheboto attack prompted local villagers to declare that they would not cooperate with the militant group, reflecting its increasing isolation. The NSCN–K's only military successes came in December, when it ambushed and killed two Assam Rifles personnel in Tirap district, in Arunachal Pradesh.

Almost half of the conflict's fatalities in 2016 occurred in the state, as it increasingly became a site of contestation between various Naga groups seeking to assert control of the Naga population there. Due to the area's isolated terrain and distance from Nagaland – where ceasefires are policed by the Ceasefire Monitoring Group – the security forces staged operations there against Naga armed groups seeking to expand their influence. On 7 July, just three months after the National Socialist Council of Nagalim–Reformation (NSCN–R) renewed its ceasefire with the government, Assam Rifles personnel killed four militants from the group in Tirap district. The activities of the NSCN–IM led to another four fatalities in Manipur State in 2016. In one of these incidents, the group reportedly killed a civilian in Sibilong, in Tamenglong district, for assisting the NSCN–K.

Troubled peace process

There were positive signs for the prospective political settlement between the central government and the NSCN–IM in early 2016, as the parties' previously informal meetings and bi-weekly dialogue were effectively institutionalised on 20 January. Furthermore, members of other armed groups increasingly flocked to the NSCN–IM: on 13 March, Khole Konyak, president of the National Socialist Council of Nagalim–Khole Kitovi – which had merged in 2011 with the remnants of the Unification faction of the council (NSCN–U) to form the NSCN–U – defected to the NSCN–IM with a large contingent of followers. He took part in peace talks in New Delhi two days later. More grounds for optimism came with the revelation that, despite this defection, the NSCN–U had renewed its ceasefire with the government on 18 April, as had the NSCN–R.

There were fears that the death from illness of Isak Chisi Swu, a founding member of the NSCN–IM and its general secretary, on 28 June would spark a power struggle among tribes within the group. Instead, his death led to an outpouring of respect and solidarity that transcended political boundaries. The federal and

Nagaland governments quickly vowed to continue the peace process and build on Swu's work, while the NSCN–R suggested on 3 July that Nagaland had achieved unity in grief, and that the group had no quarrel with the NSCN–IM. However, these announcements resulted in little substantive progress. Throughout the year, various parties to the conflict stated that a solution would be thrashed out 'soon', but on 28 July Nagaland Chief Minister T.R. Zeliang hinted that the peace process was being delayed by divisions between Naga organisations. Tensions between the groups persisted as they attacked and detained militants from rival organisations, while engaging in intensifying competition for control of Tirap, Changlang and Longding districts. There was a clash between NSCN–IM and NSCN–K fighters in Tirap district on 15 April, while on 7 May the NSCN–U allegedly attacked an NSCN–R camp in Nagaland. On 27 June, the newly founded Eastern Naga National Government condemned the NSCN–IM's extortion activities in the state. Large-scale defections between the groups continued: 54 members of the NSCN–R defected to the NSCN–IM on 8 August, while 77 militants from various organisations defected to the NSCN–U on 1 July. The NSCN–R, the NSCN–U and four smaller factions of the Naga National Council formed a working group that conspicuously lacked representatives of the NSCN–IM or the NSCN–K. This suggested that, far from moving towards unity, Naga political groups had instead begun to cluster in rival power centres.

Tension between Naga political organisations and Manipur's Meitei-dominated government further dampened optimism about the Naga peace process. The United Naga Council (UNC) imposed an economic blockade on Manipur on 1 November, in response to the state government's proposal to create two new districts that would bifurcate what it deemed to be Naga 'ancestral lands'. The UNC's resolve only hardened on 8 December, when the Manipur government reacted to the blockade by creating seven new administrative districts that cut into these territories. Such moves are at odds with the NSCN–IM's desire to establish 'Greater Nagalim', which would incorporate Naga-dominated territories in Manipur, Assam and Arunachal Pradesh as part of a proposed political settlement.

The group stepped up its military activities on 24 October, when it was reportedly involved in an armed attack targeting Manipur Chief Minister Okram Ibobi Singh in Ukhrul district. It has also been implicated in a series of incidents on 15 December, which included a raid on an India Reserve Battalion post in Noney district and two separate skirmishes that together resulted in the deaths of three police officers in Tengnoupal district. New Delhi responded to the political turmoil by deploying federal paramilitary forces to Manipur; by 26 December, the number of paramilitary personnel deployed there had reached 17,500. This led the NSCN–IM to question the sincerity of the central government's participation in the peace process. Two weeks later, the group added that it would defend its ancestral lands 'at all costs'. These events formed part of a pattern in which the NSCN–IM appeared to indulge in a series of low-intensity but politically significant acts of violence, presenting a serious challenge to policymakers working to reduce political tension in Manipur and secure peace with the group.

Despite the overall reduction in violence, there remained a lack of law and order in most areas in which Naga armed groups operated. On 12 September, the National Investigation Agency reported that the NSCN–K had intensified its extortion operations, imposing a 24% 'tax' on salaries and funds across 24 state-government departments. However, civil-society organisation Against Corruption and Unabated Taxation continued its 'One Government, One Tax' drive against extortion by armed groups, launching on 21 August a campaign against the NSCN–IM's Wokha district commander. Indeed, the organisation drew the ire of the NSCN–IM on several occasions. Despite this increased pressure, Naga armed groups largely retained control of extortion in their respective territories across Nagaland throughout 2016.

The competing groups found a significant number of opportunities to extort civilians in Tirap, Changlang and Longding districts. On 21 October, the Arunachal Naga Students Federation urged the NSCN–IM to refrain from such activities there, referring to the deleterious impact of extortion payments on educational institutions. Beyond extortion, it was reported in August that the

NSCN–IM recruited five or six child soldiers from Naga-inhabited districts in the state each year, demonstrating the conflict's wider effect on civilians.

India–Pakistan (Kashmir)

Indian-administered Jammu and Kashmir experienced an upsurge in militant attacks and violent protests in 2016, as relations between India and Pakistan became increasingly antagonistic. Indian Prime Minister Narendra Modi's government gained

Key statistics	2015	2016
Conflict intensity:	Low	Medium
Fatalities:	200	400
New IDPs:		
New refugees:		

control of the deteriorating security environment in the state, but at a high cost in casualties among civilians and military personnel. Pakistan-backed militant groups such as Lashkar-e-Taiba, Jaysh-e-Mohammad and Hizbul Mujahideen took advantage of the charged security climate to step up their attacks on Indian security forces, sustaining pressure on the Indian government and drawing international attention to Pakistan's territorial claims in Kashmir. Both New Delhi and Islamabad used international forums to articulate their claims on the area and denounce each other, resulting in some of their most forceful statements on the dispute for many years.

Militant attack in Pathankot

The decline in New Delhi's relations with Islamabad intensified on 2 January 2016, when militants attacked the Indian Air Force base at Pathankot, in Punjab, around 50 kilometres from the border with Pakistan. The assault killed seven Indian security personnel, one civilian and four militants. It surprised many observers as, on 25 December 2015, Prime Minister Modi had unexpectedly met with his Pakistani counterpart, Nawaz Sharif, in Lahore. This perceived diplomatic breakthrough seemed to signify mutual willingness to make progress in a comprehensive bilateral dialogue on Kashmir. In the aftermath of the Pathankot attack, the sides indefinitely suspended talks between their foreign secretaries planned for mid-January. Nonetheless, they maintained back-channel contact with each other through their national-security advisers. Although Sharif stated on 5 February that he was willing to resume formal talks soon, India maintained its position that the discussions would only begin again after Pakistan took concerted action against the perpetrators of the assault. New Delhi alleged that Jaysh-e-Mohammad, led by Maulana Masood Azhar, was behind the attack and, in an unprecedented move, granted a five-member Joint Investigation Team from Pakistan access to the air base. After the Pakistani authorities rejected a reciprocal request and prevented Indian investigators from interviewing Azhar, relations between the sides deteriorated in a manner that made it difficult for them to restart the comprehensive bilateral dialogue. The Pakistani military appeared to value groups such as Jaysh-e-Mohammad for their capacity to carry out attacks on important targets, a capability that was useful for signalling purposes. In this context, the Pathankot operation was likely designed to end the dialogue, and the normalisation of ties, between India and Pakistan.

The countries' foreign secretaries met in New Delhi on 26 April, but made little progress in resolving the dispute. Reportedly, India pressed Pakistan to take action against Azhar while Pakistan raised the issue of India's alleged interference in Balochistan. The Pakistani delegation was said to have highlighted the arrest in the province of Kulbhushan Yadav, a former Indian naval officer who Islamabad accused of being an operative of India's foreign-intelligence agency, the Research and Analysis Wing. Pakistan accused Yadav and the agency of funding and training Baloch separatists, a charge that India strongly denied.

As the Indian government made steady progress in its diplomatic efforts to corner Azhar, there were reports that Pakistan was actively shielding him. On 3 May, the *Indian Express* published claims by Indian intelligence officers that Pakistan's clandestine intelligence agency, Inter-Services Intelligence, had moved Azhar and Jaysh-e-Mohammad's base from Maujgarh Fort to the town of Bahawalpur, in Punjab province (the location of the group's madrassa). As Bahawalpur was considered to be a Jaysh-e-Mohammad stronghold under the protection of Inter-Services Intelligence, the move made it harder for India to gain intelligence on his activities. Indian officials suspected that Jaysh-e-Mohammad militants had been trained to launch cross-border attacks at Maujgarh Fort. On 17 May, Interpol issued a red notice for Azhar and his brother, Abdul Rauf Asghar, in connection with the attack in Pathankot, after India's National Investigation Agency (NIA) provided an Indian court with evidence of a telephone conversation linking them to the operation. Pakistan did not act on the notice or others previously issued by Interpol. Yet on 18 May, Punjab's provincial law minister, Rana Sanaullah, stated that the Pakistani government had been involved with some terrorist groups and therefore could not take action against them.

The following week, the *Hindu* reported that Indian forensic teams had confirmed that statements signed by the Afzal Guru faction of Jaysh-e-Mohammad claiming responsibility for the Pathankot attack were genuine. Found in a stolen police vehicle used by the attackers, as well as on the body of one of them, the documents claimed that the attack had been carried out in response to India's execution of Guru, a Kashmiri man convicted of planning the December 2001 assault on the Indian parliament. However, there was reason to doubt the claim: the Pathankot attack took place almost three years after Guru's execution but shortly after Modi's visit to Lahore.

By December, the NIA had gathered sufficient evidence to file a charge sheet against Jaysh-e-Mohammad at a court in the city of Mohali, in Punjab. The NIA charge sheet presented evidence on the role of Azhar and Asghar, as well Shahid Latif and Kashif Jan (also based in Pakistan), in the Pathankot attack. Pakistan did not attempt to hold the perpetrators of the Pathankot attack to account, leaving Azhar and other militant leaders to become increasingly vocal in their threats against Indian security forces in Jammu and Kashmir.

Deteriorating India–Pakistan relations

Following the death of Hizbul Mujahideen commander Burhan Muzaffar Wani in a battle with Indian security forces on 8 July, Kashmir experienced protests that were larger and more violent than any since summer 2010. The vast civil unrest affected several districts in the Kashmir Valley and Jammu region until the arrival of winter in December. As the Indian state struggled to restore order, Pakistan used the opportunity to demand 'freedom' for Kashmiris, essentially ending attempts to resume the comprehensive bilateral dialogue.

Almost daily exchanges of artillery and small-arms fire across the Line of Control, along with escalating rhetoric from New Delhi and Islamabad, continued to undermine the 2003 ceasefire agreement between the sides. The situation grew worse on 18 September, when Jaysh-e-Mohammad militants killed 19 Indian soldiers at an army base in Uri, west of the Jammu and Kashmir city of Srinagar. Having warned that it would retaliate at the time and place of its choosing, New Delhi announced on 29 September that Indian special forces had launched what it called a 'surgical strike', killing terrorists at their bases on the other side of the Line of Control. Pakistan denied that such an assault had taken place; perhaps concerned about the possibility of further escalation, neither the Indian military nor the government produced any details to validate their claims. A BBC News investigation found on 23 October that Pakistani police officials had privately conceded that some form of a covert ground assault took place.

The rhetoric of Modi and Sharif reflected the growing tension between the sides. On 27 May, Modi stated that 'Pakistan's failure to take effective action in punishing the perpetrators of terror attacks limits the forward progress in [India–Pakistan] ties'. Similarly, on a state visit to the United States in June, he told a joint sitting of Congress that terrorism was being 'incubated in India's neighbourhood'. Modi added that global counter-terrorism cooperation should be based on a policy that 'isolates those who harbour, support and sponsor terrorists' – a reference to Pakistan.

Pakistan highlighted India's attempts to suppress protests in Jammu and Kashmir with the aim of calling for a plebiscite on independence for the state. Sharif established his position in the crisis by immediately condemning Wani's killing. Taking a similar line, Pakistan's Ministry of Foreign Affairs described the incident as an extrajudicial killing. On 12 July, Aizaz Chaudhry, then Pakistani foreign secretary, raised the upsurge in violence in Jammu and Kashmir with the ambassadors of the permanent members of the UN Security Council. He urged them to diplomatically pressure India to demilitarise Kashmir, allow international observers to report on human-rights abuses there and accept international mediation on the conflict. He also condemned the loss of civilian life in Kashmir, alleging that the Indian security forces were responsible for killing innocent people. On 15 July, Sharif referred to Wani as a 'martyr' and declared 19 July as a 'Black Day' to express solidarity with the people of Jammu and Kashmir. The statement drew a sharp response from the Indian Ministry of External Affairs, which criticised Pakistan for continuing to glorify members of internationally proscribed terrorist organisations. In another effort to internationalise the dispute, Pakistan urged the UN Human Rights Council to investigate alleged human-rights violations by Indian security forces in Jammu and Kashmir. On 5 August, Sartaj Aziz, Sharif's adviser on foreign affairs, asked Médecins Sans Frontières to provide assistance to protesters injured in clashes with Indian security forces. In another push, Prime Minister Sharif wrote to Ban Ki-moon, then UN secretary-general, and the UN High Commissioner for Human Rights to ask them to end human-rights violations in Kashmir.

Modi used similar tactics to draw attention to human-rights violations in Pakistan. His allusions and direct references to alleged human-rights abuses in Balochistan by the Pakistani security forces signalled a departure from the approach of his predecessors, who had rarely mentioned the issue. On 13 August, Modi stated that these alleged violations needed to be highlighted to the international community. Two days later, in a speech to mark India's 70th Independence Day, Modi again raised the issue of human-rights violations in Balochistan, drawing a swift rebuke from Islamabad. A day earlier, Abdul

Basit, Pakistan's high commissioner to India, had dedicated his country's Independence Day to the 'freedom of Kashmir'.

This cycle of condemnation and response continued for the remainder of the year. Although it was difficult to say whether India or Pakistan gained the upper hand diplomatically, the significant dangers of escalation in Kashmir came into focus in September, when Pakistani Minister of Defence Khawaja Asif stated that his country was open to using nuclear weapons against India in the event of war.

Unrest in Indian-administered Kashmir

Militant attacks in Jammu and Kashmir intensified as India–Pakistan relations grew more hostile, particularly after the death of Wani. On 24 June, the *Daily Excelsior* reported claims by the Indian Army that there were approximately 220 active militants – around 100 of them locals – in the Kashmir Valley. The report also quoted a Multi Agency Centre estimate that 36 militants infiltrated India by crossing the Line of Control between January and May 2016, compared to 35 in all of 2015. Separately, the Indian Army reported that Pakistan contained more than 40 camps in which militants received specialised training from the Pakistani military in navigation, armed combat and other skills – most of them in Pakistan-administered Kashmir.

Anticipating protests in the wake of Wani's death, the Jammu and Kashmir authorities took precautionary measures such as suspending mobile-phone and internet services, delaying the annual Hindu Amarnath Yatra pilgrimage and cancelling train services between Baramulla and Banihal, in the Kashmir Valley and Jammu respectively. The authorities also placed separatist leaders Syed Ali Shah Geelani and Mirwaiz Umar Farooq under house arrest for inciting protesters to breach the recently imposed curfew. Moreover, the security forces arrested Jammu Kashmir Liberation Front chairman Mohammad Yasin Malik for planning to carry out a protest march.

As news of Wani's death spread, violent protests erupted immediately in the southern districts of the Kashmir Valley. The first 24 hours of violent clashes between protesters and Indian security forces claimed the lives of 15 civilians and injured nearly 200 other people, including 90 security personnel. The violence declined incrementally as the year wore on, with 48 fatalities in July, 19 in August, 13 in September, two in October and one in December. Many of these people died when stone-throwing protesters defied curfews and clashed with security personnel, who retaliated with lethal and non-lethal force to break up the demonstrations. As the clashes wore on, 22 policemen and 52 Indian Army soldiers died fighting militants attempting to cross the Line of Control. The Indian security forces killed 15 militants in July, 12 in August, 28 in September, ten in October, eight in November and five in December, indicating a correlation between militant-related activity and attempts at cross-border infiltration.

Militant groups such as Hizbul Mujahideen and Lashkar-e-Taiba issued statements threatening more violence against the Indian security forces in Kashmir as the hostilities continued. For instance, addressing an anti-India rally in Muzaffarabad on 11 August, Hizbul Mujahideen leader Syed Salahuddin warned that his followers would cross the Line of Control unless India ended its violence against Kashmiris. The following month, he asserted that he would turn Kashmir into a 'graveyard' for the Indian Army. On 27 September, Lashkar-e-Taiba warned Jammu and Kashmir Education Minister Naeem Akhtar against continuing to encourage demonstrators to return to work. On 10 August, the NIA broadcast a video of a 21-year-old man confessing that he had received three rounds of training at a Lashkar-e-Taiba camp and had been instructed to mingle with the local population to avoid suspicion, so that he could carry out attacks. As Lashkar-e-Taiba stepped up its threats, the group's charitable wings, Jamaat-ud-Dawa and the Falah-e-Insaniat Foundation, intensified their fundraising campaigns in Pakistan. Although Indian security forces had largely regained control of the restive districts in the Kashmir Valley by the end of the year, the Indian Ministry of Home Affairs made preparations for a winter deployment of additional units to manage the unrest.

Disrupted civilian life

The growing animosity between India and Pakistan threatened civilians living in Jammu and Kashmir. The curfew that the Indian authorities imposed on many districts in the Kashmir Valley caused significant disruption to the economy, education and social activity there. While the measure was suspended between 29 August and 2 September, it was quickly reimposed due to security concerns in most towns in the ten districts of the Kashmir Valley, including Badgam, Bandipora, Ganderbal and Srinagar. According to figures released on 25 September by Kashmir's business chambers in Indian-administered Kashmir, since the unrest began the area had experienced financial losses of more than US$1.4 billion (including an estimated US$450 million in tourism and US$140m in the fruit industry).

On 25 August, the *Hindu* published official data showing that 5,871 civilians had been injured in the first 45 days of the demonstrations. The security forces had wounded approximately 3,000 of these people with pellet ammunition, and another 122 with conventional bullets. Around 3,219 of the injuries occurred in four districts of southern Kashmir: Anantnag, Kulgam, Pulwama and Shopian. Violent confrontations with the security forces led to the injury of 831 civilians in Baramulla, 517 in Bandipora, 786 in Kupwara, 246 in Budgam, 126 in Ganderbal and 146 in Srinagar. An estimated 2,600 security personnel were wounded in the clashes. Furthermore, the Jammu and Kashmir police reported that mobs had been involved in 1,018 violent incidents, destroying 29 buildings and damaging 51 others.

For the first time since 1990 – when an uprising threatened Indian rule in Jammu and Kashmir – the authorities maintained the curfew on all ten districts in the Kashmir Valley during Eid al-Adha. Along with the shutdown of communications services in the area, the move prompted complaints from some members of the Jammu and Kashmir Peoples Democratic Party. One parliamentarian resigned from the party and the federal government in protest at the handling of the unrest in Kashmir. In another measure that drew widespread criticism and demonstrated the authorities' misreading of the situation, the state banned on 2 October the publication of local newspaper the *Kashmir Reader*, accusing its editors of inciting violence. Although the Kashmir Editors Guild protested against the ban the next day, stating that it went against the 'basic spirit of democracy as well as the freedom of the press', it remained in place until 28 December. The fact that the paper regained its publication rights at this point suggested a calculation that there was a reduced chance of protests during winter, implying that the ban might be reinstated if the demonstrations resumed in warmer weather.

Cross-border shelling displaced civilians on both sides of the Line of Control. The Indian authorities reported in October that 26,000 civilians had been displaced in the Jammu region due to an increase in such shelling, which they believed to be Pakistan's response to covert Indian operations in Pakistan-administered Kashmir. Due to the intensity of the shelling, the Jammu and Kashmir government temporarily closed 174 schools in Jammu district. Nonetheless, state officials reported that an estimated 98% of registered students sat exams in November, indicating that young Kashmiris had adapted to a dangerous environment.

The attack in Pathankot likely made India–Pakistan cooperation on Kashmir unworkable for the remainder of the year. Thereafter, the comprehensive bilateral dialogue remained at a standstill despite both sides' efforts to revive it. The death of Wani and the subsequent killing of 19 Indian soldiers all but guaranteed that relations between New Delhi and Islamabad would descend into further acrimony. The widespread unrest sparked by the former incident demonstrated especially clearly how the character of militancy in Indian-administered Jammu and Kashmir had changed. Although the Indian government dealt with the protests firmly, it was unable to address the political issues driving the violence. New Delhi bought itself time during the winter months, but needed to formulate a defined strategy to address the concerns of Kashmiris before spring 2017. If the government was unable to do this, there was a risk that the confrontation and loss of life would spiral further out of control, and that it would be forced to contend with greater international involvement in the conflict.

Pakistan*

Key statistics	2015	2016
Conflict intensity:	High	High
Fatalities:	3,000	1,250
New IDPs:		
New refugees:	8,500	

Over the course of 2016, Pakistan endured domestic political challenges, various crises involving its neighbours and large-scale terrorist attacks, as well as battles against militants in its tribal belt and separatists in Balochistan. Despite these pressures, the country maintained its democratic orientation. One of the Pakistani government's key priorities was to implement its counter-terrorism strategy, the National Action Plan (NAP). Another was to reach a political agreement on a merger between Khyber Pakhtunkhwa and the Federally Administered Tribal Areas (FATA). Although the government made progress in both of these efforts, Prime Minister Nawaz Sharif lost political capital when documents leaked from a Panamanian law firm revealed that his children owned undeclared offshore companies and assets. Opposition parties from across the spectrum, including Imran Khan's Pakistan Tehreek-e-Insaf and the Pakistan Peoples Party, demanded Sharif's resignation or called on the military to remove the government from office.

Implementation of the National Action Plan

The authorities continued to combat militancy and terrorism in urban and rural areas throughout Pakistan. An apparent improvement in domestic security indicated that the NAP was working to a certain extent. Nonetheless, high-casualty attacks on civilian targets became more commonplace. The majority of high-casualty attacks occurred in urban areas; a large number of them were sectarian in nature, while others were directed against the security forces. Tehrik-e-Taliban Pakistan (TTP) continued to target cities, an approach con-

---

* *The Armed Conflict Survey* classifies Pakistan's militancy in the northwest, insurgency in Balochistan and sectarian violence as a single conflict, thereby reducing the number of active conflicts assessed in the book to 36.

sistent with its main objective of weakening the Pakistani state as a means to impose a sharia system of governance.

Jamaat-ul-Ahrar, a faction of the TTP, claimed responsibility for a suicide bombing that targeted Christians celebrating Easter in Gulshan-e-Iqbal Park in Lahore, killing 70 people and wounding more than 300 others. Another suicide bomber struck the Civil Hospital in Quetta on 8 August, killing 74 people and injuring more than 100 others. Most victims of the attack were lawyers and journalists who had convened at the state-run hospital to protest against the targeted killing of the Balochistan Bar Association president, Bilal Anwar Kasi, who had been murdered earlier the same day. Both the Islamic State, also known as ISIS or ISIL, and Jamaat-ul-Ahrar claimed responsibility for the attacks. On 2 September, Jamaat-ul-Ahrar killed 14 people and injured 52 others in a suicide bombing outside a courthouse in Mardan district. On the same day, the group stated that it was behind the attack on a Christian neighbourhood on the outskirts of Peshawar known as Christian Colony. The security forces' quick response resulted in the deaths of four heavily armed terrorists. One private security guard died in the attack. On 24 October, ISIS stated that it had been behind an attack on Balochistan Police College in the Sariab area of Quetta that had killed at least 61 police cadets and guards, and injured more than 100 other people, most of them cadets. However, officials blamed the Al-Alimi faction of Lashkar-e-Jhangvi, an affiliate of the TTP, for the assault. In another attack claimed by ISIS, 52 civilians died and another 102 were injured by a bomb detonated at the Sufi shrine of Shah Norani in Balochistan's Khuzdar district.

The TTP predominantly targeted religious minorities. On 7 May, the group's Hakimullah Mehsud faction claimed responsibility for killing two Shia men, a religious scholar and a human-rights activist, in Karachi. The faction also stated that it had carried out the assassination of renowned Sufi singer Amjad Sabri after a concert in the city on 22 June, claiming that his music was 'blasphemous'. In early August, Jamaat-ul-Ahrar said it killed two Hazara men in Quetta. Many other members of minority groups fell victim to targeted killings in the year; often, no organisation claimed to have been behind the attack, but officials stated that the TTP or affiliated groups were responsible. The regularity of targeted killings and the authorities' apparent inability to stop them amplified fears among Pakistan's minority communities. The government also paid increasing attention to ISIS. Appearing before the Senate in February, the director-general of the Intelligence Bureau acknowledged that the group posed a serious threat to the country's security. He noted that organisations such as the TTP, Lashkar-e-Jhangvi and Sipah-e-Sahaba had a 'soft corner', and were recruiting, for ISIS. Pakistani officials doubted that there were operational links between ISIS in Pakistan and the group's leadership in Syria or Iraq. However, given that hundreds of Pakistanis (including entire families) had reportedly left the country to fight for ISIS, it seemed that the group was gaining traction.

The federal government and its provincial counterparts drew criticism for allegedly neglecting to implement the NAP in a coherent and rigorous manner. The lack of detail in the 20-point plan meant that the authorities applied some elements of it but ignored others; struggled to identify the implementation techniques for each point; and lacked substantive policy guidelines, as well as defined and targetable enemies. Another obstacle was federal and provincial leaders' inability to obtain the military's approval for countering terrorist organisations the security establishment regarded as strategic assets, such as Lashkar-e-Taiba and Jaysh-e-Mohammad. The patchy implementation of the NAP prompted the government of Balochistan to express concern that Islamabad was not providing sufficient resources for counter-terrorism operations. Simultaneously, there was uncertainty about the NAP's division of authority between Islamabad and provincial governments. The latter were reluctant to relinquish control over policing and law-enforcement activities to either the federal government or the military. Due to the ongoing tension between Islamabad and provincial governments, the military hinted that it was best placed to lead the implementation of the NAP, suggesting that the civilian leadership could play a supporting role. However, federal and provincial officials pushed back against the suggestion, hoping to prevent further military encroachment into civilian affairs.

The federal government took steps to speed up the implementation of the plan. In August, Prime Minister Sharif appointed Lieutenant-General (Retd) Nasser Khan Janjua, Pakistan's national security adviser, as head of a high-level committee of bureaucrats, military leaders and police officers that would monitor and evaluate the implementation of the NAP. In October, Sharif convened a meeting of all provincial chief ministers, the head of the army and the directors of the military and civilian intelligence agencies, among others, to address concerns about the plan. The Pakistani media reported that the chief ministers of Balochistan, Khyber Pakhtunkhwa and Sindh raised the lack of funding and material resources from the federal government, deficiencies in intelligence-sharing and -coordination, and the drawn-out process for obtaining approval of measures to enact the NAP.

Nonetheless, provincial governments had some success in implementing the NAP, despite its shortcomings. For instance, various local news outlets reported in early February that 182 madrassas, or religious seminaries, had been closed under the NAP, and that the State Bank of Pakistan had frozen 1 billion rupees (US$9.5 million) in accounts linked to militant organisations. The same month, the Pakistani military reported that it had arrested 97 TTP, al-Qaeda and Lashkar-e-Jhangvi militants in Karachi. The detainees included Farooq Bhatti, deputy chief of al-Qaeda in the Indian Subcontinent (AQIS), and those allegedly behind the June 2014 assault on Karachi airport. Meanwhile, Punjab's law minister announced that security forces had arrested 42 suspected ISIS sympathisers. Between April and June, the Balochistan government stated several times that it would register all madrassas operating in the province with the provincial education department. Concurrently, the Sindh police force announced that it would start tracking the finances of madrassas to identify those with links to militant groups. Officials from Khyber Pakhtunkhwa gathered information on 1,208 of the 3,208 madrassas in the province, stating that they were working with intelligence and law-enforcement agencies to track the location and attendance of the seminaries. A progress report released in July noted that since the NAP's adoption in December 2014, the state had executed 411 terrorism convicts – including Mumtaz Qadri, a bodyguard who had killed his employer, then

Punjab governor Salman Taseer, for attempting to reform Pakistan's blasphemy laws. The report also stated that the security forces had arrested 5,611 alleged terrorists, and carried out 4,230 intelligence operations and 122,722 search-and-clear operations, in the period.

Although provincial governments employed measures to check the rise of extremism, as well as to monitor, delist and close seminaries suspected of inciting violence, they admitted that there were many obstacles to tackling extremism. Provincial officials noted that in addition to countering radicalisation, they also had to deal with students who funded and otherwise facilitated terrorist activities. Furthermore, law-enforcement agencies acknowledged the danger they encountered in searching madrassas, operations that required the deployment of heavily armed paramilitary units.

The proliferation of madrassas had created significant challenges for Islamabad and provincial governments. Pakistani newspaper *Dawn* reported on 29 April that each of the five seminary boards reported that they oversaw growing numbers of these institutions and the students enrolled in them. According to its report, the Deobandi school of thought had the largest network in Pakistan, educating 2m students in 18,600 madrassas countrywide. The government lacked the resources to effectively monitor so many institutions, likely harming its efforts to counter terrorism under the NAP.

Two developments in late 2016 highlighted the gap between NAP strategy and the reality on the ground. In October, former Mangla Corps commander Lieutenant-General Tariq Hayat Khan remarked that deradicalisation centres established in Swat did not 'bear fruit', as they failed to reorient TTP militants. Khan subsequently questioned the logic of sending militants and their sympathisers through these centres when, according to him, the larger problem was the radicalisation of society itself. In December, *Dawn* reported that a confidential interior-ministry report conceded the government had made little progress in blocking the re-establishment of proscribed terrorist organisations or restricting their financing. However, the interior ministry denied the existence of the report.

Impasse in FATA reforms

The proposed merger of the FATA and Khyber Pakhtunkhwa created political tensions due to uncertainty about the range of reforms under consideration. In January 2016, members of the All FATA Political Alliance, a multi-party coalition of lawmakers who sought to integrate FATA into Khyber Pakhtunkhwa, objected to their exclusion from the FATA Reforms Committee, a group Prime Minister Sharif established to consider issues related to the FATA's constitutional status. These issues centred on residual British colonial law that deprived FATA residents of legal protection such as due process and the separation of police, judicial and executive authority. Lacking this protection, some residents of the tribal belt had been subjected to collective punishment and unfair trials.

As the federal government pushed forward the consultation process, the All FATA Political Alliance threatened to protest if it was left out of future discussions. The alliance asserted that the FATA Reforms Committee lacked legitimacy because Islamabad was only holding negotiations with tribal leaders who supported the government position. The federal government's missteps and selective consultation process created significant anxiety and resistance among FATA's residents, political organisations and civil-society groups, which perceived the reforms as an imposition by an elite that had always been disconnected from the realities of life in the tribal belt.

Headed by Sartaj Aziz, the prime minister's foreign-affairs adviser (de facto foreign minister), the FATA Reforms Committee published its proposal in August. The committee recommended a five-year merger plan, but anticipated that related, wide-ranging financial, reconstruction and development activities would take ten years to implement. The committee also suggested eliminating collective punishment and extending the jurisdiction of the Supreme Court and the Peshawar High Court to the FATA. However, soon after the federal government revealed its merger plans, critical differences between the parties emerged, with the Khyber Pakhtunkhwa government demanding that the merger take place before the 2018 general elections. Khyber Pakhtunkhwa officials argued that this would enable the tribal agencies to gain representation at the federal and provincial levels of government, which had been denied to them since Pakistan gained independence. Khyber Pakhtunkhwa Chief Minister Pervez Khattak contended that a five-year road map for a merger could jeopardise the process and generate greater uncertainty in the FATA. Khattak was also critical of the committee's recommendation for a hybrid judicial system known as the Tribal Areas Riwaj Act, calling for the province to have one government and one legal system. Human-rights campaigners argued that the proposed changes aimed to retain some of the most repressive legal arrangements in the FATA. For example, critics pointed out that even the name of the proposed law, 'Riwaj', referred to customary laws – suggesting that *jirgas* (tribal councils) could adjudicate on the application and implementation of laws, in contravention of the Pakistan Penal Code.

By October, Jamiat Ulema-e-Islam–Fazl (JUI–F) had joined the protests against the federal government's merger plans. Maulana Fazlur Rehman, the organisation's head and an ally of the government, stated that people living in the FATA should decide its future. Together with the Pakhtunkhwa Milli Awami Party, the JUI–F publicly called for a referendum, to no avail. As opposition to the merger grew, Prime Minister Sharif deferred the approval of the FATA Reforms Committee proposals to address the concerns of his allies. The political impasse persisted for the remainder of the year, as the JUI–F maintained that the FATA should become a separate province.

As political groups vied with one another to determine the future of the FATA, the military establishment exploited the area's lawlessness to support militants waging a costly insurgency in Afghanistan. However, if Islamabad brought the tribal areas under Pakistan's constitutional framework, the presence of FATA-based militant groups such as Lashkar-e-Taiba and Jaysh-e-Mohammad would underscore the fact that even governed spaces were being used to create and support lethal terrorist networks.

Militancy and regional economic development

The impact of the China–Pakistan Economic Corridor (CPEC), the centrepiece of the countries' strategic relationship, on Pakistani security become more apparent as the project developed. A key concern of Pakistani authorities remained the threat to the initiative posed by separatist groups in Balochistan. These organisations included the Baloch Republican Army, the Baloch Liberation Army and the Baloch Liberation Front, which attacked Chinese workers, Pakistani security personnel and civilians, gas pipelines, railroad tracks and government installations, while blocking highways and occupying coal mines. Regarding Balochistan as being under illegal Pakistani occupation, they mobilised against the government's perceived unequal allocation of resources, the threat of integration to local language and culture, and the military's repressive behaviour and violation of human rights. However, Islamabad dismissed these claims, blaming Kabul and New Delhi for destabilising Balochistan. In one case, Islamabad alleged that India had reopened consulates in the Afghan cities of Jalalabad and Kandahar as covert bases from which to support Baloch insurgents and the TTP.

The arrest of Indian national Kulbhushan Yadav in Balochistan in March cemented Pakistan's narrative that India was interfering in its domestic affairs. Balochistan Home Minister Sarfaraz Ahmed Bugti accused Yadav of working for Indian foreign-intelligence agency the Research and Analysis Wing, adding that the organisation was funding and training members of the Baloch Republican Army. India rejected Pakistan's allegations but confirmed Yadav's Indian citizenship, stating that he had taken early retirement from the Indian Navy and no longer worked for the government. Yadav's capture also created tension between Pakistan and Iran. Several Pakistani military officials claimed that he had entered Balochistan from Iran and alleged that the Research and Analysis Wing had established networks along the Balochistan–Iran border, thereby hinting at Tehran's complicity. Lieutenant-General Raheel Sharif, then Pakistan's army chief, reportedly raised the issue of Indian interference in Balochistan during his meeting with Iranian President Hassan Rouhani in late March. However, Rouhani criticised 'Pakistani elements' for attempting to undermine improving Iran–Pakistan relations and objected to the assertion that his country was involved in the Yadav incident.

Pakistan's security concerns about CPEC and Balochistan enabled the military to play a greater role in the project, despite careful opposition from the civilian leadership. The Pakistani political elite worried that this would further entrench and legitimise the role of the military in the country's political and economic affairs. There was a possibility that the trend could compromise the legitimacy of future civilian governments – which would be portrayed as corrupt and incompetent, in contrast to a more stable military capable of delivering goods and services to the public. Nonetheless, Prime Minister Sharif begrudgingly ceded some political ground to allow the military to play a more active part in CPEC. He appeared to be responding to Beijing's call for Islamabad to rely more heavily on the military's help in pursuing the project. Indeed, China expressed concern over the security situation on 21 July, following an increase in attacks on CPEC workers by separatist groups. And, in a rare public statement that demonstrated Beijing's growing unease about the lack of a political consensus on CPEC in Pakistan, the Chinese Embassy in Islamabad urged political leaders to address their differences for the benefit of the project. The combined impact of a deteriorating security environment in Balochistan, pressure from China and political disagreements between Islamabad and provincial governments threatened to cast Sharif in a negative light, likely informing his decision to give the military more sway on CPEC.

To address growing insecurity in the province and domestic criticism of Prime Minister Sharif's approach to handling threats to CPEC, the government announced in August that it would create a dedicated security service for the project, the Special Security Division (SSD). Composed of 9,000 soldiers and 6,000 paramilitary personnel, the new force was tasked with providing security for Chinese nationals and projects under the CPEC umbrella across the country. This development was deemed necessary as a spokesperson for military-run construction company the Frontier Works Organisation revealed in September that militants had killed 44 Pakistani workers and wounded more than 100 others in separate attacks on CPEC initiatives since 2014. The official also noted that 19 of these fatalities had occurred since November 2015, indicating a growing trend. In December 2016, the Pakistan Navy established Task Force-88, which was deployed to protect Gwadar Port from both 'conventional and non-conventional threats'.

Arguing that the Sharif government was denying most provinces their share of CPEC's benefits, Pakistani senators demanded in late July that the agreement establishing the project be made public. The provincial governments of Balochistan, Sindh and Khyber Pakhtunkhwa cautioned that discrimination against them could 'pose a threat to the federation', alleging that Punjab province was the principal beneficiary of CPEC. Although the federal government denied them, such claims had become widespread due to the perception that Sharif favoured Punjab due to his political roots and his party's strong support in the province (holding 312 out of 371 seats in the provincial assembly). The fact that Shahbaz Sharif, the prime minister's brother, was chief minister of Punjab province only deepened mistrust of the federal government and spurred accusations of favouritism. Such concerns led to the creation of the Corridor Front, an organisation comprising political parties and civil-society activists that advocated equally sharing CPEC's benefits among all provinces.

On 4 October, former Balochistan senator Sanaullah Baloch pointed out that despite having provided Pakistan with most of its extractive resources for decades, his province had received almost nothing from CPEC. 'How will a meagre share in the CPEC – $600 million out of $46 billion [sic] – bring miracles in the life of the Baloch?', he asked. He also pointed to the government's SSD as another example of discrimination against the Baloch people, saying that non-Baloch Pakistanis were joining the force. Sardar Akhtar Mengal, leader of the Balochistan National Party, stated in October that 'since the emergence of Pakistan in 1947, [Punjabis] have proved at every step that only Punjab is Pakistan ... they only need our resources, and that is their real concern because they see themselves as the real owners.' Thus, there was a possibility that the government's approach to CPEC could exacerbate separatism in Balochistan.

Continued operations against Tehrik-e-Taliban Pakistan

The Pakistani security forces continued their offensive against the TTP and affiliated militant groups in North Waziristan as part of *Operation Zarb-e-Azb*, increasingly targeting their opponents with airstrikes. The military claimed to have killed at least 130 militants, some of them affiliated with the TTP and Lashkar-e-Islam, in Datta Khel and Shawal between January and April. However, these figures were difficult to verify, as the military enforced a media blackout in its area of operations. The military's official communications wing, Inter-Services Public Relations (ISPR), stated in March that the Pakistan Army had eliminated at least 252 militants and cleared 640 square kilometres of the Shawal Valley in North Waziristan in the final phase of *Operation Zarb-e-Azb* (which began on 25 February). On 18 April, despite ongoing violence, the military declared the operation a success, claiming that all terrorists in North Waziristan had been killed or driven out. On the same day, Lieutenant-General Sharif announced that internally displaced persons (IDPs) would be resettled following final search-and-clear operations, signalling the end of the military offensive. Yet the Pakistan Army stated on 20 May that it had destroyed the last militant camp and removed more than 2,000 terrorists from the Shawal Valley, following a 90-day operation that led to the deaths of six soldiers and more than 120 militants.

To mark the conclusion of *Operation Zarb-e-Azb*, Lieutenant-General Sharif inaugurated a variety of development projects, including a new hospital in South Waziristan, as well as a 72km-long dual carriageway that formed part of CPEC. The ISPR announced that the military had undertaken 567 development projects in North Waziristan – spanning communications, power and transport infrastructure – as part of a 'post-operation comprehensive rehabilitation' plan for the FATA.

The United States had helped Pakistani security forces isolate and target militant groups. On 25 May, the US designated the Tariq Gidar Group (TGG) and Jamaat-ul-Dawa al-Quran (JDQ), both linked to the TTP, as 'specially designated global terrorists'. Based in Darra Adam Khel, the TGG was responsible for multiple high-fatality attacks, including the massacre of 132 schoolchildren and nine members of the teaching staff at the Army Public School in Peshawar on 16 December 2014. The TGG's leader, Umar Mansoor, also planned the 20 January 2016 attack on Bacha Khan University in Charsadda, which killed 20 people and injured almost 60 others. The JDQ, a Salafist group based in Peshawar and eastern Afghanistan, pledged allegiance to the late Taliban leader Mullah Omar, and had long-standing ties to al-Qaeda and Lashkar-e-Taiba. On 30 June, the US State Department formally designated AQIS a foreign terrorist organisation, allowing officials to freeze AQIS assets in any US jurisdiction and criminalising interactions between US citizens and AQIS members or affiliates. The US used unmanned aerial vehicles to kill Mansoor on 9 July and Lashkar-e-Islam leader Mangal Bagh, along with two of his associates, on 23 July.

In parallel with *Operation Zarb-e-Azb*, the Pakistani military conducted smaller campaigns in Khyber Agency. Launched in mid-August and centring on the Tirah Valley, *Operation Khyber* III was designed to reinforce troops positioned near Afghanistan and prevent cross-border infiltration by militant groups. Failing to rule out the possibility of a sustained deployment in the area, the military killed close to 40 suspected militants in a series of airstrikes in the Tirah Valley between September and December. Pakistan's military spokesperson, Lieutenant-General Asim Bajwa, stated in September that since the start of *Operation Zarb-e-Azb*, the operation had claimed the lives of 516 army personnel and more than 3,500 militants.

Lieutenant-General Qamar Javed Bajwa became chief of the army on 30 November, making Lieutenant-General Sharif the first incumbent of the post in more than 20 years to have stepped down without seeking another three-year term. Although Sharif maintained that he would retire as stated, the Pakistani media speculated whether he might be granted an extension or promoted to field marshal in recognition of his military successes against the TTP and Baloch separatists. Pledging to sustain these security gains, Bajwa approved in December the death sentences of 21 men whom military courts had convicted of killing civilians, including Ismailis in Karachi, human-rights activist Sabeen Mahmud and security personnel. One of the men was Muslim Khan, a senior TTP leader known as the 'butcher of Swat' who had been convicted of killing 31 people.

Extrajudicial killings, blasphemy laws and displacement

Throughout 2016, there were reports of the Pakistani security forces' involvement in extrajudicial killings and other human-rights abuses, particularly in Balochistan. Despite the reporting restrictions enforced by the Pakistani state, Zohra Yusuf, head of the Human Rights Commission of Pakistan (HRCP), publicly condemned the security forces for their alleged human-rights violations. She highlighted reports of numerous civilian casualties in the districts of Kalat and Mastung, cautioning the authorities that a disproportionate use of force would further alienate locals and work to the detriment of the NAP.

Leaked official data obtained by Pakistani television channel Geo News appeared to vindicate these concerns. The channel revealed on 1 July that nearly 1,000 bullet-riddled bodies had been found across Balochistan in the past six years. The highest numbers of bodies were discovered in Quetta and Kalat – 346 and 268 respectively. However, the Pakistani authorities denied that state agencies had a hand in enforced disappearances or extrajudicial killings, saying that criminals, separatists or extremists were responsible for the abductions and killings.

The HRCP's annual report, published on 1 April, noted a reduction in the enforcement of blasphemy laws but maintained that they were still disproportionately used against religious minorities. According to the United States Commission on International Religious Freedom, Pakistan had detained more people on blasphemy charges than any other country: the authorities continued to hold at least 38 people who had been incarcerated for their religious beliefs.

The presence of Afghan refugees in Pakistan became an increasingly significant source of international tension. Although around 1.5m registered and 1m undocumented refugees had resided in the country since the 1980s, Islamabad announced that they would be repatriated by 31 December 2016 and 31 March 2017 respectively. In an apparent effort to signal the government's commitment,

Aziz declared on 20 June that camps for Afghan refugees in Pakistan posed a threat because terrorists could use them as safe havens. On 28 June, Khyber Pakhtunkhwa police arrested more than 2,000 Afghan refugees in Peshawar and declared that all Afghan refugees would have to register with the authorities over the next two days or be declared illegal settlers. Is response, Afghan officials suggested that Islamabad exploited the vulnerability of Afghan refugees whenever relations between Pakistan and Afghanistan became strained, using the threat of repatriation to pressure Kabul. However, Pakistan's chief commissioner for refugees rejected the assertion that his government had politicised the issue, reiterating that refugees posed security and economic risks to the country. However, a September 2016 report by the Khyber Pakhtunkhwa government showed otherwise, finding that Afghan refugees were implicated in only 134 of the 10,549 major crimes recorded in the province between September 2014 and September 2016. This figure sharply contradicted the oft-quoted statistic that Afghans were responsible for 15–30% of crimes in Khyber Pakhtunkhwa. Khalid Aziz, former head of the Regional Institute of Policy Research and Training in Pakistan, condemned the mischaracterisation of Afghan refugees: 'this is a classic case of the term "post-truth" … truth is created to justify a certain action'.

Yet most national and provincial political parties supported Prime Minister Sharif's position on the refugees. One of the few exceptions was JUI–F chief Rehman, who openly opposed what he called their 'forced repatriation'. Rehman urged Islamabad to reconsider its position, arguing that expelling refugees could seriously harm Afghan–Pakistani diplomatic and people-to-people relations. The Pakhtunkhwa Milli Awami Party also actively opposed the move. Zulfiqar Ahmed Bhutta, a lawyer acting in a personal capacity, filed on 20 August a petition in the Supreme Court to extend the repatriation deadline by one year. Bhutta's petition claimed that Khyber Pakhtunkhwa officials were misusing their positions to buy Afghan properties and assets at significantly reduced prices. Despite this opposition, the government insisted that it would forcibly repatriate all Afghans who had not left the country by March 2017, declaring 2016 'the year of returns'.

According to the UN High Commissioner for Refugees, more than 380,000 registered Afghan refugees returned from Pakistan in 2016, the highest such number since 2007. The UN Office for the Coordination of Humanitarian Affairs (OCHA) reported that 114,511 families returned to the FATA in 2016. But it added that 76,507 families would remain displaced: 29,360 in North Waziristan, 23,879 in South Waziristan, 9,524 in Khyber Agency, 7,965 in Orakzai, 5,457 in Kurram and 322 in Tank. The OCHA also noted that due to security operations, natural disasters and other factors, an estimated 71% of houses in the FATA had been 'fully damaged', while the remainder had been partially damaged.

There were also other indicators of the enormous task facing the authorities as they tried to rebuild the lives of IDPs. Published on 29 December, the OCHA's survey of the humanitarian needs of returnees in Khyber Pakhtunkhwa and the FATA showed that 69% of girls and 29% of boys did not attend school. The survey also found that 81% of respondents lacked job opportunities and basic facilities, while 58% had no access to clean drinking water. Furthermore, 54% of women said that they lacked access to female doctors, 58% that primary healthcare was unavailable to them and 66% that they had no access to reproductive-health services. In addition, 74% of respondents reported 'market destruction', 68% damage to roads they used and 64% damage to local irrigation facilities. The OCHA classified 23% of surveyed families as 'food insecure'. Thus, Pakistan struggled to resettle IDPs while conducting security operations in parts of the FATA, placing tremendous strain on returnees in the neglected region. There was a possibility that the returnees would fall deeper into poverty if they did not receive sustained, adequate support.

The shortfalls in state support for returnees formed part of a broader government programme that continued to produce mixed results. Although CPEC projects gained momentum in 2016, they appeared to exacerbate historical disputes between the provinces. Unless Sharif's government distributed the benefits of CPEC in an equitable manner, it appeared likely that Pakistan would not realise the full potential of the undertaking. The security situation seemed to improve under the NAP – as suggested by the decline in recorded civilian and military fatalities – but the frequency with which the TTP and affiliated groups carried out high-fatality attacks, as well as targeted killings, demonstrated that they were still a potent threat. As a consequence, it appeared that the Pakistani government needed to rework the plan to effectively disrupt and defeat such groups, partly by sustaining the cooperation of the military and persuading the security services not to work with terrorist groups in pursuit of their strategic interests. The FATA reforms had considerable potential. Yet Islamabad could not afford to disenfranchise civilians in the tribal areas in the process, as any measure they perceived as an imposition by the federal government was likely to fail.

Chapter Six

Asia-Pacific

China (Xinjiang)

Key statistics	2015	2016
Conflict intensity:	Low	Low
Fatalities:	200	9
New IDPs:		
New refugees:		

The conflict in the western Chinese province of Xinjiang saw little change on the ground in 2016. Recorded fatalities fell to just nine – from around 200 in 2015 and 400 in 2014. This apparent trend could have reflected a significant reduction in the threat to civilians from terrorist activity and armed clashes. Yet even as the conflict seemed to stabilise, the authorities continued to strengthen their counter-terrorism policy on the province. A document released on 22 January at a political conference in Beijing called on them to 'firmly curb terrorist activities in [Xinjiang], prevent these activities from spreading inland, and prevent violent terrorist attacks in large and medium-sized cities'. This concern was echoed in a November speech to the National Congress of the Chinese Islamic Association by Wang Zuoan, the head of the State Administration for Religious Affairs, who said that religious extremism was spreading eastwards from Xinjiang to 'inland provincial areas'.

The Xinjiang Uighur Autonomous Region became the first Chinese province to introduce its own interpretation of a national counter-terrorism law that came into effect on 1 January, providing more detailed regulations on how the

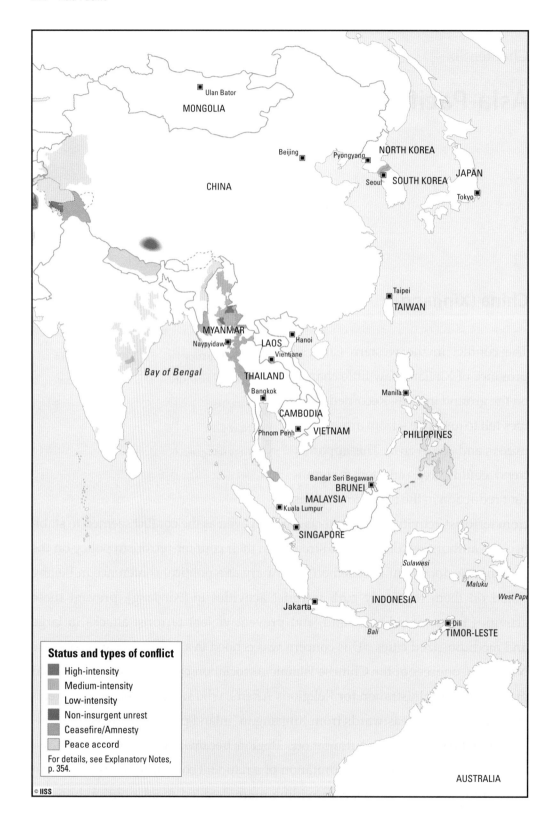

security forces could deal with suspected terrorist offences. The law banned the dissemination of information on terrorist attacks, perhaps explaining why only two fatal incidents in Xinjiang became public knowledge in 2016.

The first of these events occurred on 10 September, when a deputy county police chief died in an explosion during a raid on the home of suspected extremists in Pishan County, in the prefecture of Hotan. Three other officers were critically injured in the blast, with local sources reporting that they died on the way to hospital. It was unclear whether the explosion had been caused by an improvised explosive device, a suicide bomb or something else. By 19 September, at least 17 people had been arrested in connection with the raid. The other reported incident came on 28 December, when three militants detonated a bomb – a suicide bomb, according to some reports – at Communist Party offices in Moyu County, in Hotan, before attacking people there with knives. Two people were killed and three others injured before the attackers were shot dead.

Other developments suggested that there had been other, unreported security incidents in the conflict. The Xinjiang regional government announced on 11 April that it would pay rewards of between 200,000 renminbi (RMB) and RMB5 million (around US$29,000 and US$728,000 respectively) for tip-offs on terrorist activity. The authorities also completed a complex of seven new prisons near Urumqi in September.

On 29 August, Chen Quanguo replaced Zhang Chunxian as head of the Communist Party in Xinjiang. Hong Kong-based press outlets stated that Zhang had been replaced due to a perceived failure to stem terrorist attacks, despite his success in achieving high rates of economic growth in the region. In December, the Xinjiang government was reportedly setting up a team of 30 special prosecutors to handle terrorism cases in Kashgar, Hotan, Aksu, Ili, Bayingolin and Urumqi prefectures – again suggesting that a large number of security incidents did not become publicly known. Thus, it was unlikely that the reduction in reported incidents and fatalities came about as a result of a discernible de-escalation of the conflict; rather, it stemmed from an increasingly effective crackdown on reporting from the region.

China continued diplomatic activity linked to counter-terrorism and regional economic growth in Xinjiang. These efforts gained extra impetus on 30 August, when a suicide bomber drove a car through the gates of the Chinese Embassy in Bishkek, in Kyrgyzstan, killing himself and injuring five other people. No Chinese diplomats were hurt in the incident. According to Kyrgyzstan's authorities, the suicide bomber was an ethnic Uighur with a Tajikistan passport, and a member of the East Turkestan Islamic Movement (ETIM). Kyrgyzstan said the attack was ordered by Syria-based Uighur militants who were allied with Jabhat al-Nusra. The authorities detained five Kyrgyzstan citizens and sought another two people, one of them a Turkish citizen, in connection with the incident. The Russian Federal Security Service stated on 13 September that Russian, Tajikistan and Chinese citizens were involved in the attack.

The suicide bombing underlined the growing threat to Chinese interests in Central Asia as the country pursued its Belt and Road Initiative, a vast economic project covering infrastructure, energy and trade. Xinjiang-based companies signed in April deals with Pakistan worth an estimated US$2 billion, following a 2015 agreement to set up the China–Pakistan Economic Corridor as part of the initiative. Kazakhstan and firms in the Chinese province signed in early May deals worth a similar amount.

Uighurs as foreign fighters

A Chinese counter-terrorism official said in February that 'domestic and overseas "East Turkistan" forces are stepping up their instigating efforts, and there's a growing tendency for activities that are masterminded overseas, organized online and implemented within the country'. Indeed, Chinese Uighur militants were likely to have sympathisers in ethnically and linguistically similar communities in Central Asia, which could provide them with housing and other assistance.

Addressing reports of Uighur militants being killed in Indonesia in early 2016, Beijing stated that it was willing to improve counter-terrorism cooperation with Jakarta, adding that the ETIM had opened a 'channel of personnel transfer' with Indonesian terrorist groups. China would also have been concerned about the activities of the Turkestan Islamic Party in the war in Syria, in which the group formed part of the Jaysh al-Fateh rebel alliance.

Chinese President Xi Jinping and Turkish President Recep Tayyip Erdogan spoke on the sidelines of the G20 meeting in Hangzhou in September, agreeing to deepen counter-terrorism cooperation between their countries. This was a significant development as many Chinese Uighurs had travelled to Turkey – most to join the country's sizeable Uighur community, but some on their way to fight in the conflict in Syria. Soon after the attack in Bishkek, Kyrgyzstan hosted a counter-terrorism exercise with other members of the Shanghai Cooperation Organisation: China, Kazakhstan, Russia and Tajikistan.

In July, China's defence minister, Chang Wanquan, thanked visiting Afghan chief of the army staff, General Qadam Shah Shahim for his country's work in combating the ETIM and supporting core Chinese interests. The Afghan authorities had reportedly notified Beijing whenever Uighur militants were arrested in Afghanistan. Soon after Shahim's visit, Chinese state media reported that China, Afghanistan, Pakistan and Tajikistan would set up a counter-terrorism alliance. China's engagement with the peace process in Afghanistan was partly motivated by the conflict in Xinjiang, which borders northeastern Afghanistan. During a visit to Beijing in May, Afghan Chief Executive Abdullah Abdullah reaffirmed his country's support for Chinese efforts to defeat the ETIM.

Development efforts and strict security measures

Recognising that poverty was one driver of insurgency and radicalisation in Xinjiang, Beijing continued its efforts to promote economic development in the region. The provincial authorities announced that they would spend RMB110bn (US$17bn) on 100 projects for 'improving livelihoods' in Xinjiang. Premier Li Keqiang said on 10 March that the region had an 'especially important strategic position', adding that Xinjiang officials should ensure young people had 'something to do and money to earn'. In contrast to these positive initiatives, the government's strict security and cultural policies in Xinjiang appeared to be counterproductive – and were often held up as another key driver of recruitment by groups such as the ETIM.

State media reported in March that restrictions on movement would be relaxed, and the passport application reworked, as part of the 'year of ethnic unity progress'. While vowing to maintain a strong security posture in Xinjiang, Zhang called on companies there to employ more people from ethnic minorities and to show greater respect for their culture. However, Radio Free Asia reported in late March that the authorities had detained around 41 Uighurs in Yining County, in Ili Kazakh Autonomous Prefecture, allegedly for failing to attend the funeral of a local Communist Party official. The authorities described the detainees as religious extremists.

In June, China published a White Paper on religious freedom in Xinjiang, which stated that 'no citizen suffers discrimination or unfair treatment for believing in, or not believing in, any religion', and that 'religious feelings and needs are fully respected'. This assertion was not borne out by events. The police reportedly arrested Uighur bloggers and other writers in the run-up to Ramadan, in an effort to prevent criticism of restrictions on religious activity during the Muslim holy month. On 18 May, 98 Chinese Uighurs were detained at Istanbul airport after using fake Kyrgyz passports to attempt the hajj, a pilgrimage that China allowed only a restricted number of citizens to participate in. The World Uyghur Congress, an activist organisation based in Germany, said on 6 June that 17 people had been arrested in Xinjiang for encouraging the observance of Ramadan.

Beijing strengthened its policies that appeared to infringe on religious freedom. New education rules in Xinjiang, stipulating that parents and guardians could not 'organise, lure or force minors into attending religious activities', came into force on 1 November. This legislation was in line with a previous government policy that sought to prevent children from attending mosque or following a religion.

Municipal police in Shihezi announced on 19 October that all residents of Xinjiang would be required to hand their passports to the authorities by 16

February 2017, and to apply to the police to have the documents returned. An official explained that the policy was designed to 'maintain social order'. Yet the police stated on 25 November that they were only holding the passports of residents with suspected links to terrorism, and not those of ordinary citizens. Meanwhile, China reportedly demolished around 5,000 mosques across Xinjiang between September and December. This 'mosque rectification' campaign purported to be in the interests of public safety, but was seen by locals as a further measure to restrict religious freedom.

Myanmar

Key statistics	2015	2016
Conflict intensity:	Medium	Medium
Fatalities:	600	300
New IDPs:		
New refugees:	10,500	

Myanmar's democratically elected parliament, dominated by Aung San Suu Kyi's National League for Democracy (NLD), held its first session on 1 February 2016, setting high expectations for further democratisation of the country's politics and progress in a peace process involving more than a dozen ethnic armed organisations (EAOs). The United States officially lifted its sanctions on Myanmar on 7 October, in recognition of the country's steps towards democracy. The US Department of the Treasury removed from its blacklist senior military figures such as former dictator Than Shwe, as well as military-run businesses such as the Union of Myanmar Economic Holdings and the Myanmar Economic Corporation. Washington also lifted its ban on imports of jadeite from Myanmar and restrictions on banking and financial transactions involving the country, reinstating preferential trade tariffs. However, a dramatic increase in violence in Rakhine and Shan states had severe domestic and international repercussions, suggesting that the decision to provide sanctions relief had been premature. The emergence of a new Rohingya insurgency in Rakhine State, following years of discrimination against the ethnic group, prompted a heavy-handed security crackdown that damaged Myanmar's international reputation and relations with its neighbours.

The Network for Human Rights Documentation–Burma, a non-governmental organisation, reported a 'stark increase' in human-rights violations by both the Tatmadaw (Myanmar armed forces) and EAOs in 2016, with 98 confirmed incidents of abuse in the first six months of the year, compared to 84 in all of 2015.

Drawn-out peace process

The transfer of power to the new civilian government went smoothly, with Myanmar's first civilian president for 54 years, Htin Kyaw, sworn in on 30 March. His former rivals for the presidency, military candidate Myint Swe (a one-time intelligence chief included on a US sanctions list during the elections) and ethnic Chin candidate Henry Van Thio, both became vice-presidents. Suu Kyi initially took on four ministerial posts, including that of foreign minister – which provided her with a seat on the National Defense and Security Council. The Tatmadaw controlled six of 11 seats on the council, the highest-ranking government body as per the constitution. There continued to be tension between the NLD and the military establishment, particularly due to Suu Kyi's exclusion from the presidency. In a unilateral solution to the issue, she was appointed as state counsellor – a new post that gave her wide-ranging powers similar to those of a prime minister – on 5 April. Suu Kyi later relinquished her ministerial portfolios in energy and education, but remained foreign minister, a minister in the president's office and head of the government in all but name.

The five-day Union Peace Conference, the first stage of the political dialogue, began in Naypyidaw on 12 January as scheduled. Suu Kyi admitted that the conference was 'just a token recognising the accomplishments of the Nationwide Ceasefire Agreement', or procedural sleight of hand to keep within the stated time frame for the process. Although EAOs had welcomed the change in government, the talks excluded several non-signatories to the agreement.

The next stage of the peace process – the 21st Century Panglong Conference, held between 31 August and 3 September – began with considerable fanfare. Suu Kyi called for a 'durable peace' and 'a future infused with light' as she opened the talks; Ban Ki-moon, then the UN secretary-general, also spoke at the opening of the event. Although it had been made clear that no significant decisions would be made at the conference, the proceedings were symbolically important due to the participation of around 1,800 people from the government, the Tatmadaw, political parties and 17 of 21 EAOs. China played a major role in bringing these groups together, allowing EAO leaders to use Chinese territory to travel to a pre-Panglong militant conference at Mai Ja Yang, in Kachin State, in July. Moreover, Sun Guoxiang, China's special envoy on Asian affairs, met the United Wa State Army (UWSA) and the National Democratic Alliance Army in Shan State in August to encourage the groups to attend the Panglong conference. However, the UWSA walked out of the event on its second day, ostensibly due to a perceived slight.

The Tatmadaw prevented three allied EAOs – the Arakan Army, the Myanmar National Democratic Alliance Army (MNDAA) and the Ta'ang National Liberation Army (TNLA) – from attending the conference after they refused to sign a pledge to abandon their armed struggle at an unspecified point in the future. At the same time, the Tatmadaw stepped up its operations against EAOs in the run-up to the event, perhaps to pressure groups that had not signed the ceasefire. Nevertheless, the United Nationalities Federal Council, an alliance of nine non-signatory EAOs, attended the conference and was permitted to present a paper calling for radical changes to the governance of both the country and the armed forces. Senior General Min Aung Hlaing, the commander-in-chief, reaffirmed at the event that the Tatmadaw would support the peace process, although he warned against 'localism' – a seeming reference to local autonomy – and insisted on six principles for peace that EAOs had rejected during previous talks. He added that the Nationwide Ceasefire Agreement would be used as a baseline for the talks. This meant that only signatories to the ceasefire would be allowed to participate in discussions to finalise the framework for the political dialogue, as well as the dialogue itself, effectively barring at least ten groups from meaningful involvement in the peace process. Reflecting the drawn-out nature of the process, the conference was scheduled to reconvene every six months.

Escalation in the north

The change in government may have altered the political calculations of the Tatmadaw, which escalated its operations against some rebel groups. Clashes between the military and EAOs such as the Kachin Independence Army (KIA), the Shan State Army–North and TNLA continued at low to medium intensity throughout the year, primarily in northern Shan State. The battles displaced thousands of civilians.

The Tatmadaw stepped up its operations against the KIA, one of Myanmar's largest rebel groups, in April, launching sustained offensives near the group's headquarters at Laiza, in Kachin State. Several KIA outposts fell to the Tatmadaw. In May, there was an outbreak of violence in the jade-rich area of Hpakant, in central Kachin State; the military accused the KIA of orchestrating a spate of bombings targeting mining operations and the police in the township. The clashes prompted several mining companies to suspend their operations. The Hpakant offensive followed the military's seizure in early May of two logging camps – of the kind armed groups often used to harvest valuable resources such as hardwood timber – in KIA territory in Mansi township. The KIA and the Tatmadaw also repeatedly clashed in Hpakant and Tanai townships in August. The military resumed its artillery strikes on KIA outposts near Laiza on 18–23 August.

In November, there was a major outbreak of violence in Muse, the main crossing point on the border with China, and Kutkai. This was a marked setback for the peace process that once again displayed the divide between the Suu Kyi-led government and the Tatmadaw. It also suggested that the process would continue to be ineffective unless they were unable to reconcile their differences. The Northern Alliance–Burma (NA–B) – comprising the KIA, the TNLA, the MNDAA and the Arakan Army – launched an attack on security installations and bridges near Muse and Kutkai on 20 November. With the Tatmadaw accusing it of having positioned fighters in China prior to the assault, the NA–B seized control of the town of Mong Koe, near Muse, eight days later, leading to the closure of the border crossing and trading zone there. The KIA described the attack as a 'limited offensive' in response to pressure from the Tatmadaw. But the scale and coordination of the assault suggested careful planning. According to the government, the assault resulted in the deaths of five security personnel and three civilians, as well as the injury of 18 civilians and 11 security personnel.

The Tatmadaw recaptured Mong Koe on 5 December. As the fighting began to subside, the NA–B claimed to have withdrawn from the town to protect civilians from Tatmadaw airstrikes, calling on the military to halt its offensive. The following day, the bodies of nine police officers were discovered in Mong Koe. State media reported on 9 December that at least 30 people had been killed near Mong Koe since 20 November (the report could not be verified). On 20 December, a mass grave containing the bodies of 18 civilians was discovered near Mong Koe. According to locals, the victims had been detained by the Tatmadaw.

China called for an immediate ceasefire as large numbers of people fled across the border; a Yunnan official said on 29 November that around 14,000 Myanmar citizens had sought refuge in the country in recent weeks. The NA–B called on Beijing to mediate between the sides in the conflict, but a meeting with government representatives in Kunming, scheduled for 1 December, fell apart due to disagreements about its format. Lieutenant-General Sein Win, Myanmar's defence minister, proposed on 2 December in the lower house of parliament that members of the NA–B be classified as terrorist groups. Although the proposal failed to pass, it further complicated reconciliation efforts. The Shan State parliament – controlled by the Tatmadaw and the Union Solidarity and Development Party – voted to designate NA–B groups as terrorist organisations. The government's National Reconciliation and Peace Center invited the NA–B to talks on 20 December, as the fighting continued.

On 16 December, the Tatmadaw seized a strategically important KIA outpost in Gidon, in Kachin State, as part of an operation that had begun in September.

A former Tatmadaw officer estimated that 400–500 soldiers and 20–30 KIA troops died in the prolonged battle, with the disparity in numbers due to terrain that favoured the rebels (the statistics could not be independently verified). The Tatmadaw had invested heavily in the operation as Gidon is near the KIA's Laiza headquarters. The loss of the outpost increased the vulnerability of the headquarters and thus presumably reduced the group's bargaining power in peace talks. On 27 December, the Tatmadaw captured another KIA battalion headquarters in Waingmaw township.

Unsettled by the military's activities, the UWSA, Myanmar's largest armed group, reinforced two hilltop positions that it had seized from erstwhile ally the National Democratic Alliance Army on 28 September. The UWSA claimed to have captured the positions because the latter group's deployments there were 'too weak' to withstand a Tatmadaw offensive. Meanwhile, the military, which had begun a force build-up in the area, demanded that the UWSA withdraw.

Persecution of the Rohingya community

Rohingya Muslims living in Rakhine State continued to face a dire situation. According to the United Nations, instability in the state had displaced around 120,000 people, most of them Rohingya, since an outbreak of intercommunal violence in 2012. On 1 July, following sporadic attacks on Rohingya and other Muslims in the first half of the year, a mob comprising around 500 people burnt down a mosque in Lone Khin, in Kachin State. Although the incident occurred far from Rakhine State, it demonstrated the persistent risk of escalating intercommunal violence. During a visit to Sittwe, the capital of Rakhine, on 14 July, Senior General Min Aung Hlaing urged locals to 'prevent any misunderstanding between national races'. The government formed the Emergency Management Central Committee to prevent religious violence, and appointed former UN secretary-general Kofi Annan to lead the Advisory Commission on Rakhine State, an organisation with the declared mission of 'finding lasting solutions to the complex and delicate issues in the Rakhine State'.

However, the emergence of Rohingya insurgent group Harakah al-Yaqin (Faith Movement) undermined this incipient progress. The organisation killed nine officers in three night-time raids on checkpoints and the border-police headquarters in Maungdaw and Rathedaung townships on 9 October. The incidents involved at least 250 attackers armed with bladed weapons and firearms, eight of whom were killed, while another two were arrested. They seized 62 firearms and 10,000 rounds of ammunition, according to the police.

A disproportionate response by the security forces spurred around 69,000 people to flee to Bangladesh and killed more than 100 Rohingya – although unconfirmed eyewitness accounts suggested that the number of fatalities was much higher. These 'clearance operations', as the Tatmadaw called them, also led to the deaths of 30 security personnel and the arrest of more than 500 Rohingya, at least six of whom reportedly died in custody.

The government denied that the security forces had committed any atrocities or other crimes, but its claims that they had 'found the bodies' of Rohingya in villages they searched stretched credulity. In a similar vein, on 12 October the Tatmadaw accused armed attackers of setting fire to 25 houses in a failed assault on the border-police station that had been attacked three days prior. Yet using satellite data, Human Rights Watch estimated that by 7 December the security forces had burned down more than 1,500 houses in Maungdaw township. The Tatmadaw's 'True News Information Team' denied this, saying that the 'attackers' were responsible.

Eyewitnesses reported that the security forces shot and killed around 70 Rohingya as they attempted to cross the Naf River into Bangladesh on 16 October. Those who reached the border faced an uncertain fate: Bangladeshi border guards reportedly turned back 1,000 people in the last two weeks of November. The Tatmadaw stated that 69 'violent attackers' were killed and 234 others arrested in Maungdaw township during 9–12 November. Independent reports suggested, and unverified video footage showed, that Tatmadaw helicopters had attacked a village on 12 November. On 16 December, UN High Commissioner for Human Rights Zeid Ra'ad al-Hussein said that his organisation was receiving daily reports of rapes and killings in Rakhine State. Yet these reports could not be verified due to severe access restrictions on media outlets and independent observers. The authorities also prevented non-governmental organisations that provided healthcare to Rohingya from entering affected areas, and suspended food aid for four weeks. Aid organisations were allowed to assist people who had been displaced from Maungdaw to Buthidaung township for the first time on 23 November.

These developments led members of the international community to warn of an attempt at ethnic cleansing, crimes against humanity or even genocide targeting Rohingya. Suu Kyi faced mounting criticism for having defended the crackdown. According to an official quoted by Reuters on 1 November, the Tatmadaw had not responded to the government's 20 October request for information on the alleged killings, arrests, looting and destruction of property. This suggested that the military was in sole control of the operations and intent on keeping information from the government.

Myanmar formed a state-level commission to investigate the alleged abuses. Headed by Vice-President Myint Swe, the commission reported on 14 December that the security forces 'followed the law and acted legally in their response to the attackers'.

As these crises unfolded, Senior General Min Aung Hlaing on two occasions publicly discussed the constitutional provisions for the military to declare a state of emergency and seize power. On 28 November, 13 political parties called on the National Defense and Security Council to intervene in the crises facing Myanmar and 'effectively counter domestic and overseas threats', arguing that the current government was unable to act.

Philippines (ASG)

The threat from the Abu Sayyaf Group (ASG) increased in 2016. The group intensified its kidnapping activities while coordinating militarily with other insurgent groups in the Philippines and strengthening its links with the Islamic State, also known as ISIS or ISIL. There were no substantive changes in the government's approach to dealing with the ASG after Rodrigo Duterte's inauguration as president on 30 June. Manila combated the group by stepping up its security operations and signing a trilateral agreement with Indonesia and Malaysia to carry out joint patrols in the Sulu Sea.

Key statistics	2015	2016
Conflict intensity:	Low	Low
Fatalities:	200	250
New IDPs:		
New refugees:		

Kidnapping spree

According to data published by the Philippine police and military, the ASG made around 353 million Philippine pesos (US$7.3m) from ransom payments, thought to be its main source of income, in the first six months of 2016. The Armed Forces of the Philippines (AFP) said on 13 December that the ASG was holding 18 foreigners and five Filipinos as hostages. The group abducted four fishermen off the coast of Sulu one week later. Jolo Mayor Hussin Amin stated in June that soldiers had aided the ASG in the kidnappings in return for a portion of the ransom money – a claim that others had also made – but the AFP denied the allegation.

The authorities attempted to resolve a crisis involving two Canadians, a Norwegian and a Filipina whom the ASG had kidnapped in eastern Mindanao and taken to Sulu on 21 September 2015. After issuing several warnings that it had not received ransom payments, the group beheaded both Canadians: John Ridsdel's remains were found in Jolo on 25 April, while Robert Hall died on 13 June after another deadline passed. The ASG released Filipina hostage Marites Flor on 23 June. The ASG released Kjartan Sekkingstad, the remaining hostage, on 16 September. He was handed over to the authorities by Nur Misuari, leader of a Moro National Liberation Front (MNLF) splinter group, in Indanan, in Sulu. Misuari also received three Indonesian hostages abducted in July following the payment of their ransoms. President Duterte revealed that the government paid PHP50m (US$1m) for Sekkingstad, and PHP20m (US$400,300) for the three Indonesians.

The ASG significantly intensified its kidnapping spree in the second half of the year despite military offensives and other efforts to contain the threat. The group abducted at least 17 people, including Malaysian and Indonesian sailors, as well as Filipinos – one of whom it beheaded after discovering that his family could not afford the ransom. Jurgen Kantner, a German citizen, became an ASG hostage on 6 November. Having killed his partner, the group demanded a ransom payment of PHP500m (US$10m) for Kantner.

Intensifying security operations

The military campaign against the ASG led to the deployment of an additional battalion to Sulu in February and two extra companies to Basilan in March. However, this did not change the dynamics of the conflict, as the military continued operations against the group in its jungle hideouts. The AFP claimed that the militants' detonation of improvised explosive devices in several urban locations – including a port terminal and a police station in Lamitan, in Basilan – were designed to distract the security forces from the offensive.

The ASG escalated the conflict in Basilan province on 9 April, ambushing government troops in Tipo-Tipo. A spokesman for the AFP's Western Mindanao Command said that around 120 militants participated in the attack, which caused the deaths of at least 18 soldiers and nine members of the ASG. Another 56 soldiers were injured in the incident. Moroccan national Mohammad Khattab was among the dead, indicating what AFP chief Lieutenant-General Eduardo Año said were links between the group and ISIS. Indeed, ISIS claimed responsibility for the attack through its Amaq news agency – although Manila dismissed the claim.

The AFP launched a large-scale operation to pursue ASG fighters who had been involved in the ambush, subsequently declaring that it had killed dozens of the group's members. It said on 16 April that 37 ASG members had been killed in Tipo-Tipo, and two days later that it had captured three ASG camps in

Basilan. Meanwhile, the military portrayed a bombing on 14 April that injured the Lamitan police chief as an attempt to divert attention away from these operations. Around 1,000 families were displaced by the military operation.

The AFP's difficulties in countering the ASG had been particularly apparent ten days earlier, when it replaced Brigadier-General Alan Arrojado with Major-General Gerry Barrientos as commander of Joint Task Group Sulu. Three weeks later, Arrojado resigned from his other position as Sulu army commander due to disagreements about strategy. The military deployed an additional infantry battalion to Sulu on 12 June, following the relocation in May of a marine battalion from Zamboanga City to Tawi-Tawi to enhance coastal security. Another major clash between the AFP and the ASG began on 21 June in Patikul, in Sulu, resulting in the injury of 18 soldiers, as well as the deaths of around ten of the 200 ASG fighters involved.

In a video released on the same day, ISIS announced the appointment of Isnilon Hapilon, head of the ASG in Basilan, as leader of the Middle Eastern group in Southeast Asia. While this was a significant step, it did not constitute the establishment of an official ISIS *wilayat* (province). In the video, a Filipino, a Malaysian and an Indonesian thought to be members of Katibah Nusantara, the Bahasa-language ISIS battalion in Syria, called on their countrymen to carry out attacks in Southeast Asia. President Duterte warned that the Philippines risked being 'contaminated by the ISIS disease' in three to seven years unless the ASG was defeated. He also stated that ASG fighters had been radicalised due to a lack of governance in their home provinces.

The AFP began major offensives against the ASG in Sulu and Basilan in early July, acting on top commanders' pledge to eliminate the group by the end of the year. General Ricardo Visaya, the AFP's chief of staff, stated in June that the ASG had '500 or a thousand or more maybe' members – significantly more than the military's previous estimate of 400. Around 10,000 soldiers were deployed to Sulu and Basilan during the offensives. The Philippine Navy blockaded some areas of Sulu to trap ASG groups on the island and prevent them from receiving reinforcements. The BRP *Tarlac*, the navy's newest and largest vessel, took part in the effort. The AFP also seized around 200 speedboats in the provinces of Sulu, Basilan, Tawi-Tawi and Zamboanga to limit the ASG's mobility.

In Basilan, the military offensive concentrated on Tipo-Tipo, beginning with air and artillery barrages there on 6 July. The following day, the AFP launched operations in Patikul. By 15 August, the military had captured the group's last stronghold in Tipo-Tipo, a hilltop base in Silangkum that reportedly included ten bunkers, four tunnels and several foxholes. On 29 August, approximately 120 ASG militants killed 15 soldiers and injured ten others in an ambush in Patikul. The clashes between the sides became less intense after the first week of September: although the AFP continued its operations across both island provinces, the ASG split into smaller groups to evade the security forces. The AFP claimed that operations led to the deaths of more than 100 militants and an estimated 28 soldiers and members of the Barangay Peacekeeping Action Teams, a local government-led militia.

Militants killed 15 people and injured around 70 others by bombing a market in Davao City on 2 September, in an attack that appeared to involve unusually close cooperation between insurgent groups. An ASG representative initially admitted that his group had been behind the operation, before retracting the statement and naming ISIS as responsible. The attack was likely carried out by members of the ISIS-linked Maute Group to aid its ASG allies by distracting security forces from their campaigns in Sulu and Basilan.

Visaya said on 5 October that more than 50% of the AFP's air and naval assets had been deployed in Mindanao to combat the ASG. Manila undertook further operations to limit the group's cross-border kidnapping activities. On 27–28 September, the military offensive in Sulu killed four notorious ASG kidnappers: Nelson Muktadil and his brother, Braun Muktadil, as well as two members of the Alhabsy Misaya sub-group. On 8 December, the Malaysian police killed the alleged leader of the group responsible for the kidnapping of Flor, Hall, Ridsdel and Sekkingstad, along with two other ASG members, in a firefight off the coast of Semporna, in Sabah. The police arrested three other militants and freed a hostage.

The military operations in Basilan and Sulu had a significant humanitarian impact, particularly due to the AFP's use of artillery and airpower. Basilan Governor Jim Saliman declared on 11 July a 'state of calamity' – defined by the government as 'a condition of mass casualty and/or major damages to property, disruption of means of livelihoods, roads and normal way of life' – in Tipo-Tipo, Ungkaya Pukan and Al-Barka. By 19 September, the campaign had displaced 18,783 people in Basilan and 23,920 people in Sulu, according to the Department of Social Welfare and Development.

Insufficient political and development efforts

Alongside its military operations, the Philippine state needed to engage in political and developmental initiatives to create the conditions for defeating or marginalising the ASG. Although such efforts were under way, they appeared to be insufficient. Visaya said in June that the AFP would 'separate the terrorists from the civilian communities in order to cut off their logistics and contain them in areas conducive for battle', 'bring social services to the people' and spur economic development. On 21 March, then-president Benigno Aquino inaugurated a 138-kilometre paved circumferential road in Basilan. The government hoped that by improving opportunities for commerce and island-wide travel, it could reduce the flow of recruits to the ASG and facilitate the security forces' pursuit of those who did join the group.

President Duterte said on 25 November that he was ready to talk about opening borders in the south. Cross-border trade is a centuries-old way of life in the southwestern Philippines, but modern nation-states regard it as smuggling. Easing restrictions on such activity had the potential to help alleviate poverty, criminality and terrorism in the Sulu Sea region in the long term. Thus, Malaysia's announcement in April that it would suspend barter trade in Sabah ports – a direct response to the kidnapping of Malaysian citizens – was likely to have been counterproductive. The policy was meant to prevent smuggling and make it harder for the ASG to cross the border, but by late May the prices of rice, cooking oil and sugar appeared to have nearly doubled in Tawi-Tawi and Sulu. There was a possibility that these communities, already among the poorest areas in the Philippines, would turn towards the ASG as a source of funding and employment. Indeed, the group often distributed ransom payments in local communities.

One of the reasons that the government struggled to defeat the ASG related to the group's connections with local powerbrokers and members of the security services. On 15 September, a senior police officer in Basilan claimed that the majority of security incidents in the province were not instigated by the ASG, but by members of the armed forces who had links with criminal groups. Three days later, the authorities arrested a municipal official and alleged member of the ASG at a checkpoint in Zamboanga City, confiscating a grenade and around 100 rifle rounds in his possession. The police announced on 27 September that they had arrested three men suspected of selling firearms to the ASG and other armed groups in San Juan. The weapons recovered during the arrest were marked with the logo of the Department of National Defense, but it was unclear how the suspects acquired them. One of the detainees had run in local elections in Sulu in May.

The ASG's cross-border kidnapping activities prompted Southeast Asian governments to take concerted action against the group. In May, the foreign ministers and defence chiefs of Indonesia, Malaysia and the Philippines jointly called for intensified efforts to improve regional maritime security and assist vessels under threat, including through patrols in the Sulu Sea. The resulting formal agreement, reached on 2 August, included the right to chase suspects across borders in 'hot pursuit'. At a meeting in November, the parties discussed the possibility of deepening cooperation through joint exercises, coordinated maritime patrols and joint air patrols.

Philippines (MILF)

The ceasefire between the government of the Philippines and the Moro Islamic Liberation Front (MILF) held throughout 2016, with no serious security incidents involving both parties. But congress's failure to pass the Bangsamoro Basic Law (BBL) – a legal instrument designed to establish an autonomous region as part of the 2014 peace agreement with the MILF – before the end of 2015 increased instability in Mindanao.

Key statistics	2015	2016
Conflict intensity:	Low	Low
Fatalities:	300	90
New IDPs:		
New refugees:		

The delay also stymied attempts to decommission the MILF's weapons and demobilise its fighters. At the same time, MILF splinter group the Bangsamoro Islamic Freedom Fighters (BIFF) continued to seek independence for the region by launching small-scale attacks.

The MILF said that by failing to pass the BBL, congress had made reconciliation 'very hard', accusing its members of seeing Moros as 'sub-human'. Expressing 'deep disappointment and grave dismay', the MILF urged its armed wing to 'uphold the primacy of the peace process' and 'strictly follow the [MILF Central Committee's instructions]'. The group's leaders also spoke of 'growing restlessness and frustration' in the ranks and claimed that individuals with links to the Islamic State, also known as ISIS or ISIL, were recruiting young people in central Mindanao. In response, the group formed a task force of Muslim preachers who aimed to counteract distortions of Islam by ISIS and prevent radicalisation. The Armed Forces of the Philippines (AFP) planned for the contingency in which elements of the MILF returned to violence. This was a prescient move: there were several minor clashes between MILF fighters and the security forces, which had few repercussions but reflected growing unrest within the group.

A government soldier and a member of the MILF died on 10 February, after security personnel pursuing the BIFF entered a MILF base in Datu Saudi-Ampatuan, in Maguindanao. Both the MILF and the army described the incident as an unfortunate accident, stating that they had since strengthened the ceasefire mechanism in the area. By the end of February, the group had repositioned more than 1,700 of its fighters in Maguindanao to help the military's anti-BIFF operations and prevent further such accidents. The MILF also contributed more than 30 fighters to a government operation targeting the Abu Sayyaf Group in Basilan in late March.

Efforts to rejuvenate the peace process began to take shape soon after, as the sides extended their bilateral ceasefire to March 2017. Candidates for the 9 May presidential elections sought to gain the support of the MILF and its followers, to no avail. Presidential candidate Rodrigo Duterte visited the MILF's headquarters at Camp Darapanan, in Maguindanao, on 27 February, stating that he would incorporate the BBL into his vision for a federal system of government. Having won the elections, Duterte vowed at his swearing-in ceremony on 30 June to 'implement all signed peace agreements in step with constitutional and legal reforms'.

Although Murad Ebrahim, chairman of the MILF, warmly congratulated Duterte on a 'historic victory … [for] a true son of Mindanao in whose veins Moro blood runs', the new president and the group took different approaches to the peace process. Duterte wanted to accommodate the political ambitions of Nur Misuari – founding chairman of the Moro National Liberation Front (MNLF) and a former governor of the Autonomous Region in Muslim Mindanao – to stabilise the southern Philippines. Duterte repeatedly spoke of Misuari in positive terms and promised to visit him in Sulu, where he lived in hiding. Having long opposed the MILF peace deal, Misuari endorsed Duterte and his vice-presidential running mate Senator Ferdinand 'Bongbong' Marcos Jr, who led the movement to block the BBL in congress. As such, there was a possibility that Duterte's appointment as president would have a significant impact on the Bangsamoro peace process. Yet by the end of 2016, the president had focused his initial peacebuilding efforts on negotiations with communist insurgent group the New People's Army (NPA), while waging a controversial 'war on drugs'.

The Duterte administration produced in July a road map to address all ongoing peace negotiations in the Philippines. Under the plan, a Moro commission would draft a 'more inclusive' replacement for the BBL that aimed to combine the government's separate agreements with the MILF and the MNLF.

On 11 July, Muslimin Sema, leader of the largest faction of the MNLF, compared the federalism plan to 'keeping an activated bomb at the backburner', adding that it might not be possible to keep the Bangsamoro deal on hold indefinitely. The MILF and the MNLF's Sema faction began work on 'harmonising' their respective peace agreements on 30 July. Duterte said on 24 August that for constitutional reasons he could not establish a Bangsamoro military or police force as part of any deal – one of the issues that met with opposition in congress.

Representatives of the MILF met with government officials over two days in Kuala Lumpur in mid-August. There, they formally launched the implementation phase of the 2014 Comprehensive Agreement on Bangsamoro, alleviating concern about the future of the peace process. Ebrahim publicly welcomed the inclusion of Misuari's faction in the process, adding that the MILF and MNLF had formed a group to investigate the 'possible infiltration of [ISIS] sympathisers' into their organisations, and that Misuari had deployed men to Sulu to contain the ASG.

Duterte suspended the arrest warrant for Misuari for six months, allowing him to travel to Manila for talks on 3 November. The Bangsamoro Transition Commission, tasked with drafting a law to establish the entity, was reconstituted on 10 November following a delay that led MILF leaders to warn of 'jitters' among their subordinates. Misuari refused to join the commission or a joint panel, stating that his faction would instead form its own peace commission to engage with the government. Duterte and Misuari spoke in the presidential palace for a second time on 28 November, shortly before the president reiterated his support for a new BBL that did not have 'constitutionally sensitive' provisions in a meeting with Ebrahim.

Ongoing operations by the Bangsamoro Islamic Freedom Fighters

The BIFF continued its efforts to derail the peace process, often targeting infrastructure projects in its attacks. Clashes involving a bridge-construction project near Datu Salibo, in Maguindanao, began on 5 February and evolved into the largest military operation against the group in a year. The BIFF opposed the construction of transport links that would allow the military to deploy its armoured vehicles more easily. Using heavy weaponry, the army killed at least seven members of an unusually large BIFF force near Datu Salibo, forcing the group out of much of its territory.

The BIFF had planted a large number of improvised explosive devices (IEDs) in the area, reflecting its importance to the group. On 20 February, the army claimed to have defused around 100 of these bombs in Barangay Tee, in Datu Salibo, during a week of operations. The reason for the group's extensive use of IEDs became apparent three days later, when the AFP captured what it called a 'major' BIFF camp in Barangay Tee. The group attacked army outposts in Datu Piang and Datu Salibo on 9 March and 22 March respectively, but was pushed back on both occasions. Civilians in Datu Salibo faced a serious threat from IEDs, and the fighting temporarily displaced around 1,500 of them.

Members of the MILF appeared to be involved in a large-scale clash between the AFP and the ISIS- and ASG-linked Maute Group in Butig, in Lanao del Sur, in February – a battle that resulted in dozens of fatalities. It seemed probable that individual MILF members from nearby camps had fought against the AFP because they had family links with members of the Maute Group, although the MILF leadership claimed that they had also been motivated by congress's failure to pass the BBL. Secretary of National Defense Delfin Lorenzana said on 30 December that the BIFF and the Maute Group had likely formed a 'tactical alliance', a development with the potential to present a serious challenge to the security forces.

The MILF stood aside as the AFP carried out operations against the BIFF in Maguindanao in July–September. Further expanding their cooperation, the MILF and the government signed on 12 July an agreement to jointly combat drug trafficking in areas of Mindanao under the group's control. The MILF ordered more than 700 of its fighters to assist the AFP in tracking suspected drug traffickers.

On 12–15 July, clashes between government forces and BIFF militants in Datu Unsay and Shariff Aguak, in Maguindanao, led to the deaths of 33 insurgents and the injury of seven soldiers, according to the AFP (the insurgents disputed the figures). The military also said that it had gained control of two villages used as BIFF bases in the fighting, which displaced around 6,600 people.

The BIFF split into two factions on 22 July. Abu Amir, spokesman for the splinter group, said that Imam Minimbang – also known as Ustadz Karialan – had formed the new organisation after being unable to accept BIFF leader Commander Bungos's allegiance to ISIS. However, BIFF spokesman Abu Mama Misry denied that there had been a split.

Government forces carried out operations against the Karialan group in Shariff Saydona Mustapha, in Maguindanao, on 26–28 July. Six militants and a soldier died, while another ten militants and four soldiers were injured, in the operations. Fighters from the MILF's 105th and 118th base commands complicated the offensive by assisting the BIFF – perhaps motivated to do so by the fact that Wahid Tundok, commander of the 105th, was reportedly Karialan's cousin.

On 12 August, a counter-narcotics operation in Esperanza, in Sultan Kudarat, resulted in the death of a BIFF commander and the arrest of 11 of his men. A raid in Liguasan Marsh, in Cotabato, two days later was less successful, leading to the deaths of four security personnel and the injury of eight others, but no arrests. Following the operations, the BIFF largely restricted its activities to hit-and-run attacks and the detonation of IEDs. Although the group also attacked army and police facilities in Datu Salibo, Guindulungan and Shariff Aguak on 14–15 December, it was repelled by the security forces. On 24 December, BIFF members blockaded the Maguindanao highway in Guindulungan for three hours, before escaping when the security forces approached.

Continued risk to civilians

Civilians in Mindanao continued to be at relatively high risk from the instability caused by the BIFF's activities, military operations and broader insecurity. The group's fighters reportedly burnt down six houses in a village in Midsayap, in Cotabato, following a clash with government forces on 20 October. Eleven days later, a violent territorial dispute between local MNLF and MILF groups in Makilala, in North Cotabato, forced 140 families to temporarily flee their homes.

Other MILF activities also threatened public safety. On 15 December, a clash between rival MILF commanders in Ampatuan, in Maguindanao, injured three fighters. On 31 December, the authorities arrested four MILF members in Marantao, Lanao del Sur, on kidnapping-for-ransom charges – an activity usually associated with the ASG. These individuals likely acted independently, as the MILF leadership condemned kidnapping and appeared to remain committed to the peace process.

Philippines (NPA)

Philippine President Rodrigo Duterte came to power on 30 June 2016 promising to reach a 'comprehensive peace agreement' with insurgent group the New People's Army (NPA) that included social, political and economic reforms. He immediately set about the task. By the end of September, following productive and cordial official talks, the government and the Communist Party of the Philippines (CPP), the NPA's parent organisation, had implemented indefinite ceasefires – but not a joint ceasefire. The measure created a significant change in the dynamics of the conflict, all but ending clashes between the sides and reducing the average monthly fatality rate from 18 to one. The initial effectiveness of the process reflected the importance Duterte attached to peace with the rebels, and the degree to which they trusted him as a self-described 'leftist' and a former pupil of exiled CPP founder José Maria Sison.

Key statistics	2015	2016
Conflict intensity:	Low	Low
Fatalities:	250	150
New IDPs:		
New refugees:		

The speed of the process likely surpassed even the CPP's high expectations. In the two weeks following the presidential election, and before he took office, Duterte spoke with Sison via Skype to discuss a potential ceasefire and the release of political prisoners, including Benito and Wilma Tiamzon, the CPP's highest-ranking leaders. Duterte also offered cabinet positions to members of the National Democratic Front of the Philippines (NDFP), a leftist umbrella organisation controlled by the CPP. The incoming administration named Rafael Mariano, leader of the farmers' militant group Kilusang Magbubukid ng Pilipinas, as head of the Department of Agrarian Reform, and Judy Taguiwalo, an academic and activist, as head of the Department of Social Welfare and Development. By 26 May, Duterte's negotiating team and the NDFP had agreed on a framework for resuming peace talks.

They met for two days of exploratory discussions in Oslo on 15 June, leading to the release of a joint communiqué in which they agreed to recommend to Duterte that NDFP 'peace consultants' be freed from prison. By 19 August, the Tiamzons and around 17 of their allies had been released on bail to allow them to participate in the talks. The sides declared an indefinite ceasefire on 21 August, marking the beginning of six days of talks in Oslo. The government panel also agreed to recommend that Duterte grant amnesty to detained rebels, a key demand of the CPP.

In addition to the interim ceasefires, the agenda covered measures to affirm previous agreements; accelerate the negotiations (partly by establishing a timeline for completing various aspects of the talks); implement the Joint Agreement on Safety and Immunity Guarantees, which granted immunity to members of the CPP who were involved in the peace process; and release all incarcerated rebels, subject to congressional approval. Following the talks, presidential peace adviser Jesus Dureza asked the United States and the European Union to remove the CPP and the NPA from their lists of international terrorist groups (by the end of the year, neither the US nor the EU had done so).

On 9 October, having completed a second round of talks, the sides declared that they had created a framework for deals on social, economic, political and constitutional reform, as well as for ending hostilities and agreeing on the disposition of forces. However, they were unable to agree on a draft joint ceasefire – as opposed to separate indefinite ceasefires – by the initial deadline of 26 October. Although a government negotiator said that they were likely to sign a ceasefire in late November or early December, this did not happen. Officials said in late November that the sides had been unable to agree on several points, including the definition of a hostile act (in relation to practices such as the NPA's collection of 'revolutionary taxes', or protection money); buffer zones between forces; and the make-up of the ceasefire-monitoring team. Luis Jalandoni, the CPP's chief negotiator, said on 29 September that the NPA would neither give up its arms nor cease to be a functional organisation as part of a peace deal – likely another sticking point.

The rebels increased pressure on the government in November and December. At a 23 November press conference at a jungle encampment in the Sierra Madre mountains, a regional NPA commander stated that his group had 'no reason to enter into a friendship or alliance' with President Duterte while the Philippines remained a US ally and continued to host American troops. The group also demanded the release of 434 insurgents by January 2017. The CPP probably believed that the government might lose interest in political reform once it had achieved its primary goal of ending the violence. This belief may have partly explained why the sides did not declare a joint ceasefire. Duterte reportedly demanded that the CPP sign such an agreement before he would release any insurgents.

Although it was unclear how they viewed such swift rapprochement with the rebels, the Armed Forces of the Philippines (AFP) did not voice any opposition during the initial advances of the peace process under Duterte. However, by late 2016 both the military and the NPA had reportedly grown restless and begun to object to each other's activities. The insurgents complained that what the army called its 'peace and development operations', as well as 'medical missions', in some areas were in fact 'offensive' counter-insurgency operations. Yet given that the NPA remained active in many of these areas, the criticism appeared to be self-serving.

The CPP also accused the security forces of using *Operation Tokhang*, a counter-narcotics effort, to detain suspected NPA members by including their names on lists of suspected drug traffickers. Moreover, the CPP demanded that the government end *Operation Bayanihan*, a counter-insurgency mission, by January 2017 as a condition for extending its ceasefire. The insurgents had long condemned the operation as a cause of human-rights violations. Although Secretary of National Defense Delfin Lorenzana rejected demands for the security forces to withdraw from rural communities in which the insurgents had influence, AFP chief Lieutenant-General Eduardo Año announced on 20 December that *Operation Bayanihan* would be replaced by another initiative in January 2017. In a statement to mark its 48th anniversary, the CPP declared on 26 December that 'there are bound to be armed skirmishes as the AFP conducts armed provocations. Thus, the termination of the CPP's unilateral ceasefire declaration becomes inevitable.'

Marked security improvements

Before the mutual ceasefires, the sides engaged in several battles – albeit none that were strategically important. These clashes tended to follow an established pattern: the NPA used small arms and improvised explosive devices to ambush the security forces, while the latter conducted military operations against the rebels. The southern island of Mindanao experienced the worst of the violence.

On 12 February, the security forces prevented the NPA from attacking a police station and militia base in Carmen, in Surigao del Sur province, by battling the group in nearby Cortes. Several rebels reportedly died in the clash. Four days later, the NPA killed six police officers and injured up to 15 others while reportedly targeting a construction company in Baggao, in Cagayan province. Further small-scale battles between the sides broke out across the Philippines.

Duterte's initial unilateral ceasefire, declared on 25 July, held for only five days before the NPA attacked an army patrol. But the indefinite ceasefires declared by both parties in late August largely held, easing tension between the sides. On 22 September, Duterte ordered the AFP to prevent Magahat–Bagani and other paramilitary groups with links to the armed forces from operating during the peace talks.

From 21 August onwards, the indefinite ceasefires significantly improved the lives of civilians in conflict-affected areas. However, as indicated by unconfirmed reports of harassment of rural indigenous communities by army personnel and extortion and kidnappings by the NPA, there was a possibility that this progress would be quickly reversed.

The NPA continued its campaign of assassinations targeting former members of the security forces, alleged informers, suspected criminals and fighters who had left the group. In addition to intimidating voters in some areas – an activity that local militants also engaged in – the NPA assassinated several political figures in the run-up to the 9 May general elections. On 5 April, the group abducted and killed a municipal-council candidate in Malibcong, in Abra province, who had reportedly refused to pay the rebels for a 'permit' to campaign. The NPA also reportedly demanded that some candidates buy a permit to win.

On 18 May, two NPA child soldiers died in a battle in San Isidro, in Davao del Norte. In contrast to the Moro Islamic Liberation Front, which had signed a UN action plan to end the recruitment and deployment of children, the NPA continued to use minors in its ranks. As a consequence, there were several other violent incidents involving children recruited by the NPA. Having vowed in February to step up its collection of revolutionary taxes from mining, construction and agricultural businesses, the group engaged in this activity throughout the year – despite the ceasefires.

Southern Thailand

The military dynamics of the conflict in southern Thailand changed little in 2016, as insurgents there continued to conduct frequent, small-scale attacks on civilians and the government. Most of these incidents were either drive-by shootings or bombings targeting civilians or the security forces. The insurgents persisted with their campaign against civilians whom they regarded as agents of the state, such as teachers.

A lack of progress in peace negotiations, opposition to the talks from some rebels and two anniversaries important to the insurgents led to a surge in violence in the first three months of the year. Coordinated attacks using several improvised explosive devices (IEDs) became increasingly common in southern Thailand. Yet the Internal Security Operations Command (ISOC) announced that there were 301 violent incidents in fiscal year 2016 (ending in September), the lowest number since 2004. In July, the organisation said that 80% of violent incidents in the region were so-called criminal 'side issues', only some of which were thought to be linked to insurgent groups. However, there was likely a significant overlap between the memberships of insurgent and criminal groups. Deep South Watch, a non-governmental research organisation based in the south, stated that around 300 people died in insurgency-related incidents in southern Thailand in 2016, an annual increase of around 20%.

Key statistics	2015	2016
Conflict intensity:	Medium	Medium
Fatalities:	250	300
New IDPs:		
New refugees:		

Delayed negotiations

A team of peace negotiators from the Thai government visited Malaysia in January to draw up a framework for talks with Mara Patani, an umbrella group representing the insurgents. By 24 February, the sides had settled on '95%' of the terms of reference for the peace dialogue, according to ISOC official Lieutenant-General Nakrob Bunbuathong. The talks had proceeded despite five coordinated bombings by the insurgents (which caused no casualties) on 12 February, the anniversary of an attack on a marine base in Bacho district, in Narathiwat. Other attacks on the same day in Pattani and Yala killed one soldier.

The detonation of a 100-kilogram car bomb in Pattani province on 27 February, one day before the third anniversary of the ongoing round of talks, seemed to be a message of opposition to the negotiations from elements within the insurgency. The blast injured seven police officers and five civilians. The following day, Malaysian peace talks facilitator Ahmad Zamzamin Hashim and Mara Patani Chairman Awang Jabat called on civil-society groups to contribute to the peace process. Meanwhile, Bangkok ordered ISOC to absorb the Southern Border Provinces Administrative Centre for 'integration' purposes, thereby strengthening the armed forces' control of the south (the centre's remit included social, political and economic issues).

Barisan Revolusi Nasional (BRN), the main insurgent group in southern Thailand, marked its 56th anniversary, on 13 March, by carrying out 17 attacks. Three of the assaults injured seven soldiers and security volunteers in Cho-airong district, in Narathiwat, as the BRN stormed a local hospital before using the facility as a base to target security forces stationed nearby. During the attacks, which may have been carried out by members of the group who opposed the peace process, the insurgents destroyed medical equipment and endangered patients – acts that, as Human Rights Watch noted at the time, constitute a war crime. By the end of the month, at least ten people had been arrested in connection with the incidents. Another complex attack took place on the night of 31 March, when seven bombs exploded in succession in Pattani's Yaring district, killing two civilians and wounding five police officers.

In April, the Thai cabinet approved plans to set up a locally recruited, 2,000-person naval regiment of 'para-marines' in Tak Bai district, in Narathiwat. Marines deployed to the south from other provinces were scheduled to be withdrawn following the establishment of the regiment. The process mirrored an army initiative in which soldiers would be replaced by locally recruited soldiers and security volunteers over an indeterminate period of up to three years. However, some observers expressed concern that these projects would foster overreliance on local militias. Prime Minister Prayuth Chan-ocha said on 27 July that all troops from outside the south would be withdrawn by October.

Bangkok and Mara Patani held another round of talks in Kuala Lumpur on 27 April, but were unable to reach an agreement. Two days later, Prayuth said that Thailand could not negotiate with anyone who violated the law and that he did not recognise Mara Patani. The group's official status had long been a sticking point for the government side, which remained wary of implying equality between itself and the insurgents. To circumvent this problem, the government referred to itself as 'Party A' and Mara Patani as 'Party B' in the terms of reference – but the formulation evidently failed to allay Prayuth's concerns.

Six days before the meeting, Prayuth had dismissed Nakrob as secretary of the government's negotiating team and ISOC deputy director due to 'policy differences'. Nakrob, an experienced negotiator who had taken part in the intermittent talks since 2013, reportedly accepted the terms of reference. Narong Sabaiporn replaced him. Although Prayuth insisted that the talks would continue, Nakrob's removal slowed the process. Mara Patani reportedly refused to discuss the situation on the ground in detail, perhaps because it did not represent all insurgent groups and thus had limited influence on front-line fighters. Lead government negotiator General Aksara Kerdphol said that Bangkok would move the process forward only after the agreement of a limited ceasefire.

Insurgents bombed railway stations in Chana and Thepha districts, in Songkhla province, in April. There was speculation that these locations had been targeted in protest against the government's development push in the south, which some perceived as weakening southern cultural identity by largely excluding locals from new projects. On 25 July, the government announced the 'Triangle of Security, Sustainability and Wealth', an economic-development plan for Pattani, Yala and Narathiwat set to run from 2017 to 2022. The authorities stated that the plan – which included incentives for investors, infrastructure improvements and agricultural projects – would be discussed with representatives of the insurgents.

As in previous years, there was an increase in attacks in the lead-up to, and during, Ramadan, which began on 7 June. On 1 June, the security forces overran a camp established by Runda Kumpulan Kecil, a BRN combat unit, in a mountainous area of Chanae district, in Narathiwat. The operation led to the deaths of four militants, two of whom had participated in the 13 March attack on a

Cho-airong hospital. One soldier also died in the operation, while other militants were thought to have escaped. ISOC announced that the security forces captured nine insurgent camps in June, seizing almost 20 firearms. By the end of Ramadan, 32 people had been killed in violent incidents. Meanwhile, ISOC filed defamation charges against members of three non-governmental organisations that published in January a report alleging the systematic use of torture by the security forces. The report contended that the practice undermined support for the state and had the potential to act as a recruitment tool for the insurgents.

Political tension in Thailand rose following a 7 August referendum on a constitution that was drafted following the country's 2014 military coup. The constitution passed with around 61.4% of the national vote. But voters in the three southernmost provinces rejected the charter, with low rates of 'yes' votes in Yala (41%), Narathiwat (38%) and Pattani (35%). Officials claimed that southern voters had rejected the constitution due to intimidation from insurgent groups rather than disapproval of the military. It was possible that bombings conducted in the week leading up to the referendum reduced voter turnout. But distrust of the government and fear that the constitution would further undermine southern cultural identity – the document emphasised the promotion of Buddhism – may also have been significant factors.

On 11–12 August, insurgents carried out a nationwide bombing spree that killed four people and injured 35 others, reaffirming their willingness to engage in large-scale operations outside the south. Some of the 13 bombs they detonated had been planted in the resort town of Hua Hin, in Prachuap Khiri Khan province, where one woman died and at least 19 other people were injured; in Muang district, in Trang province, where one person died and five others were injured; and Muang district, in Surat Thani province, where one person died. The insurgents also detonated devices in Phang Nga province and in the tourist resort of Patong, on Phuket. Another bombing in Hua Hin killed one person and wounded three others.

Officials initially claimed that there were no links between these attacks and the southern insurgency, hinting that political opponents of the government were responsible. The police announced that the IEDs used in the attacks resembled those employed in the south, and stated on 22 August that at least 20 people, most of them Muslims from southern Thailand, had been involved in the bombings. But it took until 6 September for a senior government official to announce that southern insurgents were indeed behind the atrocity. According to a self-styled spokesman for the BRN, the assaults, along with two other bombings in September, were retaliation for the government's 'lack of sincerity' in the peace talks.

The government and Mara Patani took part in another round of unofficial talks in Kuala Lumpur on 2 September, but again failed to reach a final agreement on any substantive issues. The insurgents' negotiating group pledged to consider the government's proposed 'safety zones', and called on civil-society organisations to submit proposals for them. A Mara Patani spokesman said that the two parties agreed on the terms of reference, as did the BRN. However, the government failed to sign off on the terms, with Prawit Wongsuwan, the defence minister and a deputy prime minister, reportedly saying that security must come first.

The insurgents continued to signal their dissatisfaction with, or wholesale rejection of, the peace process through the use of violence. On 6 September, a motorcycle bomb killed three civilians and injured nine other people in Tak Bai district. Three days earlier, a bomb exploded on a train in Pattani province. Although insurgents rarely claimed responsibility for such attacks, a member of Runda Kumpulan Kecil announced that his unit had conducted these assaults because the government was not 'showing sincerity' in the peace talks. Outgoing army commander General Teerachai Nakwanich rejected the accusation.

The government established on 30 September a 13-member special delegation it described as a 'front-line cabinet' for the southern provinces, led by General Udomdej Sitabutr, the deputy defence minister. The body was designed to streamline and coordinate projects and policy, while promoting reconciliation, in the south. However, the delegation was unlikely to create a political solution to the conflict as nearly all of its members were army generals. Nonetheless, two of them were natives of Narathiwat and Pattani respectively: Lieutenant-General Wiwat Pathompak and Panu Uthairat, outgoing secretary-general of the Southern Border Provinces Administrative Centre. The delegation visited Pattani on 2 November. On the same day, insurgents conducted at least 12 attacks across southern Thailand, killing three people. Although officials claimed that the assaults were unrelated to the visit, the violence could easily be interpreted as a protest against the government.

In response to the attacks, the authorities tightened their security measures in Songkhla province, most of which is outside the insurgents' usual area of activity. As part of the effort, the authorities deployed around 400 security personnel to checkpoints in Hat Yai district.

There was little political progress after 13 October, when the death of King Bhumibol Adulyadej, who had spent 70 years on the throne, sank Thailand into a state of mourning. He was succeeded by his son, Maha Vajiralongkorn. However, insurgents continued to use violence as a form of political communication. On 24 October, a bombing outside a noodle shop in Muang Pattani, in Pattani province, killed one woman and injured 18 other civilians. The attack took place on the 12th anniversary of the Tak Bai incident, in which the security forces killed 85 Muslims by packing them into trucks after breaking up a demonstration. This incident directly reignited the southern insurgency. In a separate attack on the anniversary, insurgents killed two more people elsewhere in Pattani.

The authorities detained 44 men, most of them southern Muslims, in Bangkok for three days in October, due to suspicions that they intended to carry out a bombing later in the month. On 17 October, the security forces arrested another man suspected of planning to detonate car bombs in the city. On 3 November, the alleged plots prompted the government to strengthen its security measures in Bangkok.

Insurgents killed a teacher and injured another civilian in a drive-by shooting in Mayo district, in Pattani, on 28 October. The authorities found a note nearby that read 'for you who killed Melayu [ethnic Malay] people'.

Chapter Seven

Europe and Eurasia

Armenia–Azerbaijan (Nagorno-Karabakh)

Key statistics	2015	2016
Conflict intensity:	Low	Low
Fatalities:	60	200
New IDPs:		
New refugees:	1,000	

The overall dynamics of the conflict in Nagorno-Karabakh remained largely unchanged in 2016. Negotiations on conflict resolution were deadlocked. In the first three months of the year, Armenia and Azerbaijan continued to regularly accuse each other of using heavy weaponry and violating the ceasefire along the Line of Contact. In a dramatic escalation, on 2 April fighting broke out in Aghdara, Tartar, Agdam, Khojavend and Fuzuli – territory controlled by the Nagorno-Karabakh Defense Army – before spreading to other areas (the sides blamed each other, but there was little verifiable information on who was responsible). Deteriorating security, armed clashes and exchanges of fire were still a constant feature of the conflict. However, three aspects of the April escalation, which became known as the Four-Day War, were particularly disturbing: the sides' use of more artillery, tanks and aircraft than at any time since the 1994 ceasefire; shifts in patterns of territorial control; and the number of casualties. The US Department of State estimated that around 200 people were killed during this phase of the escalation. Other sources suggested there were more than 300 fatalities, although these

316 Europe and Eurasia

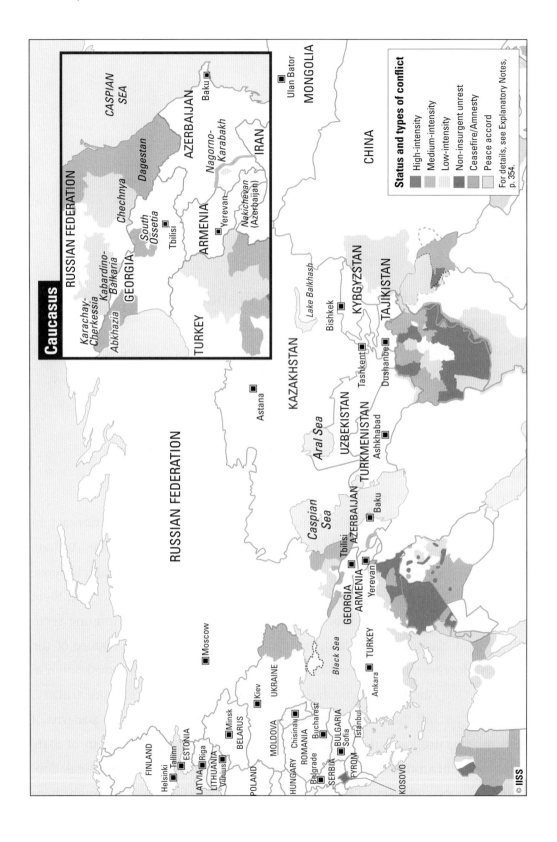

figures could not be independently verified. The hostilities reportedly involved the use of *Grad* multiple-launch rocket systems. In addition, territory changed hands for the first time since 1994. Azerbaijan gained control of an estimated 800–2,000 hectares along the Line of Contact. Although these gains were not strategically important, they had symbolic significance. Azeri officials claimed a tactical victory – as the first tangible shift in more than two decades, this was a crucial change to emphasise to the Azerbaijani population, especially given that Azerbaijan's national identity had theretofore been largely defined by defeat.

The escalation reflected both domestic and international politics. The fall in oil prices and the devaluation of Azerbaijan's currency had deepened the country's socio-economic problems and increased popular dissatisfaction with political elites. Baku may have therefore used the 'Karabakh factor' to justify extensive defence spending and to rally public opinion behind combating the 'Armenian enemy' as the most readily available unifying idea. The escalation of violence appeared to be an attempt to address internal difficulties by distracting attention away from social and economic hardship. Rising prices and discontent at growing unemployment had prompted protests in several locations across Azerbaijan in 2016. Nonetheless, these considerations did not explain the timing of the outbreak of violence. The fact that the flare-up coincided with a key visit of the Armenian and Azerbaijani presidents to the United States suggested that the attention of international audiences was also important. The deterioration of relations between Russia and Turkey appeared to have been a contributing rather than determining factor in the escalation. Turkish President Recep Tayyip Erdogan openly reaffirmed his support for Azerbaijan, stating in early April that Nagorno-Karabakh would 'inevitably return to Azerbaijan'. He may have intended to exert pressure on Russia by further destabilising an already tense situation.

Although the fighting seemed to threaten to escalate into full-scale war, it was unlikely that Azerbaijan would risk such a confrontation given the unpredictability of its outcome. At the state level, the need to be prepared for all-out war had made militarisation and increased defence spending a dominant process not only in Azerbaijan but also in Armenia and Nagorno-Karabakh. Although Azerbaijan had an overwhelming military advantage, Armenia and Nagorno-Karabakh controlled strategically important high ground and had Russia's support. Thus, Azerbaijan could not count on the success of military operations, and it was possible that there would be a repeat of the outcome of the 1992–94 war, in which the country lost 20% of its territory. Any major use of military force in the region also ran counter to Moscow's interests. Russia regards Armenia as a key ally and Azerbaijan as an important strategic partner. Moscow remained a major guarantor of security in Nagorno-Karabakh due both to its mediation efforts as a co-chair of the Minsk Group, part of the Organization for Security and Co-operation (OSCE), and to the Russian military base in Armenia. Russia had affirmed its responsibility and preparedness to protect Armenia from military aggression. Indeed, the Russian military presence was widely regarded in Armenia as a key component of the country's military strategy. Armenia's membership of the Eurasian Economic Union had further strengthened this security alliance. Russia's rearmament of both sides attracted criticism from Armenia for running contrary to this alliance, as well as from Azerbaijan. In February, Russia agreed to provide Armenia with a US$200-million loan to buy new Russian weapons. According to the agreement's appendix, these armaments included *Smerch* rocket launchers, *Igla*-S surface-to-air missile systems, *Avtobaza*-M ground-based radar-jamming and electronic-warfare systems, *TOS*-1A heavy flame-throwing systems, 9M113M guided missiles, RPG-26 grenade launchers, *Dragunov* sniper rifles and *Tigr* armoured vehicles, as well as communication systems. In response to the arms sale, Baku sent a note of protest to Moscow.

Meanwhile, bilateral relations between Russia and Azerbaijan had extended to military, economic and humanitarian spheres. In this delicate situation, Russia continued doing everything in its power to persuade both of its partners to avoid further escalation. The prospect of a large-scale entanglement with Russia seemed to moderate the behaviour of Azerbaijan, which had tended to avoid antagonising Moscow. On 5 April, the chiefs of staff of the Armenian and

Azerbaijani armed forces agreed on a Russian-mediated ceasefire. Although the violence continued following the deal, it was less intense.

However, any momentum that international attention might have provided to the conflict-resolution process appeared to subside later in the year. Russian-led diplomatic efforts failed to produce tangible results. On 16 May, the Armenian and Azerbaijani presidents met in Vienna, along with Russian, US and French representatives. They agreed to set up mechanisms for investigating violent incidents within the OSCE's framework and to expand the presence of the organisation's Monitoring Mission, while reaffirming their commitment to the ceasefire. The presidents participated in another meeting in St Petersburg on 20 June, but despite rhetorical assurances the event did little to change the status quo and they reached no formal agreements. Towards the end of the year, Armenian and Azerbaijani officials publicly blamed the other for the outbreak of violence in April. These statements maintained the hostility and belligerent rhetoric that characterised the conflict in 2016. Defining arrangements that would adequately address disagreements and provide security guarantees to both parties proved to be as elusive as ever.

Russia (North Caucasus)

There was little change in the security situation in the North Caucasus in 2016. The fatality rate there declined, albeit minimally. Meanwhile, the Islamic State, also known as ISIS or ISIL, appeared to acquire growing influence in the region as the Caucasus Emirate, previously the dominant militant jihadist group, continued to be disorganised and unable to reassert its authority. The Caucasus Emirate

Key statistics	2015	2016
Conflict intensity:	Medium	Low
Fatalities:	200	175
New IDPs:		
New refugees:		

also continued to lose leaders, with the authorities killing several of its cell commanders. Local affiliates of ISIS claimed responsibility for the vast majority of violent attacks in the year. The Russian security services consistently used heavy-handed methods to crack down on individuals whom they believed to have connections with the group or to have sought to travel to the Middle East to fight for it.

The security forces conducted a counter-terrorism raid in Karabudakhkent district on 7 July, killing the leader of the Caucasus Emirate's Makhachkala group and eight other suspected militants. The last three months of 2016 saw a rise in violent incidents. Dagestan remained the most violent region, accounting for nearly 50% of attacks. The spike in violence was particularly apparent in regions that had theretofore been largely quiet in the year, such as Kabardino-Balkaria and Ingushetia, which experienced an increase in fatalities. As in 2015, the authorities continued to target and thus antagonise Muslim worshippers who appeared to be devout, detaining hundreds of them in the process.

On 4 December, the Russian Federal Security Service stated that it had killed a leader of a local ISIS affiliate and four of his close associates. Reportedly responsible for several attacks in Dagestan and elsewhere, he had fought for the Caucasus Emirate before becoming the first jihadist in the North Caucasus to pledge allegiance to ISIS, in 2014. The long-term effect of his death on the jihadist movement remained uncertain, but it was possible that Russia would capitalise on it to divide, weaken and further suppress the insurgency. Given the losses ISIS incurred in the Middle East, this suppression may have become easier than it was in previous years. Nonetheless, there was a strong likelihood that other leaders would seek to use the ISIS brand to establish their authority in the jihadist movement and gather support for the insurgency in the North Caucasus. With Moscow's security strategy focused almost exclusively on military force, local groups' growing ideological and operational links to transnational jihadism had significant potential to create a rise in instability and violence.

Ukraine

Key statistics	2015	2016
Conflict intensity:	High	Medium
Fatalities:	4,500	700
New IDPs:	900,000	100,000
New refugees:	155,000	

The conflict in eastern Ukraine was contained in 2016, with neither side making major territorial gains and the violence restricted to areas near the established Line of Contact. The annual fatality rate declined and, according to the Organization for Security Co-operation in Europe (OSCE) Special Monitoring Mission (SMM) and the UN High Commissioner for Human Rights, 83 civilians died in the fighting, compared to around 250 the previous year. Nonetheless, the United Nations and the SMM reported incidents involving heavy, indiscriminate shelling and the killing of civilians by both sides throughout 2016. The belligerents repeatedly violated the Minsk II ceasefire, as well as local ceasefires (such as that for Orthodox Easter).

Despite periodic surges in the violence, the fighting in Donbass showed the hallmarks of an ambivalent, constantly simmering conflict, as diplomatic efforts ended in stalemate. Indeed, the sides appeared to lack the political will to implement Minsk II. By the end of the year, they had made no substantial progress in meeting the criteria of the agreement, nor developed a clear plan for moving forward.

Repeated ceasefire violations

Intermittent skirmishes and ceasefire violations, as well as the presence of heavy artillery, continued to dominate the conflict. In contrast to the events of 2015, there were no major offensives, and the Line of Contact remained largely unchanged. The violence was at its most intense in the Donetsk airport area; Yasynuvata and Avdiivka; west and north of Horlivka; the region of the Debaltseve–Svitlodarsk road; east and northeast of Mariupol; and western Luhansk. The SMM reported around 300,000 violations of Minsk II – most of them involving exchanges of mortar, artillery or rocket fire – and described the violence as persisting at 'unacceptable' levels.

The complex environment and restrictions on access for monitors on both sides of the Line of Contact made it difficult to verify which party initiated battles most often. The heaviest fighting occurred in warmer months: the UN Human Rights Monitoring Mission in Ukraine (UNHRMMU) reported a 66% increase in civilian casualties between May and August compared to earlier in the year, and documented 28 civilian fatalities – many of them from shelling and landmines – in summer. The OSCE warned in May of 'worrying levels' of armed violence and shelling by separatists, especially as the fighting escalated on the outskirts of government-held Mariupol and Avdiivka.

On 29 April, the Trilateral Contact Group – comprising representatives from Russia, Ukraine and the OSCE – reached a ceasefire agreement out of respect for the Orthodox Easter period. But the Easter ceasefire was violated days later by separatist shelling. The fighting intensified in the Donetsk and Luhansk regions, especially in July, with the Ukrainian defence ministry reporting that 27 soldiers were killed and 123 injured in that month alone. The UN deemed it the bloodiest month since August 2015. Between mid-August and mid-November 2016, the conflict killed at least 32 civilians and injured another 132.

Posturing by both the Ukrainian army and pro-separatist forces – including irregular Russian military exercises along the Ukrainian border and, allegedly, incursions by Ukrainian special forces – led to flare-ups in the conflict. The UNHRMMU documented human-rights violations by both sides, including alleged torture, kidnap for ransom, prisoner abuse and sexual assault. In July, separatists allegedly kidnapped an OSCE monitor accused of spying. The perpetrators of such crimes were rarely brought to justice. The UNHRMMU report found that the security services enjoyed widespread impunity for human-rights violations, which the authorities often attributed to the challenges of war fighting.

In late September, the Trilateral Contact Group agreed on a disengagement process intended to protect civilians in conflict-affected areas. This Framework

Decision was meant to result in a temporary local ceasefire, during which troops and weaponry would be disengaged in the areas of Petrivske, Zolote and Stanytsia Luhanska. The sides aimed to create a four-square-kilometre neutral zone around the Line of Contact in which neither would be permitted to advance. The disengagement process commenced in Zolote and Petrivske in September–October, resulting in a marked reduction in ceasefire violations.

There was a sharp rise in violence in the Svitlodarsk area of Donbass in December, with the SMM reporting heavy fighting north of Debaltseve. These clashes included shelling on both sides of the Line of Contact, some of it with large-calibre weapons banned by the Minsk II agreement, such as 122mm and 152mm artillery. Civilians and combatants also came under threat from the growing number of mines and unexploded ordnance in eastern Ukraine.

Unstable separatist movement
The leaders of the separatist Luhansk People's Republic (LNR) and the Donetsk People's Republic (DNR) engaged in infighting throughout 2016, and several of them were assassinated. In October, Arseny 'Motorola' Pavlov – a controversial, high-profile commander in the DNR – was killed by a bomb planted in his apartment building. Other DNR separatists also disappeared or died in unclear circumstances. The DNR claimed that the Ukrainian security forces were responsible for Pavlov's assassination – for which DNR 'prime minister' ovowed to take revenge.

In mid-September, a failed coup attempt against LNR leader Igor Plotnitsky led to the arrest of dozens of separatist commanders, senior officials and soldiers. Among them were alleged coup leader Gennady Tsypkalov, who had served as LNR 'prime minister' until being forced out in 2015, and LNR 'deputy defence minister' Vitaly Kiselev. Both men died in custody (Tsypkalov reportedly committed suicide).

However, the removal of high-ranking leaders had little effect on LNR and DNR combat capability, as the separatists remained well supplied – likely by Russian sympathisers and almost certainly by the Kremlin. The infighting resulted from competition for lucrative smuggling rackets and trade routes along the border, as well as for fighters and other resources in the war economy. The separatists were also aware that strengthening their presence on the battlefield could attract greater patronage from Moscow. Plotnitsky and his close advisers seemed to have secured privileged access to these resources, particularly as Tsypkalov and Kiselev were thought to have been heavily involved in the black market for products such as coal and gas. Meanwhile, the Kremlin continued to deny financing the LNR or the DNR, maintaining that it had no influence on the separatists.

Deadlock in implementing Minsk II

The slowdown in fighting in eastern Ukraine in 2016 resulted from political stagnation rather than a concerted effort to comply with the Minsk II ceasefire. Little progress was made in the effort to implement the agreement. Ukrainian President Petro Poroshenko was still unable to convince the Verkhovna Rada (parliament) to support the deal – partly because this would have involved passing a resolution that established an 'Order of Local Self-Governance in Particular Districts' in the Donetsk and Luhansk regions, clearing the way for local elections. Although he had signed Minsk II in 2015, Poroshenko was unwilling to push it through parliament as this would have likely caused the collapse of his government. One of the Rada's major objections to the deal was its provision of immunity to the separatists, a requirement that was unacceptable to Ukrainian nationalists. Moreover, under Article 9 of Minsk II, Kiev had agreed to hold the elections only after it established territorial control of the Ukraine–Russia border, a concession that the separatists had proved unwilling to make because it would have cut off their access to trade routes, supplies, personnel and funding.

Nonetheless, DNR leader Zakharchenko announced in September that the rebels were fully committed to Minsk II and saw it as the 'only solution' to the conflict. Despite such statements, the separatists were reluctant to end the fighting completely as this might cause them to lose territory if the Ukrainian government reneged on the deal. In the long term, they faced fewer threats from a frozen conflict than from allowing Kiev to reassert control of the east. As a consequence, the separatists were resistant to implementing Minsk II.

With no stable commitment to the agreement by the separatist leadership, the Kremlin or the Ukrainian government, the conflict reached a stalemate, with most battles limited to maintaining existing areas of control. Yet there was little prospect of an end to the violence, due to the close proximity of the Line of Contact, the continued presence of heavy artillery and the lack of incentives for either side to respect a ceasefire.

By the end of the year, the conflict had killed around 10,000 people, according to the UN. But there was still no clear road map for implementing Minsk II. The SMM reported that heavy weaponry and military personnel had not been withdrawn from Ukrainian territory as required under the deal. Meanwhile, the disengagement agreement appeared to have stalled. The war had some benefits for both Poroshenko and the separatists in the short term, providing the former with a justification for Ukraine's economic problems and allowing the latter to profit from black markets in the east. For instance, the trade in illegal weapons (including those sold by returning Ukrainian soldiers) flourished throughout the year. As a result, there were twice as many prosecutions for weapons offences in January–July 2016 than in January–December 2015.

In contrast, the war's detrimental effects on legal trade – between Ukraine and Russia, and between Ukraine and the European Union – pressured senior decision-makers to make substantive progress in implementing Minsk II. Poroshenko struggled to persuade Ukrainian voters and the Rada to accept the deal, as a diplomatic stalemate, an underperforming economy and internal political battles occupied his attention and weakened his support base. Russia, though not a signatory to the agreement (nor even mentioned in it), had an increasingly complex role in Ukraine due to the changing strategic environment. Following the election of Donald Trump as US president, the possibility that Washington would lift Western sanctions against Russia for its involvement in the conflict left Moscow with little incentive to act quickly. Meanwhile, Ukrainian forces were almost certainly becoming increasingly disappointed with a war against separatists who refused to relinquish power or territory.

Despite the political stagnation, the conflict settled into a pattern in which the violence could rapidly escalate at any time, as the combatants focused on securing control of territory near the Line of Contact and minimising their distance from enemy positions. The use of heavy artillery and other weaponry in or near urban areas increased the likelihood of civilian casualties, as did the growing quantity of mines and unexploded ordnance in the east. Although most ceasefires lasted only a short time, they demonstrated that the sides were capable of suspending the conflict in the right conditions. Yet the signatories to Minsk II had few reasons to fully implement it; their lack of concerted effort to do so seemed to ensure that the conflict would continue.

Chapter Eight

Latin America

Central America (Northern Triangle)

Key statistics	2015	2016
Conflict intensity:	High	High
Fatalities:	17,000	16,000
New IDPs:		
New refugees:	8,000	

The Northern Triangle countries of Honduras, Guatemala and El Salvador strengthened their institutional capacity to fight corruption in 2016, establishing new agencies and implementing reforms to reduce criminal groups' influence on politicians, business leaders and security personnel. Although the conflict there caused the deaths of around 16,000 people in the year, this was around 10% fewer than in 2015. Endemic violence and intimidation by powerful gangs combined with economic underdevelopment to drive migration northwards: the number of unaccompanied children from Northern Triangle countries apprehended at the southern US border increased by 65% in fiscal year 2016 (ending on 30 September).

Migration driven by insecurity

El Salvador experienced a 22% fall in the annual fatality rate, the largest decrease in the Northern Triangle. Nonetheless, with around 5,199 murders, 2016 was still the country's second-most-violent year since 1999. The high number of killings

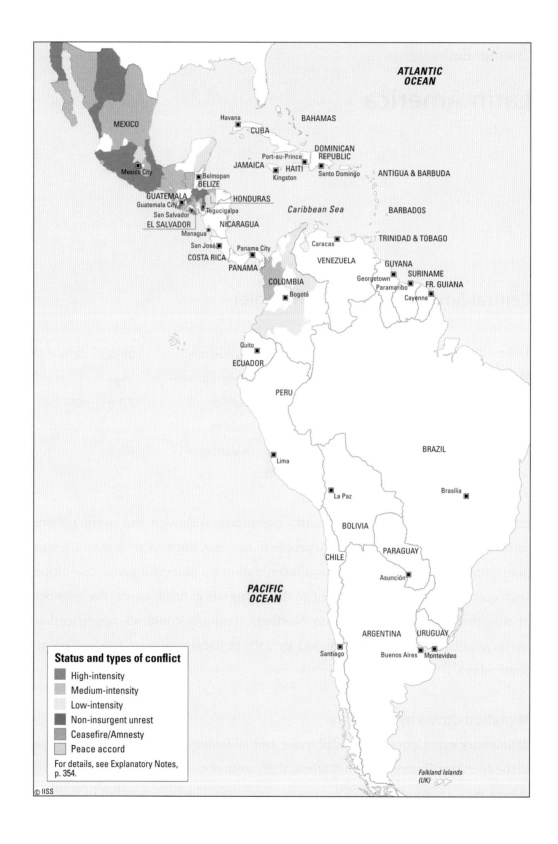

reflected the significant presence, firepower and organisational capacity of rival gangs Mara Salvatrucha (MS-13) and Barrio 18 (which remained split into two factions, Revolucionarios and Sureños). The gangs also continued their campaign against security personnel, targeting and killing 46 police officers and 27 members of the armed forces – most of them while they were off duty. In April, the government launched the Specialised Reaction Forces, a hybrid organisation comprising 600 soldiers and 400 police officers. This was the latest in a series of paramilitary security forces to be formed in the Northern Triangle in recent years, following the Military Police of Public Order in Honduras and the Special Reserve Corps for Citizen Security in Guatemala.

In Honduras, the annual number of homicides fell by just 1.7% to around 5,060, despite a 24% increase in the country's security and defence budget. There, criminal groups had an increasing impact on the economy as they expanded their traditional extortion of urban-transportation (mainly bus) companies into much larger industries. After a series of attacks on vehicles transporting food and automotive goods in August, Armando Urtecho, head of the Honduran Council of Private Enterprise, demanded 'stronger, more energetic' anti-crime measures from authorities.

Guatemala experienced a 4.5% decrease in the annual fatality rate, in line with the gradual decline in criminal violence there since 2014. Although the continuation of this trend reduced public pressure on President Jimmy Morales during his first year in office, he made no visible changes in the government's security strategy, instead spending most of his energy on the anti-corruption institutional reforms that he had promised to implement while campaigning. Judging by the few hints he gave, Morales seemed to favour a security strategy in which the military remained involved in internal security. The president confirmed in July that the armed forces would remain deployed on the streets, adding that the National Civilian Police was not yet ready to become solely responsible for fighting crime. Despite the deployment of the army and the gradual decline in homicides, Guatemala had the highest annual number of murders in the Northern Triangle: 5,459.

The 46,863 unaccompanied children from Northern Triangle countries apprehended at the US southern border in fiscal year 2016 represented an annual rise of 65%, dramatically reversing the trend in 2015. The largest number of these children came from Guatemala (18,913), followed by El Salvador (17,512). There was an even sharper increase in the number of 'family units' (individuals accompanied by at least one family member) apprehended at the US southern border: 104%. A flurry of reports on the dangers facing unaccompanied migrants transported by 'coyotes' (people smugglers) may have caused this trend. Meanwhile, the Mexican authorities apprehended 152,231 citizens of Northern Triangle countries in January–December 2016, deporting 93% of them to their countries of origin.

Children and families fled not only criminal violence but also the gangs' influence on urban peripheries, slums and rural areas that had little state law-enforcement capacity. In October, Jeh Johnson, then US secretary of homeland security, stated that 'far fewer Mexicans and single adults are attempting to cross the border without authorization, but more families and unaccompanied children are fleeing poverty and violence in Central America'. In a pointed reference to the declared migration policies of Donald Trump (then the president-elect), Johnson added: 'border security alone cannot overcome the powerful push factors of poverty and violence that exist in Central America. Walls alone cannot prevent illegal migration.'

Like Mexico, Northern Triangle countries worried that Trump would reduce the aid they received from the US government to tackle organised crime and encourage social development. Exacerbating this concern, Washington had postponed the disbursement of the Alliance for Prosperity – a US$750-million aid package approved by Congress in late 2015 – while it verified that the three countries met 16 anti-corruption, human-rights and migration requirements. The preoccupation with potential aid reductions was evident in journalists' questions to Northern Triangle foreign ministers following Trump's electoral victory in November. Salvadoran Foreign Minister Hugo Martinez said that the Alliance for Prosperity was not at risk because it had bipartisan support in Congress.

Juan Sebastian Gonzalez, US deputy assistant secretary of state for Western Hemisphere affairs, made the same argument during a visit to Guatemala on 12 December. Gonzalez said that he also expected the aid package to be disbursed because it was in the US national interest. Thus, Northern Triangle states expected to begin receiving the funds in early 2017.

Transnational anti-crime efforts

The Northern Triangle countries also adopted regional mechanisms to improve security cooperation as part of the Alliance for Prosperity, establishing on 15 November the Tri-National Force. The Salvadoran, Honduran and Guatemalan presidents launched the force – comprising around 1,500 personnel from their police and militaries, as well as border and customs agencies – with great fanfare at a ceremony in the Honduran town of Ocotepeque, which borders both Guatemala and El Salvador. But it remained unclear whether this force would be able to fulfil the wide array of missions assigned to it along a 600-kilometre shared land border. These tasks included measures to counter extortion, kidnapping, money laundering, gang violence and smuggling – each an activity for which the three countries established working groups to support the Tri-National Force. Honduran President Juan Orlando Hernández, who first proposed the initiative, said that the three countries were also strengthening the coordination of, and exchange of information between, their police and militaries.

The Tri-National Force filled a noticeable gap in regional security cooperation. Despite their recognition of major transnational links between criminal groups such as MS-13 and Barrio 18, Northern Triangle states lacked joint policies and frameworks that would allow for coordinated, efficient investigations and intelligence sharing. The first Central American security-strategy document for countering organised crime was published in 2006 and revised in 2011, but resulted in few significant measures aside from occasional joint operations. While the Central American Integration System, an established institutional framework, focused on high-level strategic meetings, the Tri-National Force appeared to provide a more pragmatic mechanism for coordinating operations.

The force had responsibility for 'intelligence and counter-intelligence', according to the Honduran government. The agreement that created the new force, signed in August 2016, stated that the three countries would enforce arrest warrants issued in any part of the Northern Triangle. Prior to the agreement, the countries' attorney generals agreed to increasingly coordinate their investigations into transnational crime, sharing information on criminal leaders and improving their communication about the arrest of suspects. The inspiration for the Tri-National Force was the Maya–Chortí Task Force, a smaller initiative that Guatemala and Honduras launched in 2015 to coordinate security operations and encourage intelligence sharing.

Collectively, Barrio 18 and MS-13, alongside smaller gangs, had between 54,000 and 85,000 members spread across urban areas in Honduras, Guatemala and El Salvador, according to estimates made by the UN Office on Drugs and Crime and the US Department of State in 2012. The Guatemalan authorities provided a reminder of the transnational nature of the gangs on 2 October 2016, when they arrested two key leaders of the Salvadoran branch of MS-13 in Guatemala City, before extraditing them to El Salvador. On 19 October, Guatemala's National Police revealed that 365 Salvadoran nationals, 329 of them from Barrio 18, had been detained on suspicion of gang-related activities in the country in the year.

Although other major criminal groups maintained drug-trafficking routes in Central America, they lacked the territorial influence and large membership of MS-13 and Barrio 18 in the region. One such group was the Atlantic Cartel, which was headquartered in Honduras but managed drug-trafficking operations in Colombia, Panama, Nicaragua, Costa Rica, Honduras and Guatemala. Investigators and security agents from the United States, Costa Rica and Honduras conducted operations against the cartel in 2016, leading to the arrest of Wilter Blanco, its leader, in Costa Rica on 22 November. A few days later, and possibly in connection with Blanco's arrest, the Costa Rican and Honduran authorities announced the arrest of 12 police officers who had allegedly been part of a transnational drug-trafficking ring.

Incomplete reforms

Alongside these modest, if important, improvements in regional security cooperation, the Northern Triangle countries persisted in their efforts to address institutional weaknesses often exploited by criminal groups. Guatemala and Honduras attempted to make crucial changes to their justice and police institutions respectively, but encountered obstacles that risked reducing the scope of proposed reforms.

Anti-corruption campaigns preserved the momentum they had gained in September 2015, when a bribery scandal led to mass protests and the resignation of the Guatemalan president, Otto Pérez Molina. The scandal set the tone of his successor's first year in office. Morales proposed an overhaul of Guatemala's customs administration and sought to introduce new anti-graft measures in its justice system. Although Morales made progress in these efforts in Congress, another corruption scandal reduced his political clout. On 15 September 2016, Julia Barrera, a spokesperson for the attorney general's office, announced that the authorities had issued an order prohibiting José Manuel Morales Marroquin and Samuel Everardo Morales – the president's son and brother respectively – from leaving the country. The president's relatives were suspected of financial improprieties relating to a supporter of his electoral campaign. The incident prompted Morales to publish a video message in which he described his family situation as 'difficult' and vowed not to interfere in the case.

Nonetheless, Morales won a small victory in his anti-corruption campaign on 19 July, when Congress approved a bill revamping Guatemala's tax authority. The main feature of the legislation was to transfer responsibility for appointing the head of the agency from the president to a panel of experts and the finance minister. The measure also established a tribunal for settling tax and customs disputes, as well as for supervising the head of the tax authority. Moreover, a reformed internal audit system would be responsible for investigating tax and customs systems and employees.

Morales met with relatively strong congressional resistance when he put forward a more ambitious bill: a reform of the justice system that involved changes to 20 articles of the constitution. The bill proposed the creation of the National Council of Justice, which would oversee careers and promotions in the judicial system and appoint judges to appellate courts. Furthermore, the council would make the system more transparent by directly selecting Supreme Court judges (revoking the right of various of external actors, including universities and the bar association, to influence the process). The draft bill was also designed to rescind the immunity from prosecution of the president, the vice-president, ministers and all members of Congress, allowing the public prosecutor to initiate an investigation into any one of them without seeking permission from the Supreme Court.

Congress voted down the reform on political immunity in November and, by the end of the year, had failed to pass the rest of the bill due to disagreements about its proposal to recognise traditional judicial practices among indigenous peoples. This initial defeat combined with corruption scandals and ongoing criminal violence to damage the image of Morales as a champion of anti-graft protesters.

Announced in January 2016, the Mission to Support the Fight against Corruption and Impunity in Honduras received congressional approval in late March. The mission had a four-year mandate and comprised Honduran and foreign legal experts tasked with fighting corruption, reforming the justice system, pursuing political reform and strengthening public security. Luis Almagro, head of the Organisation of American States, praised the Honduran government for pledging to give anti-corruption investigators full access to all of its documents.

Honduras also set up a commission to reform its police force, following reports that high- and mid-ranking National Police officers had been involved in the 2009 murder of Julian Aristides Gonzalez, head of the country's counter-narcotics office. According to *El Heraldo* and the *New York Times*, these officers had planned Gonzalez's murder in collusion with a drug trafficker. The revelations had a huge impact in Honduras and prompted immediate policy responses. On 7 April, two days after *El Heraldo* published the story,

Congress approved a law that established police vetting as an emergency measure. A few days later, President Hernández announced the formation of the Commission for the Restructuring and Reform of the Security Secretariat and the National Police, which was given 12 months to raise the standards of, and increase public support for, these institutions. Shortly thereafter, the commission suspended 25 officers on suspicion of having participated in the murder. By the end of the year, 2,091 officers had been expelled from the police following investigations and screening that uncovered various cases of corruption and gang infiltration. In September, the authorities revealed that 81 police officers had links to MS-13. The commission also passed cases involving 500 officers, including three police generals, to the public prosecutor for further investigation, indicating the extent of the challenge facing anti-corruption reformers.

Colombia

Key statistics	2015	2016
Conflict intensity:	Medium	Medium
Fatalities:	350	200
New IDPs:	200,000	90,000
New refugees:	2,500	

Although 2016 was a turbulent year for the peace process between the Colombian government and FARC, the sides reached a final deal in November – four years after talks started in Havana. The previous month, large sections of Colombia's society and political elite had received an initial agreement with scepticism, resulting in its rejection by a narrow margin in a referendum. The main opponents of the deal said that it was too lenient, allowing former guerrillas to avoid prison sentences and run in elections while offering amnesty to an excessive number of them. Despite these obstacles, however, the renegotiated peace deal paved the way for ending Colombia's 52-year war.

Search for peace

The victory of the No campaign in the referendum left many Colombians in shock, creating a crisis in President Juan Manuel Santos's administration. Opinion polls had consistently pointed to the approval of the peace deal, but No won with 50.2% of the vote. Shortly after its defeat in the plebiscite, the administration began talks with the opponents of the deal, including former president Alvaro Uribe, Santos's main political rival and the leading voice of the No campaign.

Uribe stated that the peace deal needed 'adjustments and proposals', having made too many concessions to FARC. To salvage the peace process, Santos announced that he would establish a 'broad and inclusive' commission tasked with conducting a national dialogue on potential changes to the agreement. The commission included members of Uribe's right-wing Democratic Centre party. The leaders of the many political parties that backed Yes, which had a comfortable majority in Congress, supported Santos's efforts to continue the peace process.

In the days following the referendum, there emerged a broad consensus on pursuing changes to the peace agreement rather than scrapping it. Rodrigo 'Timochenko' Londoño, leader of FARC, reiterated his commitment to the process, saying that his group maintained its 'goodwill towards peace and its disposition to use only words as a weapon to build the future'. His subordinates seemed to broadly agree with this position, as there were no recorded violations of the bilateral ceasefire in October.

The peace effort gained impetus on 7 October, just five days after the referendum, when the Norwegian Nobel Committee announced that it had awarded the 2016 Peace Prize to Santos. Nonetheless, the rejection of the deal at the ballot box reflected Colombians' persistent suspicion and resentment of FARC. Many regions with a strong guerrilla presence, such as the northeastern departments of Norte de Santander and Arauca, voted against the peace deal. The No campaign was also victorious in traditionally conservative departments such as Antioquia. However, other rural areas that had been hit hard by the war in recent years voted strongly for the peace deal, as seen in the western departments of Choco and Cauca.

The government and the opposition conducted frantic closed-door talks to draw up a list of proposed changes to the agreement, which were taken to the FARC negotiators in Havana on 25 October. To the surprise of many, the negotiating teams took less than three weeks to agree on an amended version of the peace deal. Lead government negotiator Humberto de la Calle declared that the new deal 'resolves many criticisms' directed at the original agreement. But this did not end the controversies and fierce political exchanges of the preceding months.

President Santos stated that almost all 'themes' in the original document had gone through 'changes and improvements'. The one exception was the political participation of former guerrillas, which remained unchanged despite being one of the main targets of criticism from the No campaign. Under the deal, FARC would be guaranteed five seats in the Senate and five in the Lower House, even if it did not gain a sufficient number of votes in the next two general elections. Thereafter, only elected representatives would gain seats.

The new deal failed to address another core demand of Uribe's supporters: the introduction of prison sentences for guerrillas found guilty of committing serious crimes – understood to be war crimes and crimes against humanity. Under the agreement, these actors would receive alternative sentences such as confinement to specific areas of the country and mandatory participation in demining efforts. The main change in this area was the establishment of a 15-year deadline for the work of the special tribunal tasked with receiving and judging alleged crimes by either side in the armed conflict.

Another important issue raised by the No campaign was the prospect that drug traffickers linked to FARC would be given amnesty under the deal. This concern stemmed from a clause in the document that described drug trafficking as a crime connected to the insurgency, a stipulation that the government deemed necessary to achieve peace. After all, FARC was so deeply involved with the illicit drugs trade that many (if not most) of its members would otherwise be imprisoned under normal criminal-justice procedures. Yet following heavy criticism from the No campaign, the negotiators added a significant caveat: drug trafficking would only be considered under the transitional-justice mechanism (and therefore fall under the amnesty provision) if the accused 'has not derived personal enrichment' from the activity. That is, as long as the crime was conducted exclusively by guerrillas for FARC's benefit, the crime would be handled within the transitional-justice system.

Following another tense meeting between the government's negotiating team and the opponents of the deal, it became clear that Santos's original idea for a 'national accord' for a revised peace deal would be difficult, if not impossible, to achieve. Following these discussions, the opposition said that the government had 'denied the possibility of a national accord'. After 51 days in which he seemed to be the greatest victor of the referendum, Uribe again felt left out of the decision-making process and resorted to condemning the peace deal. He said the new document was 'a slight makeover' from the one that citizens had rejected the previous month, adding that 'impunity remains practically the same as in the first deal'.

But Santos received a strong boost from members of the No campaign unconnected to the Democratic Centre. In a speech on 22 November, the president said that victims of the conflict, the business sector, the Catholic Church, governors and mayors backed the new deal.

Relying on the pro-peace majority in Congress, Santos announced that the amended agreement would not be subjected to a referendum but would instead be ratified through votes in the Lower House and the Senate. Although the new deal failed to resolve many of the doubts that Colombians expressed in the referendum, the only major voice of opposition to Santos's approach came from the Democratic Centre. As a consequence, Uribe's faction looked much more isolated than it had prior to the referendum.

There were no surprises this time and, with the Democratic Centre boycotting the votes, the peace agreement passed the Lower House with a vote of 130–0 and the Senate with a vote of 75–0 on 29–30 November. Therefore, on 30 November the peace deal began its implementation phase.

Post-conflict disputes

The ratification of the peace deal gave rise to a difficult search for the legislative, security and developmental measures needed to implement and sustain it. The first, highly symbolic steps were taken by FARC: the guerrilla group started its last march as an armed organisation, moving towards 26 transitional zones established by the government in rural areas. There, the military and UN observers would supervise the demobilisation and disarmament process. Minister of Defence Luis Carlos Villegas announced on 10 December that 'all FARC structures are now on the move towards the transitional zones'.

The march was originally scheduled to start immediately after Congress approved the deal, but the guerrillas' uncertainty about their legal status delayed it by at least ten days. They feared that the government would renege on key promises or imprison them after they surrendered their weapons. Iván Márquez, FARC's chief negotiator in the peace talks, announced on 6 December that the group would only move towards the transitional zones 'once the juridical obstructions are removed'. He said this mainly in reference to the special legislative mechanism for peace still under evaluation by the Constitutional Court, which created a road map for fast-tracking the debate and approval of laws crucial to implementing the peace agreement. The measure reduced the maximum number of congressional debates for each legislative bill from eight to four. It also prevented lawmakers from making changes to the documents, limiting them to 'yes' or 'no' votes.

A sense of urgency began to build among government authorities and FARC leaders. 'Now we need certainty, action and to move on to implementation', argued peace commissioner Sergio Jaramillo shortly after the approval of the deal. Jesús Santrich, a FARC negotiator, told the press that unless the juridical obstructions were removed, 'we will go back to the mountains'.

The Constitutional Court resolved this anxiety on 13 December, when it approved the special legislative mechanism. Shortly thereafter, Congress used the mechanism to pass the amnesty law, one of the government's key concessions and a crucial part of the juridical security that underpinned the peace agreement. The law was expected to benefit at least 5,000 FARC fighters and many more members of the armed forces, although it excluded those linked to war crimes or crimes against humanity.

The growing threat that spoiler groups would fill the vacuum left by FARC in remote rural areas also motivated the government to quickly implement the deal. Impoverished former FARC strongholds in Meta, Cauca, Norte de Santander and many other departments were focal points of armed groups' extortion rackets, as well as their activities in illicit mining and cocaine production. In the weeks following the approval of the agreement in Congress, international organisations monitoring the peace process issued warnings that criminal groups were moving into these areas. The Organisation of American States said that some of these groups 'could be related to paramilitary structures of the past' – likely a reference to right-wing paramilitary organisations that officially demobilised under a peace agreement in the 2000s. Two of Colombia's largest criminal groups, Los Urabeños (also known as Clan Usuga) and Los Rastrojos, traced their origins to former paramilitary leaders and their arsenals. But the Organisation of American States warned that a variety of smaller criminal groups were also moving into former FARC territories. The Office of the United Nations High Commissioner for Human Rights issued a similar warning, stating that 'as FARC guerrillas leave areas that are traditionally under their control, the state has not yet fully stepped in, leaving a power vacuum'.

In the uncertain days between the referendum and the signing of the amended peace agreement, the Santos administration also raised concerns about spoiler groups and the possibility that FARC would splinter. The president said that Congress needed 'urgency to make quick progress' in the analysis of the peace agreement due to threats from 'armed groups wanting to fill the spaces being left by FARC'. Villegas said in October that he was 'immensely worried about the deterioration in cohesion within FARC', citing a partial breakdown in the group's discipline and chain of command.

This statement followed the declaration in July that a FARC faction had abandoned the peace process for the first time. The 1st Front, one of the group's largest and richest factions, publicly announced that it would neither disarm nor gather in the transitional areas, prompting the FARC leadership to expel the dissidents and Villegas to name them as high-priority targets for the armed forces. Although the 1st Front argued that it had split off for political reasons – stating 'we will continue our struggle to take power for the people' – there was a strong indication that leaders of the faction had an economic motive. As the government admitted, many criminal networks were eyeing profitable enterprises in FARC territories such as illicit gold mining and cocaine production.

The 1st Front operated in the sparsely populated southeastern department of Guaviare, but lacked the capacity to do so in other regions. It received ample revenue from coca cultivation and trafficking, benefiting from the lack of state presence in Guaviare. The 1st Front was regarded as one of FARC's 'mother fronts', meaning that it was large enough to aid smaller fronts and held enough territory to host training camps. The 1st Front was thought to have 200–400 fighters, but it remained unclear whether some of them had chosen to rejoin FARC and support the peace process.

Although the group was reluctant to acknowledge dissidence within its ranks, Colombian observers reported that there was a strong possibility factions involved in illicit economies would split off. On 13 December, the group's military leadership announced in a short statement that it would expel five senior members of the Eastern Bloc, an umbrella structure that had included the 1st Front. In a bitter irony for the peace process, the most high-profile leader among those expelled was Gentil Duarte: having taken part in the Havana talks, he had been sent back to Colombia in July to lead fighters from the 1st Front who were willing to rejoin FARC.

In other indications of the economic roots of dissent, FARC also expelled the head of the 16th Front, located in the rural Vichada Department, as well as John Cuarenta, head of the 7th Front and widely regarded as the guerrillas' most prolific drug trafficker. Cuarenta had a long-standing presence in Meta and Guaviare departments, where he had orchestrated terrorist attacks and clashes with security forces. However, the FARC leadership did not reveal how many lower-ranking fighters had followed their leaders in the rebellion.

The bilateral ceasefire implemented on 29 August 2016 proved to be broadly successful. The most serious violation of the arrangement came on 16 November, when the army stated that it had killed two FARC members in a gunfight in the town of Santa Rosa, in Bolivar Department. Demonstrating the risks posed by the wide range of armed groups in the Colombian countryside, the army operation that resulted in the killings had been launched in response to reports of illegal mining by the National Liberation Army (ELN), Colombia's second-largest guerrilla organisation. Another incident linked to possible spoilers of the peace process took place on 12 November, when FARC fighters clashed with an unidentified armed group in San Pedro del Vino, in Nariño Department, resulting in the deaths of two people. The UN-backed verification mechanism stated that FARC had violated the ceasefire, leading the group to publicly acknowledge its involvement in the gunfight nearly one month later.

Prior to 29 August, FARC observed a unilateral ceasefire, with few reported violations. Yet the reduction in armed violence in Colombia in 2016, though significant, frustrated expectations that peace would prevail after the FARC agreement. The annual number of terrorist attacks decreased by slightly more than 50% in comparison to 2015. The continued defiance of the ELN was one of the main reasons that such activity continued, as the group conducted many of the 62 attacks on the country's economic infrastructure (including oil pipelines, electricity pylons, roads and bridges) in the year.

These operations by the ELN, combined with its reluctance to release kidnapped former lawmaker Odín Sánchez, caused many to doubt that the group would be willing to negotiate a peace agreement. Indeed, the ELN was also responsible for a series of attacks on military and police targets, resulting in many of the 100 fatalities among the security forces in 2016 – only 38% less than in the previous year. On 1 December, Santos postponed talks with the ELN until 2017 because of the group's failure to release Sánchez, despite its promises to do so. On 18 December, following the assassination of two soldiers in Arauca Department by the ELN's Ernesto Che Guevara unit, the head of the government's negotiating team, Juan Camilo Restrepo, said that the continued attacks pushed the possibility of a ceasefire 'further away'.

Colombia's criminal groups posed a security challenge very different to that from leftist guerrillas. These groups conducted fewer direct assaults on infrastructure, military and law-enforcement targets, but had large quantities of weapons – a legacy of the right-wing paramilitary groups from which many of them had emerged. Los Urabeños staged several bold operations against the state. On 1 April, the group imposed a strike in several parts of Colombia to mark the killing by police of José Morela Peñate (also known as Negro Sarley), one of its top leaders. This resulted in the closure of many businesses across 36 municipalities in eight departments.

Following the strike, Santos stated in a speech that 'the fight against organised crime' had become 'a priority of the Colombian state'. Villegas acknowledged that the criminal threat was complex: whereas Los Urabeños was still the most powerful criminal group, and the only one with a nationwide presence, he said that Colombia was also home to 39 medium-sized gangs and 400 small gangs, many of which worked in connection with larger criminal organisations. The most significant anti-crime measure came in May, when Villegas announced that the government no longer classified Los Urabeños, Los Puntillos and Los Pelusos as 'criminal bands' but as 'organised armed groups'. According to a Ministry of Defence directive, the move allowed the authorities to counter the groups with all the 'strength of the state, without exception', including lethal force, in accordance with international humanitarian law. Thus, the move potentially allowed for the armed forces to become more involved in internal law enforcement, perhaps by using airpower against criminal targets.

Although violence continued to threaten civilians living in most rural and many urban areas of Colombia, the government's peace process with FARC had humanitarian benefits even before it was officially approved, with an initial demobilisation of minors in the group. On 10 September, FARC released 13 of these young people to the International Committee of the Red Cross in rural areas of Meta and Antioquia. While FARC had not publicly stated the number of minors in its ranks, the Ministry of Defence reported in May that the group had recruited 170 children and adolescents, 70 of whom were girls. The demining process also proceeded, albeit at a slower pace due to the scale of the challenge. Such progress on humanitarian issues provided a first glimpse of the peace agreement's capacity to produce broad socio-economic benefits, especially in rural Colombia.

Mexico

Mexican President Enrique Peña Nieto experienced the worst period of his term in 2016. Security, a major platform of his presidential campaign, deteriorated significantly in the year, with a 22.8% rise in the annual homicide rate. Moreover, Congress diluted and stalled an effort to overhaul Mexico's police structure, one of his main institutional-reform proposals.

Key statistics	2015	2016
Conflict intensity:	High	High
Fatalities:	17,000	23,000
New IDPs:	9,000	
New refugees:	900	

Deteriorating security

One key cause of Peña Nieto's declining popularity was the widespread sense that the fight against criminal groups had regressed during his fourth year in power – a perception hard to dispute given that there were around 23,000 homicides in Mexico in 2016, compared to around 17,000 the previous year. The leap in the murder rate was particularly apparent in 22 of the country's 32 states, suggesting that his strategy of reducing the military's involvement in law enforcement had failed.

There appeared to be several factors behind the surge in violence. Some of the killing resulted from the growing ambitions of the Jalisco New Generation Cartel (CJNG), which continued to wage war against declining rivals such as the

Knights Templar. In 2016 the CJNG expanded its campaign beyond the southwestern states of Guerrero, Michoacan and Jalisco to directly challenge other gangs, including the Sinaloa Cartel, the only other group with a nationwide presence.

Early in the year, the Sinaloa Cartel used social media to announce that it had arrived in the CJNG-dominated Colima – another southwestern state – to 'cleanse' the territory (of its rivals). Although there was no conclusive evidence of a direct causal relationship between criminal ambitions and homicide numbers, Colima's annual murder rate trebled, reflecting the destabilisation of its underworld. The state with the second-largest increase in homicide numbers, Veracruz, in the southeast, was another battleground for the CJNG. There, the cartel reportedly fought remnants of Los Zetas.

Zacatecas experienced a similar territorial struggle, as the CJNG, Los Zetas and the Gulf Cartel battled for control of a strategically important drug-trafficking route in the state. In line with a trend of diversification in criminal revenue streams, the drugs involved seemed to be the synthetic products becoming more popular in United States (the cartels' primary market) rather than the more common cocaine. News reports indicated that the CJNG instigated much of the violence by establishing large laboratories for producing high-purity crystal methamphetamine in Zacatecas – perhaps prompted to do so by the destruction of one such facility in Jalisco in late 2015.

Guerrero was yet another southwestern state in which the cartels clashed over markets for drugs other than cocaine. Guerrero Governor Hectór Astudillo attributed the killing of 30 people in separate incidents on 18–21 November to 'confrontations' between criminal groups for control of the local trade in opium poppies, the main ingredient of heroin. The conflict centred on competition between La Familia Michoacána, Los Ardillos and Los Rojos.

Having traditionally avoided conflicts with other gangs, the Sinaloa Cartel came under increasing pressure as it waged war against relatively new organisations such as Los Zetas. The authorities recaptured Joaquin 'El Chapo' Guzmán, the Sinaloa Cartel's leader and founder, on 8 January, prompting the Mexican government to begin talks on his extradition to the US, where he faced a series of criminal charges. The move was widely regarded as necessary to prevent El Chapo from escaping prison once again and, perhaps more importantly, from communicating with his successors in the Sinaloa Cartel.

In its eponymous home state, the gang faced the heavily armed Beltrán Leyva Cartel, which Mexican journalist Anabel Hernández described as an ally of Los Zetas and the CJNG. According to local news reports, the Beltrán Leyva Cartel was responsible for a June attack on the town of La Tuna, home of El Chapo's mother, Consuelo Loera. In the assault, 150 armed men killed eight people and sacked Loera's house, but left her alive.

Another bold challenge to the Sinaloa Cartel was the kidnapping on 15 August of Jesus Alfredo Guzmán Salazar, El Chapo's son, from a restaurant in Puerto Vallarta, a coastal resort town in Jalisco. Eduardo Almaguer, the state's prosecutor general, said the CJNG had carried out the abduction. Jesus Guzmán Salazar was released less than a week later, reportedly following negotiations involving one of El Chapo's most trusted lieutenants, Ismael 'El Mayo' Zambada. Despite a lack of details on the negotiations, there was speculation the Sinaloa Cartel had met some of the CJNG's demands by relinquishing control of drug-trafficking routes or territories.

In late 2016, the looming extradition of El Chapo appeared to cause tension among the leaders of the Sinaloa Cartel. On 30 September, armed men ambushed a Mexican Army convoy in Bacacoragua, a town in its home state. The attack bore the hallmarks of Aureliano 'El Guano' Guzmán Loera, El Chapo's brother and one of his most violent lieutenants. Although El Guano likely carried out the ambush – particularly as Bacacoragua was thought to be directly under his control – the authorities also suspected that El Mayo had been involved. Indeed, El Mayo had been an important voice of authority during El Chapo's incarceration, and there was speculation that he would take over the leadership of the cartel. Regardless of whether El Guano had been behind the ambush, it was highly unusual for the gang to have carried out such a direct attack on the military in Sinaloa, indicating a potential breakdown in the status quo.

Yet the signs of tension did not develop into a direct confrontation between senior leaders. The Sinaloa Cartel maintained a strong position as one of the two largest gangs in Mexico, along with the CJNG. Nonetheless, in contrast to previous years, it had become unclear whether El Chapo remained the figurehead of the organisation – and, if not, who had replaced him.

Problems of command

Despite deterioration in the security landscape nationwide, Peña Nieto's *mando único* (single command) programme – a reform originally designed to place Mexico's 1,800 municipal police forces under the leadership of state governments – was only put to a vote in June 2016, more than a year and half after it had been submitted to the Senate. Along the way, the programme had undergone significant changes and been renamed *mando mixto* (mixed command). Rather than allowing state governments to take over municipal forces, the state and federal governments would only be allowed to intervene in municipal security in extraordinary cases of criminal violence.

However, the project was still the most wide-ranging reform of Mexico's law-enforcement structures in decades. Under the initiative, the Executive Secretariat of the National Public Security System would gain additional powers to supervise and regulate all police organisations in the country, improving coordination between them and the standard of their work. The secretariat would also be empowered to replace the leadership of police forces that proved ineffective or corrupt, and to enforce consistency across the police bureaucracy. Intended to address alleged corruption and human-rights violations involving the police, the bill specified that police officers would be given a minimum wage and that the rules on their promotion would be standardised. By the end of 2016, it had passed the Senate but not the lower house.

Another sweeping anti-crime institutional reform targeted the justice system. Approved by Congress in 2008 but only fully implemented on 18 June 2016 following a transitional phase, the measure introduced judgments based on oral statements during public hearings and allowed judges to hand down sentences that did not involve incarceration. With its predecessor having been based solely on written testimony, the new system was designed to increase transparency and prevent human-rights violations, as confessions would only be admissible in court if made before a judge.

On 18 July, Mexico introduced a law reforming its anti-corruption system. The new legislation required public officials to declare their assets and appointed a tribunal and prosecutor who would only investigate corruption cases involving government figures or institutions. Shortly after the law was signed, the Mexico Chamber of Commerce published a survey showing that corruption in the public sector was business leaders' top concern (followed closely by extortion and attacks on supply chains). Around 50% of respondents described the risk of corruption as 'high'.

Peña Nieto's efforts to project a positive image of Mexico abroad suffered a blow on 8 November, with the election of Donald Trump as the next US president. Trump's confrontational approach to Mexico threatened to reverse security and intelligence cooperation between the countries, and to damage their economic relationship. As a presidential candidate and president-elect, he had described Mexicans as 'rapists'; promised to build a wall on the southern US border and make Mexico pay for it; revise or scrap the North American Free Trade Agreement (NAFTA), a deal important to Mexican exports; and deport or imprison three million illegal migrants in the US, some of whom were Mexican citizens. As a consequence, the value of the Mexican peso dropped by 13% the day after Trump's electoral victory. Although the currency later rallied – closing at 19.9 per US dollar, an 8.5% decline from the start of the day – this still amounted to its largest one-day fall since January 1995.

As part of the Mexican authorities' scramble to respond to the threats posed by Trump's victory, the central bank raised interest rates by 50 basis points to prevent inflation due to a weaker peso. Felipe Acero, a senior official at the interior ministry, told Mexican lawmakers that as many as 1m people could be deported from the US to Mexico in the next four years.

The Peña Nieto administration signalled that its negotiations with the US would involve a broad range of issues beyond trade. Eduardo Sánchez, a spokesman for the president, said on 16 November that deportations and immigration would be part of any such talks. The strategy was put forward to give Mexico leverage over the White House by hinting that it could end bilateral cooperation with US authorities on issues such as immigration, crime and intelligence – all of which were important to the US national interest. Many observers expected that these issues would be linked to any US attempts to significantly alter NAFTA in ways that undermined the competitiveness of Mexican industry.

Washington channelled the majority of its security assistance to Mexico through the Merida Initiative, an aid package that covered institutional capacity-building, donations of military and security equipment, and border-security measures. By November 2016, the US had disbursed US$1.6 billion of the US$2.6bn allocated to the initiative. It seemed that Mexico would not receive the remaining amount if Trump chose to curtail foreign aid.

His election also cast doubt on the future of US–Mexico intelligence cooperation, which had reportedly led – through the Drug Enforcement Administration and other US agencies – to the capture of El Chapo in February 2014. Although the Mexican government denied US involvement in operations to capture the gang leader, it stated that the bilateral security relationship was friendly and cooperative – as did Washington.

However, judging by the events of late 2016, there was a distinct possibility that animosity between US and Mexican leaders could cause significant damage to the relationship. After provoking the ire of Mexican voters by inviting Trump to Mexico City during the presidential campaign, Peña Nieto quickly changed his approach. While Trump said that he did not discuss paying for the wall with Peña Nieto, the Mexican president had a different recollection of their conversation, posting on Twitter that he 'made clear to Donald Trump that Mexico will not pay for the wall'. The day after the meeting, with public anger at Trump rising, Peña Nieto called the then-candidate's proposals a 'threat' to Mexico. Yet Trump had already cost the Mexican government a great deal of political capital. A Consulta Mitofsky survey published on 19 November found that Peña Nieto had an approval rating of 24%, the worst of any Mexican president in his fourth year in power. In August, Peña Nieto's approval rating was 29%, according to the firm.

Ten years of the 'war on drugs'

The ten-year anniversary of the so-called 'war on drugs' came in December 2016, likely reminding Mexicans of the dangers of directly confronting powerful criminal groups by deploying the armed forces in support – and sometimes instead – of the police. Mexico's National Human Rights Commission highlighted the impact of crime on the population in the decade, reporting in May 2016 that 35,433 people had been forcefully displaced nationwide since 2007. Around 90% of these people had fled their homes because of violence.

The rise in criminal violence made Peña Nieto's campaign promise to reduce reliance on the military seem increasingly unrealistic. On 20 December, the president urged lawmakers to debate and approve an internal-security bill being drafted by the Senate's political-coordination group. The measure aimed to establish a legal framework for the military's involvement in internal security, providing a more reliable set of rules for soldiers fighting criminal groups. Peña Nieto delivered a speech defending efforts to consolidate the military's internal-security role almost exactly ten years after his predecessor announced the deployment of troops in the war on drugs.

Chapter Nine

Explanatory Notes

The Armed Conflict Survey provides analysis and data on active armed conflicts involving states and non-state armed groups across the world. The data in the current edition is accurate according to IISS assessments as at January 2017, unless specified. Inclusion of a territory, country, state or group in *The Armed Conflict Survey* does not imply legal recognition or indicate support for that entity.

General Arrangement and Contents

The introduction provides an overview of some of the key themes of the book. Next, five analytical essays focus on select transnational trends: UN peacekeeping, conflict-related sexual violence, the Islamic State's shifting narrative, the changing foundations of governance by armed groups, and rebel-to-party transitions. A graphical section follows, presenting a number of conflict maps and graphics. The sections providing data and analysis on each conflict are organised according to region. The book closes with a reference section.

The Chart of Conflict

The Chart of Conflict shows the location and intensity of armed conflicts and terrorist attacks. The theme of the 2017 edition is UN peacekeeping.

Individual conflict entries in *The Armed Conflict Survey*

For each conflict, there is an analytical essay identifying and critically assessing key developments in the political, military and humanitarian aspects of the conflict. This is complemented by an analysis of conflict intensity and data on fatalities, refugees, IDPs and returnees.

Conflict intensity

For each active conflict, an assessment is made as to the intensity of the fighting:

- High-intensity conflicts involve frequent (daily) armed clashes between governments, government forces and insurgents, or among non-state armed groups that control territory. Typically, fatalities as a direct result of conflict exceed 3,000 per year.
- Medium-intensity conflicts involve regular armed clashes between governments, government forces and insurgents, or among other non-state armed groups that control territory. Typically, fatalities as a direct result of conflict exceed 300 per year.
- Low-intensity conflicts involve occasional clashes between governments, government forces and insurgents, or among other non-state armed groups that control territory. Fatalities as a direct result of conflict are typically below 300 per year.

Fatalities

Fatality statistics in *The Armed Conflict Survey* relate to military, insurgent and civilian lives lost as a direct result of armed conflict. Some data is estimated and all results are rounded. The figures are derived from publicly available information, including international, national and local press reports, as well as the annual assessments of international organisations (where available). Care is taken to ensure that our data is as accurate and free from bias as possible. The data reflect judgments based on information available to the IISS at the time the book is compiled. It may be subject to revision.

Refugees

According to the 1951 Refugee Convention (Article 1A(2)), the term 'refugee' is applied to a person who 'owing to well-founded fear of being persecuted for reasons of race, religion, nationality, membership of a particular social group or political opinion, is outside the country of his[/her] nationality and is unable or, owing to such fear, is unwilling to avail himself[/herself] of the protection of that country; or who, not having a nationality and being outside the country of his[/her] former habitual residence as a result of such events, is unable or, owing to such fear, is unwilling to return to it.'

The refugees entry refers to the number of people that as a result of conflict moved out of the host country in the last year. It measures the flow rather than the total number, or stock, of refugees created by a conflict.

Although multiple sources are consulted as part of our analytical process, refugee data used in *The Armed Conflict Survey* is sourced from the United Nations High Commissioner for Refugees.

Internally displaced persons (IDPs)

IDPs are individuals or groups who have been forced or obliged to flee their homes as a result of, or in order to, avoid the effects of armed conflict, situations of generalised violence or violations of human rights. The IDP figures in the tables correspond to the number of people displaced that particular year, and not to the total number of IDPs since the beginning of the conflict. IDP figures and information comes from the Norwegian Refugee Council's Internal Displacement Monitoring Centre.

Attribution and acknowledgements

The IISS owes no allegiance to any government, group of governments, or any political or other organisation. Its assessments are its own, based on the material available to it from a wide variety of sources. Some data in *The Armed Conflict Survey* are estimates. Care is taken to ensure that these data are as accurate and free from bias as possible. The Director-General and Chief Executive and staff of the Institute assume full responsibility for the data and judgements in this book. Comments and suggestions on the data and textual material contained within the book, as well as on the style and presentation of data, are welcomed and should be communicated to the Director of Editorial at: IISS, 13–15 Arundel Street, London WC2R 3DX, UK; email: publications@iiss.org. Application to reproduce limited amounts of data may be made to the publisher: Taylor & Francis, 4 Park Square, Milton Park, Abingdon, Oxon, OX14 4RN; email: society.permissions@tandf.co.uk. Unauthorised use of data from *The Armed Conflict Survey* will be subject to legal action.

Key to regional maps

Status and types of conflict

High-intensity – frequent (daily) armed clashes between governments, government forces and insurgents, or among non-state armed groups that control territory.

Medium-intensity – regular armed clashes between governments, government forces and insurgents, or among other non-state armed groups that control territory.

Low-intensity – occasional clashes between governments, government forces and insurgents, or among other non-state armed groups that control territory.

Non-insurgent unrest

Ceasefire/Amnesty – agreed and/or announced by recognised leaders of parties to the conflict. Does not stand as a resolution of the conflict and does not suggest that all conflict has stopped.

Peace accord – formal resolutions of conflict ratified by recognised leaders of parties to the conflict or between intervening party and post-conflict government. In some cases, conflict may still persist.

Index

A

Abadam (Nigeria) 182
Abadi, Haider al- 91, 95–100
Abbas, Mahmoud 103, 104, 105, 106, 107
Abbassia (Egypt) 86
Abdullah, Abdullah 220, 222, 223, 283
Abra (Philippines) 308
Abu Grein (Libya) 117
Abuja (Nigeria) 184, 187, 191
Abu Sayyaf Group (Philippines) 292–302, 304
Abyan (Yemen) 160
Acero, Felipe 348
Achin (Afghanistan) 228, 231
Adamawa (Nigeria) 187
Addis Ababa (Ethiopia) 180, 181, 209–211
Aden (Yemen) 153, 155, 156, 160
Adeosun, Lamidi 184
Adnani, Abu Muhammad al- 39, 40, 42–46, 138
Afghanistan 5, 6, 7, 19, 217–232, 234, 268, 269, 273–276, 283
    Afghan National Army 224–227
    Afghan National Security Forces 217, 219, 223–229
    Bagram air base 228
    Camp Morehead 228
    Kajaki Dam 224
    Ministry of Defense 226
    National Directorate of Security 222
    Operation Khanjar 225
    Operation Qahar Selab 229
    Operation Shafaq 224, 228
    Pul-e-Charkhi prison 225
    Quadrilateral Coordination Group 219
    Supreme Court 223
African Union 7, 11, 172, 196, 209–211, 214
    High-Level Implementation Panel 209
    Mission in Somalia 11, 196–198
    Operation Antelope 197
Afrin (Syria) 139
Agadez (Niger) 184
Agdam (Nagorno-Karabakh) 315
Agha, Tayyab 221
Aghdara (Nagorno-Karabakh) 315
Agip (Italy) 192
Ahmad Wal (Pakistan) 220
Ahmar, Ali Mohsen al- 155
Ahmed, Ismail Ould Cheikh 154
Ahmed, Zannah Ibrahim 182
Ahrar al-Sham (Syria) 46, 47, 139
Ajnad al-Sham Islamic Union (Syria) 140

Akhtar, Naeem 259
Akhundzada, Haibatullah 220
Akol, Lam 206
Aksara Kerdphol 311
Aksu (China) 281
Akwa Ibom (Nigeria) 194
Alamieyeseigha, Diepreye 190
Al-Azarak (Sudan) 213
Al-Bab (Syria) 42, 136, 139, 151
Al-Barka (Philippines) 297
Aleppo (Syria) 8, 41, 42, 46, 47, 108, 111, 129–131, 133–139, 142, 143
Algeria 41, 54
Alhabsy Misaya (Philippines) 296
Aliabad (Afghanistan) 231
Allied Democratic Forces (DRC) 170, 173, 175–177, 178
Alloush, Zahran 140
Almagro, Luis 334
Almaguer, Eduardo 346
Al-Majlis al-Islami al-Suri 46
Almar (Afghanistan) 226
Al-Naba 45, 183
al-Qaeda 46, 116, 120, 132, 140, 198, 228, 230, 265, 273
al-Qaeda in the Arabian Peninsula 154, 160–162
al-Qaeda in the Indian Subcontinent 265, 273
al-Qaeda in the Islamic Maghreb 123
Al-Rai (Syria) 138, 139, 151
King Salman bin Abdulaziz Al Saud 90
al-Shabaab (Somalia) 11, 58, 196–198, 201
Al-Tal (Syria) 141
Amaechi, Rotimi 190
Amaq 294
Amin, Hussin 293
Amir, Abu 303
Amnesty International 88, 100, 112, 214
Ana (Iraq) 42, 92
Anantnag (India) 260
Anbari, Abu Ali al- 138
Anbar (Iraq) 41, 42, 91–93, 96, 100
Andhra Pradesh (India) 239–243
Angola 64
Ankara (Turkey) 147
Annan, Jeannie 31
Annan, Kofi 21, 290
Año, Eduardo 294, 307
Ansar al-Dine (Mali) 122, 123, 128
Ansar al-Sharia (Libya) 120
Ansari, Abu Duaa al- 85

Antioquia (Colombia) 336, 343
Aoun, Michel 108, 109, 111
Arab League 111
Arakan Army (Myanmar) 287, 288
Aramco (Saudi Arabia) 90
Arauca (Colombia) 336, 342
Arish (Egypt) 85
Arjona, Ana 57, 58
Armenia 315–319
Arrojado, Alan 295
Arsal (Lebanon) 110
Arunachal Pradesh (India) 236, 245, 247, 249–251
Asghar, Abdul Rauf 255
Asian Development Bank 238
Asif, Khawaja 258
Asir (Saudi Arabia) 158
Asosa (Ethiopia) 211
Assad, Bashar al- 8, 54, 90, 95, 128, 129, 132, 142, 143
Assam (India) 232–237, 245–247, 249, 250, 251
Astudillo, Hectór 345
Atlantic Cartel (Honduras) 332
Aung San Suu Kyi 285–288, 292
Avdiivka (Ukraine) 321, 322
Awan, Paul Malong 208
Ayn Issa (Syria) 42, 136
Azaria, Elor 102, 104
Azaz (Syria) 42, 136, 139
Azerbaijan 315–319
Azhar, Maulana Masood 254, 255
Aziz, Khalid 275
Aziz, Sartaj 219, 257, 267, 275
Azzi, Sejaan 112

B

Baalbek (Lebanon) 112
Bab al-Mandeb Strait 159
Bab al-Nayrab (Syria) 134
Bacacoragua (Mexico) 346
Badakhshan (Afghanistan) 222, 225
Badgam (India) 260
Badia (Syria) 41
Badibanga, Samy 172
Badi, Salah 115
Badme 181
Badreddine, Mustafa 111
Baghdadi, Abu Bakr al- 40, 41, 44, 45
Baghdad (Iraq) 58, 96, 97, 98
    Green Zone 97, 98
Baghlan (Afghanistan) 227
Bagh, Mangal 273
Baghran (Afghanistan) 225
Bahah, Khaled 155
Bahariyah (Syria) 141
Bahawalpur (Pakistan) 255
Bahrain 45, 138, 158, 159
Bahr el-Ghazal (South Sudan) 206
Baiji (Iraq) 42, 94
Bajwa, Asim 273
Bajwa, Qamar Javed 273
Balajan Tiniali (India) 235
Balkh (Afghanistan) 222, 230
Balochistan (Pakistan) 220, 254, 257, 262–265, 269–271, 274
Baloch Liberation Army (Pakistan) 269
Baloch Liberation Front (Pakistan) 269
Baloch Republican Army (Pakistan) 269
Baloch, Sanaullah 271
Bamako (Mali) 122–124, 127
Bambari (CAR) 163, 167, 169
Bamungaon (India) 236
Bandipora (India) 260
Bangkok (Thailand) 314

Bangladesh 19, 34, 233, 234, 291
Bangsamoro Islamic Freedom Fighters (Philippines) 299, 301, 302, 303
Bangsamoro (Philippines) 299, 300, 301
Bangui (CAR) 165–169
    PK5 165, 166
Bani Naim (Palestinian Territories) 101
Bani Zayd (Syria) 134
Ban Ki-moon 15, 17, 21, 131, 201, 257, 287
Baramulla (India) 258, 260
Barangay Tee (Philippines) 302
Barghathi, Mahdi al- 121
Barghouti, Marwan 104
Barisan Revolusi Nasional (Thailand) 310, 311, 313
Barnawi, Abu Musab al- 183
Barrera, Julia 333
Barrientos, Gerry 295
Barrio 18 (Central America) 6, 329, 331, 332
Barzani, Masoud 98, 99
Bashiqa (Iraq) 93, 99, 151
Bashir, Omar al- 211, 212, 213
Basilan (Philippines) 294–298, 300
Basit, Abdul 257
Basra (Iraq) 96
Bassil, Gebran 112
Bastar (India) 242
Batta, Anna 65
Bauma, Fred 171
Bayelsa (Nigeria) 190, 192, 194
Bayingolin (China) 281
Bayo (Nigeria) 182
BBC 256
Beirut (Lebanon) 108–110, 112
Bekaa Valley (Lebanon) 112
Belgium 138
Benghazi (Libya) 120, 121
Beni (DRC) 170, 171, 173, 175, 177, 178
Bhamragarh (India) 239
Bhatia, Bela 242
Bhatti, Farooq 265
King Bhumibol Adulyadej 314
Bhutta, Zulfiqar Ahmed 275
Bihar (India) 238, 241
Binali, Turki al- 45, 46
Bin Jawad (Libya) 118
Biro, Ibrahim 137
Bishkek (Kyrgyzstan) 282, 283
Bita (Nigeria) 184
Blanco, Wilter 332
Blattman, Christopher 31
BLOM Bank (Lebanon) 108, 110
Blue Nile (Sudan) 10, 209–215
Boguila (CAR) 168
Boko Haram (Nigeria) 9, 10, 181–188
Bolivar (Colombia) 342
Boma (South Sudan) 206
Borno (Nigeria) 9, 181, 182, 183, 187
    Emergency Management Agency 187
Boroh, Paul 189
Bosnia 5, 64, 66
Bosnian war 27
Bossangoa (CAR) 167
Bosso (Niger) 185
Botrous, Khalid 206
Boulduc, Donald 183
Bouri (Libya) 116
Brahimi, Lakhdar 15, 16, 20
Brass (Nigeria) 192
Brazil 18
Bria (CAR) 169
Brussels (Belgium) 138

Brussels Conference on Afghanistan 232
Budgam (India) 260
Bugti, Sarfaraz Ahmed 269
Buhari, Muhammadu 181, 182, 184, 189, 191, 192, 194, 195
Commander Bungos 303
Bunia (DRC) 18, 178
Buqa (Yemen) 159
Burkina Faso 9
Burundi 64, 65, 66, 179
Bustan al-Basha (Syria) 134
Butembo (DRC) 178
Buthidaung (Myanmar) 292

C

Cagayan (Philippines) 307
Cairo (Egypt) 86, 90
Call, Charles 68
Calle, Humberto de la 337
Cameroon 10, 166, 181, 184, 186
Canada 19, 225, 293
Cauca (Colombia) 337, 340
Caucasus Emirate 319, 320
Central African Republic 7, 9, 14, 16, 21, 163–169, 179
    Popular Front for the Renaissance of Central African Republic 165, 169
    Return, Reclamation, Rehabilitation 166
    Séléka 165, 166, 168
    Union for Peace in Central Africa 165, 169
Central American Integration System 331
Central Equatoria (South Sudan) 205
Chad 119, 166, 181, 184
Chah-e-Anjir (Afghanistan) 227
Chanae (Thailand) 311
Chana (Thailand) 311
Chandel (India) 246
Changlang (India) 249, 251, 252
Chang Wanquan 283
Charsadda (Pakistan) 273
Chaudhry, Aizaz 257
Chen Quanguo 281
Chetia, Anup 234
Chevron (US) 192
Chhattisgarh (India) 238, 239, 241, 243
Chibok (Nigeria) 182, 185
China 19, 52, 131, 204, 208, 219, 221, 222, 268–270, 279–285, 287–289
    Belt and Road Initiative 222, 282
    China–Pakistan Economic Corridor 282
    Chinese Islamic Association 279
    Communist Party 281, 284
    State Administration for Religious Affairs 279
Chin (Myanmar) 28
Choco (Colombia) 337
Cizre (Turkey) 146
Clark, Edwin 191
Cohen, Dara Kay 26, 30
Cold War 16, 22, 51, 53, 62
Colima (Mexico) 345
Collier, Paul 54
Colombia 11, 28, 57, 58, 69, 332, 335–344
    1st Front 340, 341
    Congress 336, 338–340
    Constitutional Court 339
    Democratic Centre 336, 338
    Eastern Bloc 341
    Los Pelusos 343
    Los Puntillos 343
Combat Forces Abacunguzi (DRC) 177
Communist Party of India–Maoist 237–241, 244
Congo Research Group 178
Coordination of Movements for the Azawad (Mali) 124
Cortes (Philippines) 307
Costa Rica 332
Cotabato (Philippines) 303, 304
Crimea 21
Cuarenta, John 341
Cuba 52

D

Dabiq 45
Dabiq (Syria) 139, 151
Dagestan (Russia) 320
Dahiyat al-Assad (Syria) 134
Dahlan, Mohammed 104, 105
Daily Excelsior 258
Damascus (Syria) 41, 42, 111, 129
Damaturu (Nigeria) 187
Dantewada (India) 238, 243
Daraa (Syria) 42, 140, 143
Daraya (Syria) 141
Darfur (Sudan) 7, 10, 14, 16, 209–215
Darra Adam Khel (Pakistan) 272
Datta Khel (Pakistan) 272
Datu Piang (Philippines) 302
Datu Salibo (Philippines) 301, 302, 303
Datu Unsay (Philippines) 302
Davao City (Philippines) 296
Davao del Norte (Philippines) 308
Davutoglu, Ahmet 148
Dawn 266
Debaltseve (Ukraine) 321, 323
Deir al-Asafir (Syria) 141
Deir ez-Zor (Syria) 41, 42
Delta State (Nigeria) 193, 194
Demirtas, Selahattin 149
Democratic Forces for the Liberation of Rwanda (DRC) 170, 175, 176–179
Democratic Republic of the Congo 9, 14, 16–18, 20, 27, 29, 30, 167, 169–179
    Episcopal Conference of the Democratic Republic of the Congo 172
    G7 173
    Journaliste en Danger 175
    Lumumbist Progressive Movement 174
    Operation Nyamuragira 176
    Operation Sukola 1 176
    Operation Sukola 2 176, 177
    Radio Okapi 175
    Republican Guard 174
    Struggle for Change 171, 173, 174
    Union for Democracy and Social Progress 174
    Union for the Congolese Nation 172
Democratic Union Party (Syria) 135, 137
    Asayish 135, 137
Deng Gai, Taban 204
Denmark 19
Dhaka (Bangladesh) 233
Dhi Qar (Iraq) 98
Dieng, Adama 202, 207
Diffa (Niger) 185
Digna, Babiker 215
Dima Halam Daogah (India) 234
Dimapur (India) 249
Dinesh 241
Dishu (Afghanistan) 225
Diyala (Iraq) 42, 96
Diyarbakir (Turkey) 145
Djedou, Sammy 138
Djibouti 201
Doha Document for Peace in Darfur 210
Donbass (Ukraine) 321, 323
Donetsk (Ukraine) 321–324
Donetsk People's Republic (Ukraine) 323, 324

Dostum, Abdul Rashid 222, 223
Doyle, Patrick 184
Dual, Simon Gatwech 203
Duarte, Gentil 341
Dudhnoi (India) 236
Dureza, Jesus 305
Duterte, Rodrigo 293–295, 297, 300, 301, 304–307

E

Eastern Ghouta (Syria) 140
Eastern Naga (India) 251
East Jerusalem 103, 105, 106
East Turkestan Islamic Movement 282, 283, 284
East Turkistan 282
Ebrahim, Murad 300, 301
Edo (Nigeria) 187
Egeland, Jan 142
Egypt 41, 83–91, 105, 118, 119, 131, 144
   Air Force 85
   Nadeem Centre for Rehabilitation and Victims of Violence 89
   Tahrir Square 88
Eisenkot, Gadi 104
El Heraldo 334
El Niño 201
El Salvador 6, 64, 66, 327–335
   Specialised Reaction Forces 329
Emarati, Saad al- 229
Entebbe (Uganda) 211
Erdogan, Recep Tayyip 131, 148, 149, 283, 317
Eringeti (DRC) 177
Eritrea 8, 11, 180, 181
Eritrea–Ethiopia Boundary Commission 181
Es Sider (Libya) 118, 119
Ethiopia 19, 64, 65, 66, 180–181, 201, 208, 211
   Algiers Agreement 181
Euphrates River 136, 138, 145, 150, 151
Eurasian Economic Union 318
European Union 22, 114, 127, 144, 189, 210, 214, 232, 305, 325
   Battle Groups 22

F

Falluja (Iraq) 42, 92, 94, 96
Farabundo Martí National Liberation Front (El Salvador) 66
Farah (Afghanistan) 223, 227
FARC (Colombia) 28, 57, 69, 335–343
Fardous (Syria) 134
Farooq, Mirwaiz Umar 258
Faryab (Afghanistan) 223, 226, 231
Fastaqim Union (Syria) 140
Fatah 104, 105
   Central Committee 104
Fateh Halab (Syria) 134, 139
Faylaq ar-Rahman (Syria) 140, 141
Fayulu, Martin 173
Federally Administered Tribal Areas (Pakistan) 262, 267, 268, 272, 276, 277
Fezzan (Libya) 121
Firozkoh (Afghanistan) 231
Flor, Marites 293, 296
France 18, 41, 105, 120, 131, 167, 175, 214, 319
   Operation Sangaris 167
Frangieh, Sleiman 108
Free Syrian Army 132, 139, 151
Friedman, David 107
Front for Patriotic Resistance in Ituri (DRC) 175, 177
Furqan, Abu Mohammed al- 138
Fuzuli (Nagorno-Karabakh) 315

G

G20 283
Gadchiroli (India) 238, 239, 242
Gadhafi, Muammar 116
Galkayo (Somalia) 198
Galmudug (Somalia) 198, 200
Gambella (Ethiopia) 180
Gamoru (Nigeria) 184
Ganderbal (India) 260
Gao (Mali) 123, 124, 125, 127
Garbai, Baba Kaka 182
Garmser (Afghanistan) 226
Gaza 102, 103
Gaziantep (Turkey) 147, 150
Geagea, Samir 108
Geelani, Syed Ali Shah 258
Geneva (Switzerland) 130
Geo News 274
Germany 228, 284, 294
Ghani, Ashraf 220, 222, 223, 229, 232
Ghariani, Sadiq al- 116
Ghashghar (Nigeria) 183, 185
Ghazni (Afghanistan) 225, 228
Ghor (Afghanistan) 231
Ghwell, Khalifa 115, 116
Gibril Ibrahim (Sudan) 211
Gidon (Myanmar) 289, 290
Global Witness 53, 175
Goalpara (India) 236
Gogoi, Tarun 233, 234
Golani, Abu Muhammad al- 140
Gomaa, Ali 86
Goma (DRC) 17, 177, 178
Gombo, Jean-Pierre Bemba 32
Gonzalez, Juan Sebastian 331
Gonzalez, Julian Aristides 334
Gore, Alfred Lado 203
Gourmat, Salah 138
Grand Bassam (Ivory Coast) 9
Greater Equatoria (South Sudan) 204, 205, 206
Greater Upper Nile (South Sudan) 204, 205
Greece 144
Guatemala 6, 327–335
   National Civilian Police 329
   National Council of Justice 334
   National Police 332
   Special Reserve Corps for Citizen Security 329
   Supreme Court 334
Guatemala City (Guatemala) 332
Guaviare (Colombia) 341
Guéhenno, Jean-Marie 21
Guerrero (Mexico) 345
Guindulungan (Philippines) 303
Gulf Cooperation Council 111
Guru, Afzal 255
Guterres, António 13, 21
Guzmán, Joaquin 'El Chapo' 345, 346, 347, 349
Guzmán Loera, Aureliano 'El Guano' 346
Guzmán Salazar, Jesus Alfredo 346
Gwadar (Pakistan) 271
Gwoza (Nigeria) 184

H

Haas (Syria) 142
Habibi, Abdullah Khan 222
Hadaj (Syria) 138
Hadhramaut (Yemen) 154, 160, 161
Hadi, Abd Rabbo Mansour 116, 153–156, 159
Haftar, Khalifa 114, 116, 118–121
Haiti 16, 18
Hajar, Ibn 45
Hajjah (Yemen) 157, 162
Hakam, Jaafar Abdel 211
Hakimullah Mehsud (Pakistan) 263

Hakkari (Turkey) 145
Halak (Syria) 134
Halisa (Syria) 139
Hall, Robert 293, 296
Halqi, Wael al- 132
Hamadi, Nawfal 93
Hamam, Walid 138
Hamas 64, 102, 103, 105
Hama (Syria) 41, 42, 54, 143
Hamdaniyeh (Syria) 134
Hameh (Syria) 141
Hammam al-Alil (Iraq) 93
Hammarskjöld, Dag 22
Hamra (Lebanon) 110
Handarat (Syria) 134
Hangzhou (China) 283
Hapilon, Isnilon 295
Haqqani network (Afghanistan) 220
Haqqani, Sirajuddin 220
Harakah al-Yaqin (Myanmar) 291
Harakat Ahrar al-Sham al-Islamiyya (Syria) 140
Harakat Nour al-Din al-Zenki (Syria) 47, 140
Harawa (Libya) 118
Hariri, Saad 108, 109
Hasakah (Syria) 41, 42, 135, 136
Haseeb, Abdul 229
Hashim, Ahmad Zamzamin 310
Hassm (Egypt) 83, 86, 87, 91
Hat Yai (Thailand) 314
Hatz, Sophia 67
Haute-Kotto (CAR) 166
Haut-Mbomou (CAR) 166
Havana (Cuba) 335, 337, 341
Hawija (Iraq) 42, 93
Hawsh al-Farah (Syria) 141
Hawsh Nasri (Syria) 141
Haydariyeh (Syria) 134
Haysom, Nicholas 217
Hebron (Palestinian Territories) 101, 102
Helmand (Afghanistan) 223, 224, 225, 226, 230
Hernández, Anabel 346
Hernández, Juan Orlando 331, 335
Herzog, Isaac 104
Hidme, Madkam 242, 243
Hindu 255, 260
Hiran (Somalia) 200
Hit (Iraq) 42, 92
Hizb-e-Islami (Afghanistan) 222
Hizbullah 64, 107–111, 133
Hizbul Mujahideen (India) 253, 256, 259
Hoeffler, Anke 54
Homs (Syria) 41, 42, 132
Honduras 6, 327–335
  Council of Private Enterprise 329
  Military Police of Public Order 329
  Mission to Support the Fight against Corruption and Impunity 334
  National Police 334
Hong Kong 281
Horlivka (Ukraine) 321
Hotan (China) 281
Houthi, Abdul Malik al- 156
Hpakant (Myanmar) 288
Htin Kyaw 286
Hua Hin (Thailand) 312
Human Rights Watch 28, 34, 100, 166, 175, 205, 230, 291, 310
Hussein, Zeid Ra'ad al- 152, 207, 291

I

Idil (Turkey) 146
Idlib (Syria) 46, 141–143

Ili (China) 281, 284
Ili Kazakh Autonomous Prefecture 284
Imphal (India) 245, 246, 248
Indanan (Philippines) 293
India 11, 19, 221, 222, 232–261, 265, 269
  Aam Aadmi Party 243
  Against Corruption and Unabated Taxation 252
  Andhra–Odisha Border Special Zonal Committee 240
  Armed Forces (Special Powers) Act 248
  Arunachal Naga Students Federation 252
  Asom Gana Parishad 234
  Assam Accord 234
  Assam Rifles 246, 249, 250
  Bharatiya Janata Party 233, 234, 236
  Border Security Force 241
  Ceasefire Monitoring Group 250
  Central Bureau of Investigation 243
  Congress 233, 234
  Greyhounds 242, 243
  Inner Line Permit 244, 247
  Jammu and Kashmir Peoples Democratic Party 260
  Jawaharlal Nehru Institute for Medical Sciences 248
  Joint Action Committee against Anti-Tribal Bills 245
  Joint Committee on Inner Line Permit System 244, 245, 247
  Ministry of External Affairs 257
  Ministry of Home Affairs 233, 259
  Multi Agency Centre 258
  Naga National Council 251
  National Investigation Agency 252, 255, 259
  People's Resurgence and Justice Alliance 248
  Red Corridor 238, 239, 243
  Regional Institute of Medical Sciences 248
  Research and Analysis Wing 254, 269
  United Naga Council 245–247, 251
Indian Express 254
Indonesia 282, 283, 293, 294, 295, 298
Ingushetia (Russia) 320
International Committee of the Red Cross 142, 186, 343
International Criminal Court 32, 107, 142
International Crisis Group 53
International Monetary Fund 89
International Organization for Migration 144
International Residual Mechanism for Criminal Tribunals 177
International Syria Support Group 129
Interpol 255
Irabor, Lucky 183
Iran 90, 95, 111, 129, 131, 133, 151, 155, 159, 221, 222, 229, 232, 269
  Islamic Revolutionary Guard Corps 95, 97
    Quds Force 95
Iraq 5, 6, 8, 39–45, 54, 58, 85, 90, 91–100, 137, 146, 151, 152, 264
  Council of Representatives 98
  Gorran 99
  Mishraq sulphur plant 94
  Popular Mobilisation Units 93, 95, 96, 97, 100
  Qayyarah Air Base 92
  Supreme Court 97
Ishiyama, John 65, 68
Islamabad (Pakistan) 219
Islamic Muthanna Movement (Syria) 141
Islamic State 8, 9, 27, 29, 31, 39–47, 50, 83, 85–87, 91–95, 98–100, 110, 112, 114, 116, 117–120, 129, 132, 133, 135–139, 141, 143, 146, 147, 150–152, 161, 183, 185, 188, 196, 198, 221, 222, 225, 228–231, 263–265, 293–296, 299, 301–303, 319, 320
  West Africa Province 185
Islamic State in the Greater Sahara 9
Israel 90, 100–107, 111
  Israel Defense Forces 104
  Knesset 103
  Peace Now 105
  Zionist Union 104

Istanbul (Turkey) 45, 46, 131, 147, 150
Italy 83, 88, 89, 117, 118, 120
Ivory Coast 9

J

Jaafari, Bashar al- 130
Jabal Badro (Syria) 134
Jabar (Syria) 138
Jabat, Awang 310
Jabhat al-Nusra 110, 112, 132, 139, 140, 282
Jabhat al-Shamiya (Syria) 46
Jabhat Ansar al-Din (Syria) 47
Jabhat Fateh al-Sham (Syria) 46, 47, 140
Jalalabad (Afghanistan) 221, 269
Jalandoni, Luis 306
Jalawla (Iraq) 42
Jalisco (Mexico) 344, 345, 346
Jamaat-ul-Ahrar (Pakistan) 263
Jamaat-ul-Dawa al-Quran (Pakistan) 272, 273
Jamaat-ul-Mujahadeen Bangladesh 233
Jamiat Ulema-e-Islam–Fazl (Pakistan) 268, 275
Jammu and Kashmir (India) 253, 255–261
Jammu Kashmir Liberation Front (India) 258
Jamui (India) 238
Jani Khel (Afghanistan) 226
Janjua, Nasser Khan 265
Jan, Kashif 255
Jarabulus (Syria) 42, 136, 139, 151
Jaramillo, Sergio 339
Jathran, Ibrahim 118, 119
Jaysh al-Fateh (Syria) 132, 134, 139, 283
Jaysh al-Islam (Syria) 46, 140, 141
Jaysh al-Mujahideen (Syria) 46
Jaysh al-Sunna (Syria) 47
Jaysh-e-Mohammed (India) 253–256, 264, 268
Jaysh Khalid ibn al-Walid (Syria) 141
Jebel Marra (Sudan) 10, 209, 211, 213–215
Jenin (Palestinian Territories) 104
Jerusalem 102, 107
Jharkhand (India) 239, 241
Jiribam (India) 245
Jizan (Saudi Arabia) 158
Johnson, Jeh 330
Jolo (Philippines) 293
Jonathan, Goodluck 184
Jordan 104, 117, 120, 143, 144
    Rukban camp 144
Jubaland (Somalia) 200
Juba (Somalia) 196, 197, 203, 204, 205, 207, 208
Juba (South Sudan) 18, 26
Jubeir, Adel al- 154
Jui River 239
Junabi, Abd al-Wahid al- 230
Jund al-Aqsa (Syria) 140
Justice and Equality Movement (Sudan) 209–213

K

Kabardino-Balkaria (Russia) 320
Kabare (DRC) 177
Kabila, Joseph 169, 170, 171, 172, 174, 176
Kabul (Afghanistan) 8, 224, 225, 228–231
Kachikwu, Ibe 191
Kachin Independence Army (Myanmar) 288, 289, 290
Kachin (Myanmar) 28, 31, 287–290
Kaduna (Nigeria) 193
Kafr Naseh (Syria) 135
Kafr Naya (Syria) 135
Kafr Saghir Martyrs Brigade (Syria) 139
Kaga-Bandoro (CAR) 163, 166, 168, 169
Kala-Balge (Nigeria) 182
Kalat (Pakistan) 274

Kalehe (DRC) 177
Kamerhe, Vital 170, 172
Kamwina Nsapu (DRC) 175
Kananga (DRC) 176
Kandahar (Afghanistan) 224, 269
Kangleipak Communist Party (India) 248
Kansaba (Syria) 132
Kantner, Jurgen 294
Kanu, Nnamdi 190
Kara, Abdul Rauf 115
Karabudakhkent (Russia) 320
Karachi (Pakistan) 263, 265, 273
Karbi Anglong (India) 235
Karbi People's Liberation Tigers (India) 233, 235, 236
Karialan, Ustadz 303
Karkamis (Turkey) 138
Karm al-Jabal (Syria) 134
Kasai-Central (DRC) 170, 175
Kashgar (China) 281
Kashmir Reader 261
Kasi, Bilal Anwar 263
Kata'ib Sayyid al-Shuhadaa (Iraq) 96
Katibah Nusantara (Syria) 295
Katrom (Sudan) 214
Katumbi, Moise 173
Kaung Kha (Myanmar) 32
Kazakhstan 282, 283
Keita, Ibrahim Boubacar 127, 128
Kenya 200, 201, 208
    Dadaab refugee camp 200, 201
Kerala (India) 243
Kerry, John 42, 130, 131
Khakar (Afghanistan) 225
Khaldiyeh (Syria) 134
Khaldun, Ibn 47
Khamis, Emad 132
Khan al-Assal (Syria) 132
Khan al-Shih (Syria) 141
Khanasir (Syria) 42
Khan, Hafiz Saeed 229
Khan, Imran 262
Khan, Muslim 274
Khan Neshin (Afghanistan) 224, 226
Khan, Tariq Hayat 266
Khartoum (Sudan) 210
Khattab, Mohammad 294
Khattak, Pervez 268
Khojavend (Nagorno-Karabakh) 315
Khorasan Province (Afghanistan) 221, 222, 225, 228–231
Khost (Afghanistan) 226
Khyber Agency (Pakistan) 273, 276
Khyber Pakhtunkhwa (Pakistan) 262, 265, 267, 268, 271, 275, 276
Kidal (Mali) 124, 125, 126
Kiir, Salva 26, 29, 202, 203, 204, 206, 208
Kinshasa (DRC) 169, 174
Kirkan, Haydar 140
Kirkuk (Iraq) 41, 42, 93, 99, 100
Kiryat Arba (Israel) 101
Kiselev, Vitaly 323, 324
Kobani (Syria) 42
Koch (South Sudan) 204
Kodjo, Edem 172
Kohistanat (Afghanistan) 225
Kolkata (India) 243
Konyak, Khole 250
Kosovo 8
Kot (Afghanistan) 225, 226, 227, 229
Kovacs, Mimmi Söderberg 67
Kuala Lumpur (Malaysia) 301, 310, 313
Kulgam (India) 260
Kunar (Afghanistan) 229, 230

Kunduz (Afghanistan) 223, 225, 226, 227, 231, 232
Kunming (China) 289
Kupwara (India) 260
Kurdish Democratic Party 98, 99, 100
Kurdish National Council 137
Kurdish Reform Movement 137
Kurdistan 41, 42, 93, 98, 99, 100, 129, 130, 135, 136, 137, 139, 145–151, 153
   Peshmerga 42, 93, 99
Kurdistan Freedom Falcons (Turkey) 145, 147, 152, 153
Kurdistan Regional Government 91, 95, 98, 99, 100
Kurdistan Workers' Party (Turkey) 7, 145–152
Kurnool (India) 241
Kurram (Pakistan) 276
Kutkai (Myanmar) 288, 289
Kuwait 111, 154
Kwajok, Lako Jada 205
Kwaya Kusar (Nigeria) 182
Kyrgyzstan 282, 283, 284

L

Ladsous, Hervé 214
Lahij (Yemen) 160
Lahore (Pakistan) 253, 255, 263
Lainya (South Sudan) 205
Laiza (Myanmar) 288, 290
Lake Chad 10, 181, 183, 184, 186
Lamitan (Philippines) 294, 295
Lanao del Sur (Philippines) 302, 304
Lanzer, Toby 186
Lapid, Yair 104
Lashkar-e-Islam (Pakistan) 272, 273
Lashkar-e-Jhangvi (Pakistan) 263, 264, 265
Lashkar-e-Taiba (India) 253, 259, 264, 268, 273
Lashkar Gah (Afghanistan) 223–227
Latakia (Syria) 132
Latehar (India) 241
Latif, Shahid 255
Lavrov, Sergei 130
Layramoun (Syria) 134
Lebanon 56, 64, 107–113, 144
   Ain al-Hilweh Palestinian refugee camp 112
   Amal Movement 109
   Beirut Madinati 109
   Free Patriotic Movement 109
   Future Movement 108, 109
   Hizballah International Financing Prevention Act 110
   Marada Movement 109
   March 14 alliance 108
León, Bernardino 114
Liberation Tigers of Tamil Eelam (Sri Lanka) 8, 28, 50
Liberia 64
Libya 8, 41, 113–121
   Abu Sitta naval base 115
   Al-Bunyan al-Marsous 117, 118
   Al-Jufrah Air Base 120, 121
   Benghazi Defence Brigades 120
   Central Bank of Libya 115, 119
   General National Congress 113–116, 118
   Government of National Accord 113–121
   High Council of State 114
   House of Representatives 114, 115, 118, 119
   Libyan National Army 114, 119, 120, 121
   Libyan Political Agreement 113, 114, 116
   Madkhalist 604th Infantry Battalion 117
   Ministry of Defence 115
   Misrata Military Council 120
   Misratan brigades 120
   Misratan Islamist militias 114, 115, 117, 118, 120
   National Oil Corporation 115, 118, 119
   Operation Dignity 120
   Petroleum Facilities Guard 118, 119, 121
   Presidency Council 114, 115, 116
   Presidential Guard 116
   Shura Council of Benghazi Revolutionaries 120
   Special Deterrence Force 115, 116
   Steadfastness Front 115
   Tripoli Revolutionaries' Brigade 115
Lieberman, Avigdor 103, 104
Li Keqiang 283
Line of Contact (Nagorno-Karabakh) 315, 317, 321, 322, 323, 325, 326
Line of Control (Kashmir) 256, 258, 259, 261
Liwa al-Haqq (Syria) 47
Liwa al-Thawra (Egypt) 83, 86, 87, 91
Loera, Consuelo 346
Logar (Afghanistan) 228
London (UK) 13, 23
Londoño, Rodrigo 'Timochenko' 336
Lone Khin (Myanmar) 290
Longding (India) 251, 252
Lord's Resistance Army 27, 28, 31, 166, 175, 178
Lorenzana, Delfin 302, 307
Los Rastrojos (Colombia) 340
Los Urabeños (Colombia) 340, 343
Lubala, Marcel 175
Lubero (DRC) 176, 178, 179
Luhanga (DRC) 178
Luhansk (Ukraine) 321–324
Luhansk People's Republic (Ukraine) 323, 324
Lumumba, Patrice 170
Lusaka (Zambia) 52
Lyons, Terrence 65

M

M23 (DRC) 17
Maadi (Syria) 134
Mabruk (Libya) 116
Machar, Riek 26, 29, 202–205, 208
Madkhali, Rabi bin Hadi al- 116
Magahat–Bagani (Philippines) 307
Maguindanao (Philippines) 299–304
Maharashtra (India) 238, 241
King Maha Vajiralongkorn 314
Mahdi Army (Iraq) 58
Mahdi, Sadiq al- 210
Maheen (Syria) 42
Mahmud, Sabeen 273
Maiduguri (Nigeria) 181, 182, 185, 187
Makuei Lueth, Michael 206, 208
Makwambala, Yves 171
Malakal (South Sudan) 18, 207
Malaysia 293–298, 309, 310
Mali 7, 11, 14, 16, 55, 121–128
   Algiers Accord 121, 122, 123, 124, 126, 128
   Platform alliance 124
   Self-Defence Group of Imrad Tuareg and Allies 124
Maliki, Nuri al- 98
Malik, Mohammad Yasin 258
Malkangiri (India) 238, 239, 240
Manbij (Syria) 42, 136, 138, 143, 150
Manila (Philippines) 301
Manipur (India) 244–252
Mansi (Myanmar) 288
Mansoor, Umar 273
Mansour, Mohammad Akhtar 219, 220
Mara Patani (Thailand) 309, 310, 311, 313
Mara Salvatrucha (Central America) 6, 329, 331, 332, 335
Marcos Jr, Ferdinand 'Bongbong' 300
Mardan (Pakistan) 263
Marea (Syria) 136, 139, 143
Mariano, Rafael 305

Maridi (South Sudan) 205
Mariupol (Ukraine) 321, 322
Marjah (Afghanistan) 223, 225
Marjeh (Syria) 134
Márquez, Iván 339
Marshall, Michael 68
Martinez, Hugo 330
Masaken Hanano (Syria) 134
Masri, Abu Afghan al- 140
Masri, Abu Faraj al- 140
Mastung (Pakistan) 274
Maujgarh Fort (Pakistan) 255
Maungdaw (Myanmar) 291, 292
Mauritania 126
Maute Group (Philippines) 296, 302
Maya–Chortí Task Force 332
Maydaa (Syria) 141
Mayom (South Sudan) 204
Mazar-e-Sharif (Afghanistan) 221, 228
Mbeki, Thabo 209, 210
Mbomou (CAR) 166
McGoldrick, Jamie 156
Médecins Sans Frontières 142, 161, 162, 166, 207, 257
Mediterranean Sea 144
Mengal, Sardar Akhtar 271
Meshal, Khaled 105
Meta (Colombia) 340, 341, 343
Metrojet flight 9268 85, 89
Mexico 5, 330, 344–350
  Beltrán Leyva Cartel 346
  Chamber of Commerce 348
  Congress 344, 347
  Gulf Cartel 345
  Jalisco New Generation Cartel 344–347
  Knights Templar 345
  La Familia Michoacána 345
  Los Ardillos 345
  Los Rojos 345
  Los Zetas 345, 346
  National Human Rights Commission 5, 350
  National Public Security System 347
  Senate 347, 350
  Sinaloa Cartel 345, 346, 347
Mexico City (Mexico) 349
Michoacan (Mexico) 345
Mikati, Najib 109
Min Aung Hlaing 287, 290, 292
Mindanao (Philippines) 293, 294, 296, 299, 300, 302, 303, 307
Minimbang, Imam 303
Minni Minnawi (Sudan) 209, 211
Minsk agreement 321, 323, 324, 325, 326
Minyan (Syria) 134
Misrata (Libya) 117
Misry, Abu Mama 303
de Mistura, Staffan 129, 130, 137
Misuari, Nur 293, 300, 301
Mizan (Afghanistan) 230
Mladenov, Nickolay 103
Mobbar (Nigeria) 182
Moby Group (Afghanistan) 231
Modi, Narendra 253, 255, 256, 257
Mogadishu (Somalia) 196, 197, 198, 200
Mohali (India) 255
Mohammad Rasool (Afghanistan) 220
Mohammed, Lai 186
Molina, Otto Pérez 333
Mong Koe (Myanmar) 289
Mon (India) 249
Mon (Myanmar) 28, 29
Mopti (Mali) 123
Morales, Jimmy 329, 333, 334

Morales Marroquin, José Manuel 333
Morales, Samuel Everardo 333
Morobo (South Sudan) 205
Morocco 294
Moro Islamic Liberation Front (Philippines) 298–304, 308
Moro National Liberation Front (Philippines) 293, 300, 301, 304
Morsi, Muhammad 87
Moscow (Russia) 131
Mosul (Iraq) 8, 40, 42, 43, 44, 91–95, 99, 100, 137
Movement for the Emancipation of the Niger Delta (Nigeria) 191
Moyu (China) 281
Mozambique 64, 66, 67
Mozambique National Resistance Movement 66
Muadimiyat al-Sham (Syria) 141
Muang (Thailand) 312, 314
Muhajir, Abu Hasan al- 44
Mujyambere, Leopold 177
Mukalla (Yemen) 154, 160, 161, 162
Muktadil, Braun 296
Muktadil, Nelson 296
Mullah Nanai 220
Mumbai (India) 242
Mundri (South Sudan) 205
Munger (India) 238
Munich Security Conference 130
Musa Khel (Afghanistan) 226
Musa Qala (Afghanistan) 223, 224, 225
Muse (Myanmar) 288, 289
Museveni, Yoweri 211
Muslim Brotherhood (Egypt) 83, 86, 87, 88, 91, 116
Muzaffarabad (India) 259
Mwenga (DRC) 177
Myanmar 9, 11, 28, 29, 31, 32, 246, 285–292
  21st Century Panglong Conference 287
  Advisory Commission on Rakhine State 290
  Emergency Management Central Committee 290
  Myanmar Economic Corporation 285
  Myawaddy 32
  National Defense and Security Council 286, 292
  National League for Democracy 285, 286
  National Reconciliation and Peace Center 289
  Nationwide Ceasefire Agreement 286, 287
  Network for Human Rights Documentation–Burma 286
  Tatmadaw 286, 287, 288, 289, 290, 291, 292
  Union of Myanmar Economic Holdings 285
  Union Peace Conference 286
  Union Solidarity and Development Party 289
  United Nationalities Federal Council 287
Myanmar National Democratic Alliance Army 287, 288
Myassar (Syria) 134
Myint Swe 286, 292

N

Naameh (Lebanon) 110
Nablus (Palestinian Territories) 102, 104
Nad-e-Ali (Afghanistan) 225
Naf River 291
Nagaland (India) 245–252
Nagorno-Karabakh 315, 317, 318
  Four-Day War 315
  Nagorno-Karabakh Defense Army 315
Nahr-e-Saraj (Afghanistan) 223, 225
Najran (Saudi Arabia) 158, 159
Nakrob Bunbuathong 309, 311
Namibia 52
Nangaa, Corneille 170
Nangarhar (Afghanistan) 228, 229, 232
Narathiwat (Thailand) 309, 310, 311, 312, 314
Nariño (Colombia) 342
Narong Sabaiporn 311

Nasir (South Sudan) 205
Nasrallah, Hassan 111
National Council for Renewal and Democracy (DRC) 177
National Defence Force (Syria) 135
National Democratic Alliance Army (Myanmar) 287, 290
National Democratic Front of Bodoland–Progressive (India) 234
National Democratic Front of Bodoland–Songbijit (India) 233, 234, 235, 236
National Democratic Front of the Philippines 305
National Democratic Movement (South Sudan) 206
National Liberation Army (Colombia) 342
National Socialist Council of Nagalim–Isak Muivah (India) 244–253
National Socialist Council of Nagalim–Khaplang (India) 236, 246, 249– 252
National Socialist Council of Nagalim–Khole-Kitovi (India) 250
National Socialist Council of Nagalim–Reformation (India) 250, 251
National Socialist Council of Nagalim–Unification 250, 251
NATO 219
Nawa (Afghanistan) 225
Nawah-e-Barakzai (Afghanistan) 226, 227
Nawfaliyah (Libya) 118
Naxalites (India) 237, 238, 241, 242, 243
Naypyidaw (Myanmar) 286
Ndagijimana, Laurent 177
Ndomete (CAR) 168
Nduma Defence of Congo (DRC) 178
Ndume, Mohammed 182
Nepal 64, 225
Netanyahu, Benjamin 103, 104, 106
Netherlands 19
New Delhi (India) 221, 250, 254
New People's Army (Philippines) 11, 300, 304–308
New York Times 334
Ngoy, Christopher 171
Niamey (Mali) 124
Nicaragua 332
Nicholson, John 221, 223, 228, 229, 230
Niger 126, 181, 184, 185, 186
Niger Delta Avengers (Nigeria) 188–193, 195
 Operation Red Economy 189, 195
Niger Delta Greenland Justice Mandate (Nigeria) 193
Niger Delta Justice Defence Group (Nigeria) 191
Nigeria 7, 9, 10, 181–195
 Adibawa oilfields 193
 All Progressives Congress 190
 Artisan Fishermen Association 194
 Biafra 190
 Bonga oil spill 194
 Bonny terminal 191
 Borno State Islamic Preaching Board 182
 Content Development and Monitoring Board 189
 Forcados oil-export terminal 190, 193
 Ijaw National Congress 191
 Ministerial Technical Audit Committee 195
 Ministry of Environment 189
 Ministry of Niger Delta Affairs 189, 195
 National Conference 190
 National Emergency Management Agency 182
 National Human Rights Commission 194
 Niger Delta Development Commission 189, 195
 Nigerian National Petroleum Corporation 192, 193, 194
 Nigerian Petroleum Development Company 193
 Operation Awatse 192
 Operation Crocodile Smile 193
 Operation Safe Corridor 187
 Pan Niger Delta Forum 191, 192
 Presidential Amnesty Programme 189, 190, 195
 Presidential Committee on the North East Initiative 195
 Public Procurement Act 195
 Regional Security Summit 184
 Transition and Empowerment Task Force 189
 Victims Support Fund 187
Nigerian Maritime University 190
Nimr, Nimr al- 111
9/11 220
Ninewa (Iraq) 41, 42, 91, 92, 93
Nkaissery, Joseph 200, 201
Non-Aligned Movement 52
Noney (India) 247, 248, 252
Noor, Atta Mohammad 222
Norte de Santander (Colombia) 336, 340
North American Free Trade Agreement 348, 349
Northern Alliance–Burma (Myanmar) 288, 289
Northern Ireland (UK) 64
Northern Triangle 327, 329–333
North Kivu (DRC) 170, 176, 178, 179
North Oil Company (Iraq) 94
North Sinai (Egypt) 85, 91
North Waziristan (Pakistan) 272, 276
Norway 19, 52, 210, 293, 336
 Nobel Committee 336
Now Zad (Afghanistan) 224, 225
Nsapu, Kamwina 176
Ntaganzwa, Ladislas 177
Nuba Mountains (Sudan) 213, 215
Nubl (Syria) 133
Nusaybin (Turkey) 146, 147

O

Obama, Barack 19, 106, 131, 157, 219
Obeidi, Khaled al- 98
O'Brien, Stephen 142
Ocotepeque (Honduras) 331
Odisha (India) 238–243
Ogbogoro (Nigeria) 193
Ogoniland (Nigeria) 190, 194
Ogun (Nigeria) 193
Olonisakin, Abayomi 188
Olson, Richard 220
Mullah Omar 273
Ondo (Nigeria) 194
Orakzai (Pakistan) 276
Organisation for the Prohibition of Chemical Weapons 143
Organisation Internationale de la Francophonie 175
Organisation of American States 334, 340
Organization for Security and Co-operation 318, 319, 321, 322
 Minsk Group 318
 Special Monitoring Mission 319, 321, 323, 325
Organization of African Unity 52, 53
 Coordinating Committee for the Liberation of Africa (Liberation Committee) 52, 53
Organization Stabilization Mission in the Democratic Republic of the Congo 173, 174, 175, 176, 177
Oromia (Ethiopia) 180
Oromo Liberation Front (Ethiopia) 180
Osinbajo, Yemi 186, 194
Oslo (Norway) 305
Ouham-Pendé (CAR) 166
Owaija (Syria) 134

P

Pachir Wa Agam (Afghanistan) 229
Pakistan 8, 19, 219, 282, 220, 221, 228, 229, 232, 234, 253–261, 262–277, 282, 283, 262
 All FATA Political Alliance 267
 Army Public School attack 273
 Bacha Khan University 273
 Balochistan National Party 271
 China–Pakistan Economic Corridor 268, 270, 271, 272, 276
 Corridor Front 271

FATA Reforms Committee 267, 268
Frontier Works Organisation 270
Human Rights Commission of Pakistan 274
Intelligence Bureau 264
Inter-Services Intelligence 255
Inter-Services Public Relations 272
Ministry of Foreign Affairs 257
National Action Plan 262, 264, 265, 266, 274, 276
Operation Khyber III 273
Operation Zarb-e-Azb 272, 273
Pakhtunkhwa Milli Awami Party 268, 275
Pakistan Peoples Party 262
Pakistan Tehreek-e-Insaf Party 262
Regional Institute of Policy Research and Training 275
Special Security Division 270, 271
State Bank of Pakistan 265
Supreme Court 267, 275
Task Force-88 271
Tribal Areas Riwaj Act 268
Paktia (Afghanistan) 226, 228
Paktika (Afghanistan) 228, 231
Palakkad (India) 243
Palestine Liberation Organization 52
Palestinian Authority 104, 105
Palestinian Territories 64, 100–107, 111, 112
  Center for Policy and Survey Research 102
Palmyra (Syria) 42, 132, 133
Panama 332
Panu Uthairat 314
Paris (France) 210
Paris attacks 2015 138
Paris peace conference 105
Parwan (Afghanistan) 228
Pathankot (India) 253, 254, 255, 261
Patikul (Philippines) 295, 296
Patong (Thailand) 312
Patriotic Movement for Central Africa (CAR) 168
Pattani (Thailand) 309–314
Pavlov, Arseny 'Motorola' 323
Peña Nieto, Enrique 344, 347, 348, 349, 350
Peñate, José Morela 343
Pengaree (India) 237
People's Democratic Council of Karbilongri (India) 235
People's Liberation Guerrilla Army (India) 240
People's Protection Units (Syria) 135, 145, 146, 150, 151, 153
Peshawar (Pakistan) 263, 267, 273, 275
Petrivske (Ukraine) 323
Phang Nga (Thailand) 312
Philippines 11, 292–308
  Armed Forces 293–297, 299, 302, 303, 306, 307
  Bangsamoro Basic Law 299, 300, 301, 302
  Bangsamoro Transition Commission 301
  Barangay Peacekeeping Action Teams 296
  Communist Party of the Philippines 304, 305, 306, 307
  Comprehensive Agreement on Bangsamoro 301
  Department of Agrarian Reform 305
  Department of National Defense 298
  Department of Social Welfare and Development 297, 305
  Joint Agreement on Safety and Immunity Guarantees 305
  Joint Task Group Sulu 295
  Kilusang Magbubukid ng Pilipinas 305
  Operation Bayanihan 307
  Operation Tokhang 307
Phuket (Thailand) 312
Pishan (China) 281
Plotnitsky, Igor 323, 324
Popular Resistance Movement (Egypt) 86, 87
Poroshenko, Petro 324, 325
Power, Samantha 106, 208
Prachuap Khiri Khan (Thailand) 312
Prawit Wongsuwan 313

Prayuth Chan-ocha 310, 311
Puerto Vallarta (Mexico) 346
Pulka (Nigeria) 184
Pulwama (India) 260
Punjab (India) 253, 255, 265, 266, 271
Puntland (Somalia) 196, 198, 199, 200, 201
Putin, Vladimir 131, 133

Q
Qaa (Lebanon) 110, 112
Qadi Askar (Syria) 134
Qadri, Mumtaz 265
Qahtani, Farouq al- 230
Qaim (Iraq) 42, 92
Qamenas (Syria) 143
Qamishli (Syria) 135, 137
Qandala (Somalia) 196, 198
Qandil (Iraq) 152
Qaryatayn (Syria) 42, 132
Qatar 105, 111, 210, 219, 221
Qayyarah (Iraq) 92, 94
Quartet on the Middle East 105
Qudsaya (Syria) 141
Quetta (Pakistan) 263, 274
Qusayr (Syria) 111

R
Rabaa al-Adawiya (Egypt) 87
Rabbani, Salahuddin 222, 223
Rabia (Syria) 132
Radio France Internationale 175
Radio Free Asia 284
Raia Mutomboki (DRC) 178
Rakhine (Myanmar) 285, 286, 290, 292
Ramadi (Iraq) 42, 94, 95
Ramakrishna 240
Ramallah (Palestinian Territories) 104
Ramana, Bakuri Venkata 240
Ramouseh (Syria) 134
Rao, Kudumula Venkata 240
Raqqa (Syria) 41, 42, 43, 135–138, 151
Ras Lanuf (Libya) 118, 119
Rathedaung (Myanmar) 291
Rawa (Iraq) 42, 92
Red Sea 83
Regeni, Giulio 83, 88, 89
Rehman, Maulana Fazlur 268, 275
Renk (Sudan) 215
Restrepo, Juan Camilo 343
Reuters 292
Revolutionary United Front (Sierra Leone) 56
Ridsdel, John 293, 296
Rif Aleppo (Syria) 41
Rif Dimashq (Syria) 140, 141, 143
Rifi, Ashraf 109
Rivers State (Nigeria) 193, 194
Riyadh (Saudi Arabia) 129
Rodriguez, David 182
Rohling, Andrew 224
Rouhani, Hassan 269
Rubkona (South Sudan) 204
Rumiyah 45
Runda Kumpulan Kecil 311, 313
Russia 21, 41, 47, 85, 90, 105, 119, 128–135, 137, 140, 142, 143, 151, 208, 221, 222, 282, 317, 318, 319–320, 322, 323, 324, 325
  Federal Security Service 133, 282, 320
Rustow, Dankwart 63
Rutba (Iraq) 42
Rutshuru (DRC) 170, 176, 178
Rwanda 15, 16, 17, 18, 66, 170, 179

S

Saada (Yemen) 159
Sabah (Philippines) 296, 297
Sabeel (Egypt) 85
Sabri, Amjad 263
Sadar Hills (India) 245
Sadiya (Iraq) 42
Sadr, Moqtada al- 96, 97, 98
Sakhour (Syria) 134
Salahuddin (Iraq) 41, 42, 92
Salahuddin, Syed 259
Salam, Tammam 112
Saleh, Ali Abdullah 153, 155, 156, 160
Saliheen (Syria) 134
Saliman, Jim 297
Salma (Syria) 132
Samba-Panza, Catherine 163
Sanaag (Somalia) 199
Sana'a (Yemen) 154, 155, 156, 157, 159
Sanafir (Egypt) 83, 89
Sanaullah, Rana 255
Sánchez, Eduardo 349
Sánchez, Odín 342
Sangin (Afghanistan) 223, 224
San Juan (Philippines) 298
Santos, Juan Manuel 336, 337, 338, 340, 342, 343
Santrich, Jesús 339
Saraqeb, Abu Omar 140
Sar-e-Pul (Afghanistan) 225
Sarmin (Syria) 143
Saudi Arabia 41, 83, 89, 90, 108, 111, 116, 138, 153, 154, 155, 157–160, 162
Save the Children 162
Sawyer, Ida 175
Scott, James 57
Sderot (Israel) 102
Second World War 51
Sein Win 289
Sekkingstad, Kjartan 293, 294, 296
Seko, Mobuto Sese 171
Sema, Muslimin 301
Semdinli (Turkey) 146
Semporna (Philippines) 296
Serbia 27
Serraj, Faiez al- 114, 115, 121
Shaar (Syria) 134
Shabelle (Somalia) 196, 197, 200
Shabunda (DRC) 177
Shaddadi (Syria) 42, 136
Shahim, Qadam Shah 283
Shamba, Franck Diongo 174
Shanghai Cooperation Organisation 283
Shan (Myanmar) 28, 32, 285, 287, 288, 289
Shan State Army–North (Myanmar) 288
Shariff Aguak (Philippines) 302, 303
Sharif, Nawaz 221, 228, 253, 254, 256, 257, 262, 265, 267, 268, 270, 271, 273, 275, 276
Sharif, Raheel 269, 272, 273
Sharif, Shahbaz 271
Sharmila, Irom 248
Shawal (Pakistan) 272
Shehu, Garba 186
Sheikh Issa (Syria) 135
Sheikh Khader (Syria) 134
Sheikh Lutfi (Syria) 134
Sheikh Maqsoud (Syria) 135
Sheikh Miskeen (Syria) 143
Sheikh Sayed (Syria) 134
Shekau, Abubakar 183
Shell (Netherlands–UK) 190, 191, 193, 194
Sherzad (Afghanistan) 232
Shettima, Kashim 187
Shihezi (China) 284
Shirqat (Iraq) 42, 92
Shishani, Abu Omar al- 138
Shopian (India) 260
Sibut (CAR) 167
Sidon (Lebanon) 112
Siege Watch 8, 141
Sierra Leone 16, 27, 56, 59
Silangkum (Philippines) 296
Silopi (Turkey) 146
Sinai Peninsula (Egypt) 9, 83, 85, 86, 87
Sinaloa (Mexico) 346
Sindh (Pakistan) 265, 271
Singh, Okram Ibobi 245, 247, 252
Sinjar (Iraq) 42, 151, 152
Sipah-e-Sahaba (Pakistan) 264
Sirnak (Turkey) 145, 147
Sirte (Libya) 8, 43, 116, 117, 118, 120
Sisi, Abdel-Fattah Al- 83, 85–91
Sison, José Maria 304, 305
Sittwe (Myanmar) 290
Somalia 9, 11, 15, 58, 196–201
   Independent Electoral Disputes Resolution Mechanism 200
   National Leadership Forum 199, 200
Somaliland (Somalia) 199, 200, 201
Songbijit, I.K. 235
Songkhla (Thailand) 311, 314
Sonowal, Sarbananda 233
Sooka, Yasmin 207
Sool (Somalia) 199
Sori, Soni 242
South Africa 53
South Kivu (DRC) 176, 177, 179
South Kordofan (Sudan) 10, 209–215
South Sudan 7, 8, 10, 14, 16–20, 26, 27, 29, 30, 55, 179, 180, 202–208, 215
   Ceasefire and Transitional Security Arrangements Monitoring Mechanism 205
   Joint Monitoring and Evaluation Commission 203, 204
   Protection of Civilian sites 203, 207
   Regional Protection Force 208
   Transitional Government of National Unity 203, 204, 206, 208
South Sudan Democratic Front 205
South Sudan Democratic Movement/Army–Cobra Faction 206
South Waziristan (Pakistan) 272, 276
Southwest African People's Organization (Namibia) 52
South West (Somalia) 200
Soviet Union 51, 52
Spain 131
Sri Lanka 8, 28, 29, 50, 150
Srinagar (India) 256, 260
Stanekzai, Masoom 222
Stanytsia Luhanska (Ukraine) 323
Stearns, Jason 175
St Petersburg (Russia) 319
Sudan 10, 11, 119, 208, 209–215
   Korona Peace Agreement 211
   National Congress Party 212
   National Dialogue 209, 210, 211, 212
   National Document 211, 212
   National Umma Party 210, 212
   Popular Congress Party 212
   Rapid Support Forces 213
   Roadmap Agreement 210
   Sudan Armed Forces 213, 214
   Sudan Call 210, 211, 212
Sudanese Revolutionary Front 213
Sudan Liberation Movement 211, 213, 214
Sudan Liberation Movement–General Leadership 211

Sudan Liberation Movement–Minni Minnawi 209, 210, 211, 213
Sudan People's Liberation Army 203–206, 208, 213
Sudan People's Liberation Movement 202–206, 208–213
Sudan People's Liberation Movement/Army–In Opposition (South Sudan) 202–206, 208
Sudan People's Liberation Movement–North 209–213
Sudhakaran 241
Sukhnah (Syria) 132, 133
Sukma (India) 239, 242, 243
Suleimani, Qassem 95
Sultan Kudarat (Philippines) 303
Sulu (Philippines) 293–298, 300, 301
Suluk (Syria) 42
Sulu Sea 293, 297, 298
Sun Guoxiang 287
Suqour al-Sham (Syria) 46
Surat Thani (Thailand) 312
Surigao del Sur (Philippines) 307
Sur (Turkey) 146, 147
Suwayda (Syria) 42
Svitlodarsk (Ukraine) 321, 323
Swat (Pakistan) 266, 274
Sweden 19
Switzerland 186
Swu, Isak Chisi 250, 251
Sylva, Timipre 190
Syria 5, 6, 8, 28, 39, 40, 41, 42, 44, 45, 46, 47, 54, 85, 90, 92, 95, 107, 108, 109, 111, 112, 128–144, 145, 150, 151, 153, 229, 264, 282, 283, 295
   Astana peace talks 47
   Castello Road 133
   Democratic Federal System of Rojava–Northern Syria 137
   High Negotiations Committee 129, 130
   Shaer gas field 132
   Syrian Arab Army 133
   Syrian Civil Defence 143
   Tishrin Dam 150
   Yakiti 137
   Yarmouk refugee camp 42
Syrian American Medical Society 142
Syrian Arab Red Crescent 142
Syrian Democratic Council 129, 137
Syrian Democratic Forces (Syria) 129, 135–139, 143, 151

T

Ta'ang National Liberation Army (Myanmar) 287, 288
Tabqa (Syria) 138
Taguiwalo, Judy 305
Tahrir al-Sham (Syria) 47
Tajammu Fastaqim Kama Umirt (Syria) 46
Tajikistan 282, 283
Tajouri, Haitham 115
Tak Bai (Thailand) 310, 313, 314
Tal Abyad (Syria) 42
Tal Afar (Iraq) 42, 44, 95
Talamenes (Syria) 143
Taliban (Afghanistan) 7, 47, 50, 219–231, 262, 272, 273
   Operation Omari 224
Tal Kaif (Iraq) 42
Tamenglong (India) 246, 250
Tanai (Myanmar) 288
Tanganyika (DRC) 170, 176
Tank (Pakistan) 276
Tarin Kot (Afghanistan) 225, 226, 227, 231
Tariq Gidar Group (Pakistan) 272, 273
Tartar (Nagorno-Karabakh) 315
Tartus (Syria) 133
Taseer, Salman 266
Tata Institute of Social Sciences 242
Tawila (Sudan) 214
Tawi-Tawi (Philippines) 295, 296, 297

Teerachai Nakwanich 313
Tehran (Iran) 111
Tehrik-e-Taliban Pakistan 229, 262–266, 269, 272, 273, 274, 277
Telangana (India) 241
Tel as-Saman (Syria) 138
Tel Aviv (Israel) 103, 107
Tel Kurdi (Syria) 141
Tel Rifaat (Syria) 135
Tel Sawwan (Syria) 141
Tengnoupal (India) 247, 252
Thailand 308–314
   Deep South Watch 309
   Internal Security Operations Command 309, 310, 311, 312
   Southern Border Provinces Administrative Centre 310, 314
   Triangle of Security, Sustainability and Wealth 311
Than Shwe 285
U Thant 22
Thant Myint-U 22
Thepha (Thailand) 311
Thio, Henry Van 286
Thomas-Greenfield, Linda 183
Tiamzon, Benito 305
Tiamzon, Wilma 305
Tikrit (Iraq) 42
Timbuktu (Mali) 123, 124, 125, 127
Timor-Leste 66
Tinsukia (India) 236, 237
Tipo-Tipo (Philippines) 294, 296, 297
Tirah Valley (Pakistan) 273
Tiran (Egypt) 83, 89
Tirap (India) 250, 251, 252
Tobruk (Libya) 113
Togo 172
Tora Bora mountains (Afghanistan) 229
Touadéra, Faustin-Archange 163, 165, 167
Trang (Thailand) 312
Trilateral Contact Group 322
   Framework Decision 322
Tri-National Force 331, 332
Tripoli (Lebanon) 109, 115
Tripoli (Libya) 113, 115
Trump, Donald 22, 106, 107, 325, 330, 348, 349
Tshibala, Bruno 174
Tshisekedi, Etienne 171, 172, 174
Tsypkalov, Gennady 323, 324
Tundok, Wahid 303
Turkestan Islamic Party (Syria) 283
Turkey 7, 11, 45, 95, 112, 129, 131, 134, 136, 137, 138, 139, 143, 144, 145–153, 282, 283, 317
   Democratic Regions Party 145
   Democratic Union Party 135, 145
   Operation Euphrates Shield 150
   Justice and Development Party 148
   Nationalist Action Party 148
   National Security Council 149
   Patriotic Revolutionary Youth Movement 146
   People's Democratic Party 145, 147, 148, 149
   Republican People's Party 148

U

Udomdej Sitabutr 313
Uganda 27, 31, 201, 211
Ukhrul (India) 245, 248, 252
Ukraine 11, 321–326
Ulang (South Sudan) 204
Ungkaya Pukan (Philippines) 297
Union of Patriots for the Defence of the Innocent (DRC) 178
United Arab Emirates 104, 111, 119, 120, 138, 159, 160
United Kingdom 19, 117, 120, 210, 214, 267
   Department for International Development 189
   Special Air Service 117

United Liberation Front of Asom (India) 233, 234
United Liberation Front of Asom–Independent (India) 233, 234, 236, 237
United Liberation Front of Asom–pro-talks faction (India) 233, 234
United Liberation Front of Western South East Asia (India) 236
United Nations 6, 7, 13–23, 26, 27, 30, 33, 34, 52, 53, 90, 94, 100, 101, 103, 106, 107, 111–114, 118, 122, 126, 127, 129, 130, 131, 134, 142, 143, 144, 152–157, 161, 165, 166, 167, 168, 171, 174, 176–179, 186, 189, 194, 198, 201, 202, 203, 207, 208, 210, 214, 215, 217, 219, 227, 230, 231, 232, 257, 276, 287, 290, 291, 308, 321, 322, 325, 332, 339, 342
   Assistance Mission in Afghanistan 217, 231
   Assistance Mission in Iraq 100
   Charter
      Chapter VII 16
   Commission of Inquiry on Syria 142
   Commission on Human Rights 207
   Development Programme 189
   Environmental Programme 194
   Force Intervention Brigade 17
   General Assembly 52, 131, 142
   Group of Experts on the DRC 177
   High Commissioner for Human Rights 27, 153, 257, 291, 340
   High Commissioner for Refugees 6, 126, 215, 276
   High-Level Independent Panel on UN Peace Operations 15, 18, 23
   Human Rights Council 257
   Human Rights Monitoring Mission in Ukraine 322
   Institute for Namibia 52
   Interim Multinational Emergency Force 18
   Joint Human Rights Office 174
   Joint Investigative Mechanism 143
   Leaders' Summit on Peacekeeping 19
   Mission in Darfur 214, 215
   Mission in South Sudan 17, 18, 203, 204, 207
   Multidimensional Integrated Stabilisation Mission in the Central African Republic 165, 166, 167, 168
   Multidimensional Integrated Stabilization Mission in Mali 7, 122, 126
   Office for the Coordination of Humanitarian Affairs 7, 101, 126, 215, 232, 276
   Office on Drugs and Crime 6, 219, 332
   Panel of Experts on South Sudan 207
   Peacekeeping Defence Ministerial 13
   Regional Protection Force 17
   Regional Response Plan 144
   Secretariat 15, 22, 23
   Security Council 13, 14, 16, 17, 21, 22, 90, 106, 107, 114, 118, 131, 142, 143, 167, 208, 214, 215, 217, 257
      Resolution 2259 114
      Resolution 2301 167
      Resolution 2334 106
   Special Commission on Human Rights in South Sudan 18
   Special Committee on Decolonization 52
   Stabilization Mission in the DRC 17
   UN–African Union Mission in Darfur 7
   World Food Programme 161, 202
United States 6, 8, 19, 22, 28, 30, 31, 41–44, 51, 58, 59, 89, 92, 93, 95, 105–107, 110, 111, 118, 129–131, 135–138, 140, 145, 150, 151, 153, 157–160, 162, 175, 182–184, 187, 189, 196, 197, 201, 208, 210, 212, 217, 219, 220, 221, 223–232, 256, 272, 273, 274, 285, 286, 305, 306, 315, 317, 319, 325, 327, 330, 331, 332, 345, 346, 348, 349
   Africa Command 118, 182, 184
   Agency for International Development 187, 189
   Alliance for Prosperity 330, 331
   Bradbury memos 31
   Central Command 159
   CIA 28
   Commission on International Religious Freedom 274
   Congress 106, 110, 256, 330
   Department of Defense 8, 92, 137, 228
   Department of State 6, 145, 273, 332
   Department of the Treasury 285
   Drug Enforcement Administration 349
   Intelligence Community 41
      Worldwide Threat Assessment 41
   Merida Initiative 349
   Operation Inherent Resolve 137, 138
   Operation Iraqi Freedom 43
   Operation Resolute Support 221, 224, 227, 229
   Operation Tidal Wave II 137
   Pentagon 220, 226
   Senate 28, 183
      Armed Services Committee 182
United Wa State Army (Myanmar) 287, 290
Unity (South Sudan) 203, 204
Unity State (South Sudan) 26
Upper Nile (South Sudan) 204, 207, 215
Uribe, Alvaro 336, 337, 338
Urtecho, Armando 329
Urumqi (China) 281
Uruzgan (Afghanistan) 220, 223, 225, 227
Utabi, Bilal al- 230

V
Vichada (Colombia) 341
Vienna (Austria) 319
Vietnam 30
Villegas, Luis Carlos 339, 340, 341, 343
Visakhapatnam (India) 241, 242
Visaya, Ricardo 295, 296, 297

W
Waingmaw (Myanmar) 290
Wakil, Aisha 185
Walikale (DRC) 176
Wang Zuoan 279
Wani, Burhan Muzaffar 256, 257, 258, 261
Warduj (Afghanistan) 225
Warri South West (Nigeria) 192, 194
Warsaw (Poland) 219
Washington Post 44
Wau (South Sudan) 29
West Bank 101, 102, 103, 104, 105, 106
West Bengal (India) 243
Western Ghouta (Syria) 141
West Kordofan (Sudan) 215
Wilayat Sina (Egypt) 9, 83, 85, 86, 87, 91
Wittig, Katrin 65
Wiwat Pathompak 314
Wokha (India) 252
World Bank 95, 186
World Uyghur Congress 284

X
Xi Jinping 283
Xinjiang (China) 279, 281, 282, 283, 284, 285
Xinjiang Uighur Autonomous Region (China) 279

Y
Ya'alon, Moshe 104
Yadav, Kulbhushan 254, 269
Yakhchal (Afghanistan) 225
Yala (Thailand) 309, 311, 312
Yambio (South Sudan) 205
Yamgan (Afghanistan) 225
Yarkawa, Yaga 182, 185
Yarmouk Martyrs Brigade (Syria) 141
Yasynuvata (Ukraine) 321
Yei (South Sudan) 205

Yemen 41, 108, 111, 153–162, 201
  Ansar Allah 155
  Central Bank 155
  General People's Congress 155
  Houthi Supreme Revolutionary Committee 155
  Popular Resistance 86, 87, 159, 162
  Sana'a Chamber of Commerce and Industry 156
  Supreme Political Council 155
Yildirim, Binali 148
Yobe (Nigeria) 187
Yugoslavia 15
Yüksekdag, Figen 149
Yuksekova (Turkey) 146
Yunnan (China) 289
Yusuf, Zohra 274

Z

Zabdeen (Syria) 141
Zabul (Afghanistan) 225, 230
Zacatecas (Mexico) 345
Zahra (Syria) 133
Zakharchenko, Alexander 323, 324
Zambada, Ismael 'El Mayo' 346
Zambia 52
Zamboanga City (Philippines) 295, 298
Zamboanga (Philippines) 295, 296, 298
Zebari, Hoshyar 98, 99
Zeliangrong United Front (India) 246
Zeliang, T.R. 251
Zhang Chunxian 281, 284
Zolote (Ukraine) 323
Zueitina (Libya) 118, 119
Zummar (Iraq) 42
Zunheboto (India) 249